Control Dynamics
of Robotic Manipulators

Control Dynamics
of Robotic Manipulators

J. M. Skowronski

University of Southern California
Los Angeles, California
and
University of Queensland
St. Lucia, Australia

1986

ACADEMIC PRESS, INC.
Harcourt Brace Jovanovich, Publishers
Orlando San Diego New York Austin
Boston London Sydney Tokyo Toronto

Copyright © 1986 by Academic Press, Inc.
ALL RIGHTS RESERVED.
NO PART OF THIS PUBLICATION MAY BE REPRODUCED OR
TRANSMITTED IN ANY FORM OR BY ANY MEANS, ELECTRONIC
OR MECHANICAL, INCLUDING PHOTOCOPY, RECORDING, OR
ANY INFORMATION STORAGE AND RETRIEVAL SYSTEM, WITHOUT
PERMISSION IN WRITING FROM THE PUBLISHER.

ACADEMIC PRESS, INC.
Orlando, Florida 32887

United Kingdom Edition published by
ACADEMIC PRESS INC. (LONDON) LTD.
24–28 Oval Road, London NW1 7DX

Library of Congress Cataloging in Publication Data

Skowroński, Janisław M.
   Control dynamics of robotic manipulators.

   Bibliography: p.
   Includes index.
   1. Robotics.   2. Manipulators (Mechanism)   I. Title.
TJ211.S59   1986   629.8′92   86-7996
ISBN 0–12–648130–X (alk. paper)

PRINTED IN THE UNITED STATES OF AMERICA

86 87 88 89   9 8 7 6 5 4 3 2 1

Contents

Preface   vii
Acknowledgments   ix

Chapter 1   Mechanical Models

1.1 Modular RP-Unit Manipulator	1
1.2 Multiple Mechanical System. Cartesian Model	9
1.3 Generalized Coordinates. Lagrangian Model	17
1.4 Work Regions. Target and Obstacles	24
1.5 State-Space Representation	30
1.6 The Modular Manipulator Revisited	39

Chapter 2   Force Characteristics, Energy, and Power

2.1 Potential Forces. Coupling Characteristics	44
2.2 Conservative Reference Frame. Energy Surface	48
2.3 Nonpotential Forces. Power Balance. Energy Flow	60

Chapter 3   Controllability

3.1 From Classical to Recent Programs and Objectives	70
3.2 Stability, Boundedness, and Stabilization	84
3.3 Controllability under Uncertainity	98
3.4 Strong and Adaptive Stabilization	109
3.5 Optimal Controllability	117
3.6 Stabilization Synthesis	124

Chapter 4   Reaching and Capture

4.1 Reaching and Related Objectives	129

4.2 Reaching with and without Penetration	146
4.3 Stipulated Reaching Time. Maneuvering	151
4.4 Reaching with Capture. Optimal Capture	156
4.5 Reaching without Capture. Stipulated Handling Time	170
4.6 Planned Path Tracking	175
4.7 Model Reference Adaptive Control	185

Chapter 5 Avoidance of Obstacles

5.1 Conditions for Avoidance	198
5.2 Avoidance with Stipulated Handling Time	209
5.3 Reaching with Capture While Avoiding Obstacles	217
5.4 Adaptive Avoidance	221
5.5 Avoidance of Moving Obstacles	225

Chapter 6 Adaptive Indentification

6.1 Identification of States and Parameters	231
6.2 Identification of Variable Payload	239
6.3 Constant Parameter Identification and Model Tracking	242

References 247

Index 263

Preface

In a large number of well-known universities, the mechanical sciences (applied mechanics, mechanical systems dynamics and control, aero and hydro mechanics, elasticity, etc.) have separated from mechanical engineering departments into independent teaching units or have joined the systems sciences. There are serious arguments for doing so, but there is also definitely at least one substantial negative in such an action, and that is the resulting lack of cooperation between technological-production oriented engineering and its fundamental counterpart, mechanical systems dynamics. The technologists miss out on a broader scope of options and thus on flexibility in design, and the theoreticians lose reference to real problems. The gap widens as it spreads from research to teaching and then to textbooks, and this may well impede future development.

Robotics, quite naturally developed so far by the production oriented groups, begins to suffer from the above fate, particularly in view of its growing need to be applicable almost everywhere. Not only must manipulators be automatically controlled to reach from A to B, but they must also be made to follow a stipulated smooth path, avoiding various stationary or moving obstacles. In addition they are required to do this quite accurately in a specified time and space and at high speed, in cooperation or competition with other machines or other manipulators for that matter. Often they also need to do it optimally with respect to perhaps several cost functionals.

These tasks should be achieved while working in a variable environment and subject to various, frequently uncertain, dynamic payloads and hence also to unknown dynamical structural forces. Manipulators must thus be stabilized and made robust against unpredictable conditions, as well as capable of attaining the objectives mentioned. This yields a need for feedback and adaptive controllers, as well as perhaps a self-organizing structure.

Such manipulators become a complex, strongly nonlinear and strongly coupled system with many degrees of freedom. Control based on a linearized or otherwise simplified model loses effectiveness to the degree that the real working procedure deviates from the conditions assumed for simplification. This can lead to displacements of a manipulator from the desired trajectory in spite of all control efforts. It can even make the motion of the manipulator incompatible with such a trajectory, for instance, if the equilibria of the real nonlinear system do not coincide with the single equilibrium of the linearized model.

All of the above requires delving more deeply into rather sophisticated basic research in control and systems dynamics, not only to use it but also to extend it or at least to adjust it for our purposes. It means the situation in development of manipulators has matured to fundamental studies, and there is a need to investigate which part of the wide range of control dynamical results can be applied.

There are several excellent textbooks on robotic manipulators which attempt and largely succeed in presenting the material in a unified way with some of the above in mind—for example, Paul [1], Coiffet [1], and Snyder [1]; however, they stop on "unfinished" problems currently under research or requiring research. That is exactly where we would like to start, indicating to the postgraduate student or design engineer which branches and topics in control and systems dynamics are applicable and perhaps suggesting appropriate methods. Obviously no book can pretend to do the above regarding all the topics involved either in manipulator theory or in its control system mechanical background. The selection must be a matter of what seems to be more urgently needed (judging by its research popularity) and obviously biased by the experience and preferences of the author and his circle of collaborators.

We refer briefly to various existing and possible models (Chapter 1) and energy relations needed later (Chapter 2), and introduce elementary nonlinear controllability and stabilization conditions (first part of Chapter 3). Then we discuss control under uncertainty, both "worst-case design" and adaptive, referring first to stabilization (Chapter 3), then to various elementary objectives like reaching, real-time reaching, maneuvering, capture (handling) and optimal capture of an object (target), planned path tracking, and model reference adaptive control (Chapter 4). Avoidance of stationary and moving obstacles is the next main topic (Chapter 5). We close with adaptive identification of states and parameters (Chapter 6).

Acknowledgments

The author is indebted to Professors M. Corless, H. Flasher, W. J. Grantham, G. Leitmann, W. E. Schmitendorf, R. J. Stonier, and T. L. Vincent for cooperation leading the the results included in this book, as well as to some of the above-mentioned colleagues for critical comments on the text.

The author is also indebted to Professor T. Maxworthy and the Mechanical Engineering Department at the University of Southern California for encouragement and help in teaching the course.

Chapter 1

Mechanical Models

1.1 MODULAR RP-UNIT MANIPULATOR

The robot manipulators may be presently modeled as complex (branching) open or closed chain machines, but for the sake of basic dynamic studies they still are best considered consisting of simple open kinetic chains of n material links, each link with a single degree of freedom (DOF), connected together by joints. Such a manipulator may have more than n DOF, the additional due to orientation of the base and the gripper, but the DOF of the links are fundamental to our study. The joints are either *revolute* (rotational) coded R, or *prismatic* (translational) coded P. So it seems that a modular two-link RP-unit manipulator with two DOF, one rotation and one translation, can serve a twofold purpose: as a unit for composing up a suitable machine, and as an instructive example for studying both dynamics and control in the general case. It also seems that some fundamental notions of our study are best introduced and explained first on such an example.

Figure 1.1 displays the set-up. The link 1 is posed at the *base* with underlying revolute joint 1, the link 2 is connected to the first by a prismatic joint 2. Obviously the order of joints could have been reversed without influencing the generality. The mass of each link and of the corresponding joint is lumped into the mass point m_i, $i = 1, 2$, located at the end of a link, the first mass at the fixed distance r_1 from the joint 1, the second at the variable distance r. The mass m_1 rotates about the joint 1 and the mass m_2 translates together with the link 2 respective to the link 1, the motion allowed by the joint 2. These DOF are measured in terms of either *Cartesian* or *Langrangian coordinates*, the values of which give time instantaneously the *configuration* of the manipulator arm.

We place the Cartesian reference frame $0\xi_0\eta_0\zeta_0$ of the "world" coordinates fixed to the base, as shown in Fig. 1.1, with the origin at joint 1. It will thus also be called the *inertial* or, more frequently, *base system*. We also embed a Cartesian reference frame $0_i\xi_i\eta_i\zeta_i$ in each link $i = 1, 2$. These are body coordinates of the link concerned, called briefly *link coordinates*. We let the origin 0_i be at the joint $i + 1$, while the ζ_i axis is parallel to the axis of joint i

1 MECHANICAL MODELS

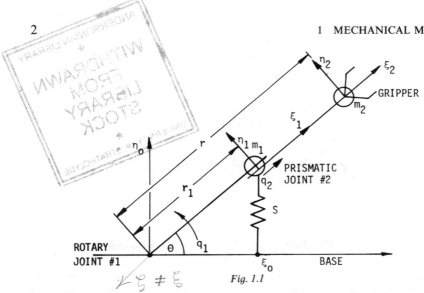

Fig. 1.1

(axis of rotation or translation). This means that 0_1 is placed at the end of link 1, that is, at joint 2.

The motion of a link is considered in terms of the motion of its link coordinates $0_i \xi_i \eta_i \zeta_i$ with respect to some reference frame. In general such motion is determined by three changes of *position* along coordinates of this frame (translation of 0_i) and three changes of *orientation* (rotation of $0_i, \xi_i, \eta_i, \zeta_i$), as well as the corresponding velocities. Obviously, the motion of link 1 is referenced to the base frame $0\xi_0\eta_0\zeta_0$. In our case link 1 rotates only about the ζ_0 axis with the position and velocities of m_1 ($\xi_{01}, \eta_{01}, \zeta_{01}$) specified by

$$\xi_{01}(t) = r_1 \cos \theta(t), \qquad \dot{\xi}_{01}(t) = -r_1 \sin \theta(t) \cdot \dot{\theta}(t),$$
$$\eta_{01}(t) = r_1 \sin \theta(t), \qquad \dot{\eta}_{01}(t) = r_1 \cos \theta(t) \cdot \dot{\theta}(t), \qquad (1.1.1)$$
$$\zeta_{01}(t) \equiv 0, \qquad \dot{\zeta}_{01}(t) \equiv 0,$$

where t is time, with $t \geq t_0$, which is an initial (reference) instant, $t_0 \in R$.

The motion of any other link $i > 1$ may be investigated by referencing it *relative* to the frame fixed at the previous link $0_{i-1} \xi_{i-1} \eta_{i-1} \zeta_{i-1}$. Such a relative reference applied sequentially leads to indirect reference to the base coordinates $0\xi_0\eta_0\zeta_0$. Alternatively, we may reference the motion of each link i *directly* to $0\xi_0\eta_0\zeta_0$.

In the case of our example, the translation of m_2 ($\xi_{12}, \eta_{12}, \zeta_{12}$) presently referred to $0_0\xi_0\eta_0\zeta_0$ gives

$$\xi_{02}(t) = r(t) \cos \theta(t),$$
$$\eta_{02}(t) = r(t) \sin \theta(t), \qquad (1.1.2)$$
$$\zeta_{02}(t) \equiv 0;$$

1.1 MODULAR RP-UNIT MANIPULATOR

and the velocities

$$\dot{\xi}_{02}(t) = \dot{r}(t)\cos\theta(t) - \dot{\theta}(t)r(t)\sin\theta(t),$$
$$\dot{\eta}_{02}(t) = \dot{r}(t)\sin\theta(t) + \dot{\theta}(t)r(t)\cos\theta(t), \qquad (1.1.3)$$
$$\dot{\zeta}_{02}(t) \equiv 0,$$

as the direct-to-base reference approach.

Note that (1.1.3) may be written as

$$\begin{bmatrix} \dot{\xi}_{02} \\ \dot{\eta}_{02} \end{bmatrix} = \begin{bmatrix} \cos\theta & -r\sin\theta \\ \sin\theta & r\cos\theta \end{bmatrix} \begin{bmatrix} \dot{r} \\ \dot{\theta} \end{bmatrix}, \qquad (1.1.4)$$

where

$$J(r,\theta) = \begin{bmatrix} \cos\theta & -r\sin\theta \\ \sin\theta & r\sin\theta \end{bmatrix} \qquad (1.1.5)$$

is the Jacobian of the transformation between the variables

$$(r,\theta) \to (\xi_{02}, \eta_{02}, \zeta_{02}).$$

More generally, the standard routine in manipulator kinematics (see Paul [1]) is to use a transformation matrix A_i, $i = 1, 2$, describing the relative translation and/or rotation between $0_i \xi_i \eta_i \zeta_i$ and $0_{i-1} \xi_{i-1} \eta_{i-1} \zeta_{i-1}$. Then the position and orientation of the link i in base coordinates are given by the matrix product $T_i = A_1 \cdot A_2 \cdot \ldots \cdot A_i$. The matrix A_i transforms the position vector $\bar{r}_{i-1}$ of the mass point m_{i-1} into $\bar{r}_i$ of m_i: $\bar{r}_i = A_i \bar{r}_{i-1}$. The matrix may include a rotation matrix (direction cosines) and/or translation matrix (components of translations vector). Depending upon the kinematic method employed, these vectors and matrices may be *either* three dimensional and 3×3, immediately generalizing (1.1.1), (1.1.2), *or* four dimensional called *homogeneous* with $\bar{r}_i = (a_i, b_i, c_i)$ represented by the matrix $[\xi_i \eta_i \zeta_i d_i]^T$, $a_i = \xi_i/d_i$, $b_i = \eta_i/d_i$, $c_i = \zeta_i/d_i$ and the transformation matrices 4×4.

In our example A_1 is a rotation matrix only, while A_2 represents only translations

$$A_1 = \begin{bmatrix} \cos\theta & -\sin\theta & 0 & 0 \\ \sin\theta & \cos\theta & 0 & 0 \\ 0 & 0 & 1 & 0 \\ 0 & 0 & 0 & 1 \end{bmatrix}, \qquad (1.1.6)$$

$$A_2 = \begin{bmatrix} 1 & 0 & 0 & r \\ 0 & 1 & 0 & 0 \\ 0 & 0 & 1 & 0 \\ 0 & 0 & 0 & 1 \end{bmatrix}. \qquad (1.1.7)$$

Then the relation to the base coordinates is given by

$$T_2 = A_1 \cdot A_2 = \begin{bmatrix} \cos\theta & -\sin\theta & 0 & r\cos\theta \\ \sin\theta & \cos\theta & 0 & r\sin\theta \\ 0 & 0 & 1 & 0 \\ 0 & 0 & 0 & 1 \end{bmatrix}. \quad (1.1.8)$$

Note here that the general homogeneous transformation illustrates the case of rotation first and then translation (see the fourth column of the T_2 matrix). We took the link coordinates positioned similarly to $0\xi_0\eta_0\zeta_0$ for illustrative purposes. A more frequently used convention is to take ζ_i along the axis of rotation or translation. Readers interested in kinematics of manipulators may look for details in Paul [1], Coiffet [1], Snyder [1], or Kooleshov and Lakota [1].

The relative reference approach is also convenient for the fast numerical simulation (see Hollerbach [1]) and thus for speeding up the on-line calculations during the work of the machine, as well as for deriving the kinetostatic relations needed for computer aided modeling (see Stepanenko and Vukobratovic [1]). In both these applications recursive kinematics is used, i.e., calculations begin with a given path of the gripper. We shall return to this topic in Sections 3.1, 4.6, and 5.5.

On the other hand the direct approach relates the *mechanical network* (see Skowronski [2]) of links and appendages (gripper, vehicle) to other objects in the world space $0\xi_0\eta_0\zeta_0$ on which the manipulator works, later called *targets*, or which it must avoid, termed *obstacles*. Once the links and appendages are referred to $0\xi_0\eta_0\zeta_0$, we consider their relation to the objects in this space or, if convenient, simplify matters by taking the space of relative distances between the elements of the network and the objects. Then the origin (zero distance) in that space is discussed as either an attracting target or repelling obstacle, see Sections 4.6 and 5.5.

Moreover representing the robot and its environment as a joint system in $0\xi_0\eta_0\zeta_0$ allows the use of the wealth of methods available in system dynamics, providing an insight into the dynamic behavior of the robot, which is much needed for synthesis, design, and optimization. Such an insight is very poor or totally missing in the numerically based kinetostatic investigation (for a review see Silver [1]).

In this section we simplify the network to the two links concerned, with the last mass m_2 also representing the gripper, and m_1 covering the vehicle as well, provided it is fixed at 0. These approximations fully suit our purpose of control-dynamical studies, and have been justified within the so-called "augmented body" approach (see Liegeois, Khalil, Dumas, and Renaud [1] and Hooker and Margulies [1]).

We turn now to the environmental object (target, obstacle) which may be represented by points in $0\xi_0\eta_0\zeta_0$. In the case of a lumped mass representation,

1.1 MODULAR RP-UNIT MANIPULATOR

it will be a single mass point; in the case of a larger body, it will be several geometric points specifying the boundary. Let us take a mass point m_3 determined by $\xi_{03}, \eta_{03}, \zeta_{03}$ as the target to be reached by the gripper, and a body determined by the band (*safety zone*)

$$\xi_{04} \leq \xi_0 \leq \xi_{05}, \quad \eta_{04} \leq \eta_0 \leq \eta_{05}, \quad \zeta_0 \neq 0 \tag{1.1.9}$$

enveloping an *antitarget* as a set to be avoided.

The objects are either *at rest* or *moving*. We assume here the first case, leaving the second to more general study in Section 1.2. We have $\xi_{03} = a$, $\eta_{03} = b, \zeta_{03} = 0, \xi_{04} = a_4, \xi_{05} = a_5, \eta_{04} = b_4, \eta_{05} = b_5$, all constants.

Our gripper m_2 will now pursue m_3 $(a, b, 0)$ while m_1, m_2 avoid the band $a_4 \leq \xi_{0i} \leq a_5, b_4 \leq \eta_{0i} \leq b_5, i = 1, 2$, with the kinematics described by (1.1.1), (1.1.2), and (1.1.4). Observe however that the motion is constrained by the fact that each joint is allowed only one DOF. Indeed, from (1.1.1), (1.1.2), and (1.1.4), we see that the motion depends only on two independent variables $\theta(t)$, $r(t)$, and their derivatives $\dot{\theta}(t) = \omega(t)$ and $\dot{r}(t)$ varying with time $t \geq t_0$. We thus choose them as the Lagrangian *joint coordinates* $q_1 = \theta(t), q_2 = r(t)$ with $\dot{q}_1 = \omega(t), \dot{q}_2 = \dot{r}(t)$. Note that the relative referenced choice $q_2 = r(t) - r_1$, $r_1 = $ const, does not change our case since all derivatives of r_1 vanish. It is easily seen that the motion is completely determined by $q_1(t), q_2(t), \dot{q}_1(t), \dot{q}_2(t)$, or by two vectors: the configuration vector $\bar{q}(t) = (q_1(t), q_2(t))^T$ in the *configuration space* R^2 and the velocity vector $\dot{\bar{q}}(t) = (\dot{q}_1(t), \dot{q}_2(t))^T$ in the corresponding tangent space of generalized velocities.

We thus arrive at the representation of motion in the *phase space* $R^{2n}, n = 2$, by the vector $\bar{x}(t) = (q_1(t), q_2(t), \dot{q}_1(t), \dot{q}_2(t))^T, t \geq t_0$, which will later be called the *state vector* of the system, while R^{2n} becomes the *state space* $R^N, N = 2n$.

The reaching ability of the joints, in particular of the gripper m_2, in both configuration and phase spaces is limited. The gripper and the joint 1 can maneuver in a bounded work space in $0\xi_0\eta_0\zeta_0$ restricted by the physical limitations of its design, while the velocities and accelerations are bounded by the bounds on the controlling forces and torques of actuators. The restrictions by design will be most frequently given directly in terms of θ, r in the configuration space. They produce a bounded set Δ_q called the *work region*. Together with the bounded set $\Delta_{\dot{q}}$ in the tangent space restricting the velocities, these two sets generate the *state work region* $\Delta = \Delta_q \times \Delta_{\dot{q}} \subset R^N$, expressed in terms of specified stop limits on $\bar{q}(t), \dot{\bar{q}}(t)$. In our case Δ_q will be bounded only by the design limitation on $q_2 = r, q_1 = \theta$ being arbitrary.

Recall that in $0\xi_0\eta_0\zeta_0$ the gripper attempted to reach the target m_3 $(a, b, 0)$ while both joints had been avoiding the obstacles. Because of the obstacles, in order to do a job on a workpiece, say m_3, it becomes important for the arm to achieve such a *configuration of all the joints* (in our case two) such as to make the reaching by the gripper possible. Since a vector $\bar{q}(t)$ in the configuration space represents the positions of all the joints, *we may reach the above-mentioned target configuration by conveniently reaching a desired stipulated point $\bar{q}$ in R^2*, which thus may be called a *configuration target*. It may be easily seen that it is different from m_3 in $0\xi_0\eta_0\zeta_0$ in that it refers to all the joint variables involved in the system, not to the gripper only. With this in mind, we still briefly call it a *target*.

Now the Cartesian target m_3 $(a,b,0)$ and the Cartesian avoidance band (1.1.6) must be transferred into the joint coordinates, obviously by inverting (1.1.1), (1.1.2), and (1.1.4). Unfortunately, this is not a unique procedure even in our very simple modular case and is obviously less so in general. Indeed the inverses of (1.1.1), (1.1.2) give

$$\begin{aligned} q_2 &= r(t) = \{[\xi_{02}(t)]^2 + [\eta_{02}(t)]^2\}^{1/2}, \\ q_1 &= \theta(t) = \arctan(\eta_{02}(t)/\xi_{02}(t)) = \arctan(\eta_{01}(t)/\xi_{01}(t)) \end{aligned} \tag{1.1.10}$$

which produces a sequence of values for $\theta(t)$. The above gives the configuration target $\mathfrak{T}_q$:

$$q_1 = \arctan(b/a), \qquad q_2 = (a^2 + b^2)^{1/2}, \tag{1.1.11}$$

and the avoidance band as $\mathscr{A}_q$:

$$\begin{aligned} \arctan(b_4/a_4) &\leq q_1 \leq \arctan(b_5/a_5), \\ (a_4^2 + b_4^2)^{1/2} &\leq q_2 \leq (a_5^2 + b_5^2)^{1/2}. \end{aligned} \tag{1.1.12}$$

In order to invert (1.1.4) we differentiate (1.1.10) obtaining

$$2r\dot{r} = 2\xi_{02}\dot{\xi}_{02} + 2\eta_{02}\dot{\eta}_{02}, \qquad \dot{\theta}\sec^2\theta = \frac{-\eta_{02}}{\xi_{02}^2}\dot{\xi}_{02} + \frac{1}{\xi_{02}}\dot{\eta}_{02}.$$

The latter yields

$$\dot{\theta} = \frac{-\eta_{02}}{\xi_{02}(r^2/\xi_{02}^2)}\dot{\xi}_{02} + \frac{1}{\xi_{02}(r^2/\xi_{02}^2)}\dot{\eta}_{02}.$$

Hence the inverse of (1.1.4) is

$$\begin{aligned} \dot{q}_2 &= \dot{r}(t) = \frac{\xi_{02}(t)}{r(t)}\dot{\xi}_{02}(t) + \frac{\eta_{02}(t)}{r(t)}\dot{\eta}_{02}(t), \\ \dot{q}_1 &= \dot{\theta}(t) = \frac{-\eta_{02}(t)}{r^2(t)}\dot{\xi}_{02}(t) + \frac{\xi_{02}(t)}{r^2(t)}\dot{\eta}_{02}, \end{aligned} \tag{1.1.13}$$

or

$$\begin{pmatrix} \dot{r} \\ \dot{\theta} \end{pmatrix} = \begin{pmatrix} \xi_{02}/r & \eta_{02}/r \\ -\eta_{02}/r^2 & \xi_{02}/r^2 \end{pmatrix} \begin{pmatrix} \dot{\xi}_{02} \\ \dot{\eta}_{02} \end{pmatrix},$$

1.1 MODULAR RP-UNIT MANIPULATOR

with the inverted Jacobian matrix

$$J^{-1}(r,\theta) = \begin{pmatrix} \xi_{02}/r & \eta_{02}/r \\ -\eta_{02}/r^2 & \xi_{02}/r^2 \end{pmatrix}, \quad (1.1.14)$$

which is a key feature of the "inverse" kinematics and dynamics described later in Chapter 3.

The relation (1.1.13) now serves to establish the corresponding target and the possible avoidance band in the space of generalized velocities to complement (1.1.11) and (1.1.12) to a *target* $\mathfrak{T}$ and an *avoidance set* $\mathscr{A}$ in the phase space. Indeed, if there were a task of attaining some velocities $\dot\xi_{02}(t) = c$, and $\dot\eta_{02}(t) = d$ by the gripper while at $\mathfrak{T}_q$, in particular to stop there ($c = 0$, $d = 0$), by (1.1.13) we would aim at $\dot q_1 = 0$, $\dot q_2 = 0$, with the target

$$\mathfrak{T}: q_1 = \arctan(b/a), \quad q_2 = (a^2 + b^2)^{1/2}, \quad \dot q_1 = 0, \quad \dot q_2 = 0. \quad (1.1.15)$$

Moreover, if one wished to avoid excessive speeds $\dot\xi_{02} = c_4, \dot\eta_{02} = d_4, i = 1, 2$ close to the obstacles, then by (1.1.13) one would have to avoid $\mathscr{A}$ specified by (1.1.12) and

$$\dot q_1 \geq \frac{a_4 c_4}{(a_4^2 + b_4^2)^{1/2}} + \frac{b_4 d_4}{(a_4^2 + b_4^2)^{1/2}},$$

$$\dot q_2 \geq \frac{-b_4 c_4}{a_4^2 + b_4^2} + \frac{a_4 d_4}{a_4^2 + b_4^2}, \quad (1.1.16)$$

when approach $\mathscr{A}_q$ from below, or (1.1.16) with a_4, b_4 replaced by a_5, b_5 when approaching $\mathscr{A}_q$ from above.

It is now of interest how $\mathfrak{T}$ and $\mathscr{A}$ relate to the work region Δ, which is convenient if made simply connected. The target $\mathfrak{T}$ (1.1.15) fits for a suitable length r of the arm, that is, when $\sqrt{a^2 + b^2} < r_{max}$; r_{max} provided by the designer. The obstacle along q_2 may obviously be excluded from Δ_q by suitable adjustment of r_{min}, r_{max}, without spoiling the connectedness. However, the angular obstacle on q_1 (1.1.12) has to be covered by Δ_q and made avoidable by means of the controlling torque rather than by exclusion. The velocity obstacles (1.1.16) can be excluded by adjusting the boundaries of $\Delta_{\dot q}$ in a similar way.

The motion of our manipulator is generated by two controlling actuators or *drives*, which may be electrical, pneumatic, hydraulic or mixed, depending on the job to be done. These actuators produce *torques* or *forces*, while acting upon the joints, rotary or prismatic, respectively, through some *transmission* that includes clutches, brakes, etc. In our case we let the control variables $u_1(t)$, $u_2(t)$ represent the input from the two actuators producing a torque $Q_1^F(\bar q, \dot{\bar q}, u_1)$ on joint 1 and a force $Q_2^F(\bar q, \dot{\bar q}, u_2)$ on joint 2, the functions $Q_1^F(\cdot), Q_2^F(\cdot)$ being called *input transmission forces*.

The actuator's job is twofold; it must be *passive* to offset the gravity of links and joints (their own weight as well as the payload) and *active* to perform a specified assignment for the manipulator. It is often convenient to replace, at least in part, the passive job by the static effort of spring forces acting on or between the lumped masses. In our case we offset the gravity $9.81m_1$ and $9.81(m_1 + m_2)$ by setting up the spring force $S = a_0 + aq_1 + bq_1^3$ (see Fig. 1.1) with the free-play $a_0 = -9.81m_1 r_1$ and a hardening effect afterward. Elementary mechanics gives the potential energy as

$$\mathscr{V}(\bar{q}) = 9.81m_1 r_1 \sin q_1 - 9.81m_1 r_1 q_1 + \tfrac{1}{2}aq_1^2 + \tfrac{1}{4}bq_1^4 + 9.81m_2 q_2 \sin q_1, \tag{1.1.17}$$

and the kinetic energy as

$$T(\bar{q}, \dot{\bar{q}}) = \tfrac{1}{2}m_1 r_1^2 \dot{q}_1^2 + \tfrac{1}{2}m_2(\dot{q}_2^2 + q_2^2 \dot{q}_1^2). \tag{1.1.18}$$

We immediately calculate the Lagrangian $L(\bar{q}, \dot{\bar{q}}) = T(\bar{q}, \dot{\bar{q}}) - \mathscr{V}(\bar{q})$. Then we assume the damping force in the joints to be $\lambda_1|\dot{q}_1|\dot{q}_1$ and $\lambda_2 \dot{q}_2$, respectively, with $\lambda_i = \text{const} > 0$, $i = 1, 2$ being the damping coefficients. With the input transmission generally described by $Q_i^F(\bar{q}, \dot{\bar{q}}, u_i)$, $i = 1, 2$, the Lagrange equations of motion (see Section 1.3) immediately give

$$(m_1 r_1^2 + m_2 q_2^2)\ddot{q}_1 + 2m_2 q_2 \dot{q}_1 \dot{q}_2 + \lambda_1|\dot{q}_1|\dot{q} + 9.81(m_1 r_1 + m_2 q_2)\cos q_1$$
$$- 9.81m_1 r_1 + aq_1 + bq_1^3 = Q_1^F, \tag{1.1.19}$$

$$m_2 \ddot{q}_2 - m_2 q_2 \dot{q}_1^2 + \lambda_2 \dot{q}_2 + 9.81 m_2 \sin q_1 = Q_2^F.$$

Now since $(m_1 r_1^2 + m_2 q_2^2) > 0$, $m_2 > 0$, we may let

$$G_1(\bar{q}) = \frac{9.81(m_1 r_1 + m_2 q_2)}{m_1 r_1^2 + m_2 q_2^2} \cos q_1,$$
$$G_2(\bar{q}) = 9.81 \sin q_1 \tag{1.1.20}$$

be *gravity characteristics*, while

$$\Psi_1(\bar{q}) = \frac{-9.81 m_1 r_1 + aq_1 + bq_1^3}{m_1 r_1^2 + m_2 q_2^2},$$
$$\Psi_2(\bar{q}) \equiv 0 \tag{1.1.21}$$

are *elastic (spring) characteristics* of the system.
Then the sum

$$\Pi_i(\bar{q}) = \Psi_i(\bar{q}) + G_i(\bar{q}), \quad i = 1, 2, \tag{1.1.22}$$

will be called the *characteristics of potential forces*. In turn, the characteristics of nonpotential forces (centrifugal, Coriolis, damping) will in our case reduce

1.2 MULTIPLE MECHANICAL SYSTEM. CARTESIAN MODEL

to the two following types: *kinetic characteristics* (centrifugal, Coriolis).

$$\Gamma_1(\bar{q},\dot{\bar{q}}) = \frac{2m_2 q_2 \dot{q}_1 \dot{q}_2}{m_1 r_1^2 + m_2 q_2^2}, \quad \Gamma_2(\bar{q},\dot{\bar{q}}) = -q_2 \dot{q}_1^2 \qquad (1.1.23)$$

and *damping characteristics*

$$D_1(\bar{q},\dot{\bar{q}}) = \frac{\lambda_1 |\dot{q}_1| \dot{q}_1}{m_1 r_1^2 + m_2 q_2^2}, \quad D_2(\bar{q},\dot{\bar{q}}) = \frac{\lambda_2}{m_2} \dot{q}_2. \qquad (1.1.24)$$

The above sums up to what may be called the nonpotential (nonconservative) characteristics

$$\Phi_i(\bar{q},\dot{\bar{q}}) = \Gamma_i(\bar{q},\dot{\bar{q}}) + D_i(\bar{q},\dot{\bar{q}}). \qquad (1.1.25)$$

Finally, denoting

$$F_1(\bar{q},\dot{\bar{q}},u_1) = \frac{Q_1^F(\bar{q},\dot{\bar{q}},u_1)}{m_1 r_1^2 + m_2 q_2^2}, \quad F_2(\bar{q},\dot{\bar{q}},u_2) = \frac{1}{m_2} Q_2^F(\bar{q},\dot{\bar{q}},u_2), \qquad (1.1.26)$$

we obtain the equations (1.1.19) in the compressive form

$$\ddot{q}_i + \Phi_i(\bar{q},\dot{\bar{q}}) + \Pi_i(\bar{q}) = F_i(\bar{q},\dot{\bar{q}},u_i), \quad i = 1,2. \qquad (1.1.27)$$

1.2 MULTIPLE MECHANICAL SYSTEM. CARTESIAN MODEL

Recent robot manipulators work at high speeds, under heavy payloads, in difficult conditions and thus are subject to nonnegligible dynamical forces, both external and internal (material). On the other hand, they must be relatively light (space structures) and highly accurate (precision to 0.05 mm is not excessive in manufacturing!), thus robust to uncertainty of all kind, in particular vibrations coming from the dynamical load. Hence, modeling of links as rigid bodies in a single simple open chain is often insufficient, particularly when accompanied by the "kinematic-only" approach that recognizes the linkage as a mechanism, not a machine, and gives steady state solutions. The modeler must now look at the manipulator as a rather sophisticated machine with complex linkage, perhaps flexible, subject to various material investigations and dynamic calculation (see Benedict [1]).

Thus, we extend our discussion of the previous section not only in number of links but also into a more general model: possibly *accommodating manipulators with several chain structures* (branching, trees, etc.) as well as the objects in the workplace like targets and obstacles, with the linkage *subject to elastic and damping forces*, both external and structural (see Liegois, Khalil, Dumas, and Renaud [1]). This will be done in an attempt to view the manipulator and its working appendages (gripper, vehicle, sensors, actuators, transmission gear, wires, compensation suspensions, etc.) as one system interacting with the objects to be reached, maneuvered, or avoided.

Here we must comment that our model is meant only as a background to control dynamical study and as such is heavily biased toward the discussion in this book. Readers interested in modeling manipulators for other purposes should look for the texts written on kinematic and dynamic modeling per se, like the already quoted basic books by Paul [1], Coiffet [1], or Snyder [1], or in the Russian literature Medvedev, Leskov, and Yushchenko [1], Popov, Vereschagin, and Zinkevich [1].

On the other hand, our later discussion on control dynamics must serve the design purposes, both present and possibly those coming in the near future, and the model must in turn support such discussion being relevant to design as well.

In order to give the designer the freedom of choice needed for attaining the control objectives mentioned, the model should be made flexible enough to accommodate the following:

(1) an arbitrary number and configuration of links with possibly multiple joints;
(2) both options: translation and rotation of each link at each particular joint concerned;
(3) the fact that the system is nonlinear and arbitrarily coupled; and
(4) full dynamic representation of the motion under internal and external forces and torques.

Note that in the last section we dealt with a subcase of an open simple chain of sequentially adjoined links, the links and joints represented by a finite sequence of n lumped masses located at the mass points $m_i, i = 1,\ldots,n$. Each m_i was positioned at the end of the corresponding link i which it represented; then either rotated or translated (not both), together with the link coordinates $0_i \xi_i \eta_i \zeta_i$, with respect to joint i positioned at the end of link $i - 1$ where the mass point representing that link was. It means that thinking of link i as being rotated or translated with respect to m_{i-1}, we may as well say that m_i was rotated or translated with respect to m_{i-1} by turning the length of link i about joint i, that is m_{i-1}, or by extending this length, respectively. The latter motion in particular is better described in terms of moving point m_i of the link i than in terms of moving link i, as the first concept is more precise. We shall use the notions of "moving link" or "moving its mass point" interchangeably depending upon which is more instructive, most of the time having rather "rotating links" and "translated mass points", than otherwise.

Because of the fact that the links and joints and thus their representing mass points were organized in a sequence, each link $i - 1$ did have only one follower "satellite" link i which moved about it. Presently, in order to comply

1.2 MULTIPLE MECHANICAL SYSTEM. CARTESIAN MODEL

with the mentioned demands (1) and (2) for the model, we shall deorganize the sequence and begin with an arbitrary configuration of the assembly of n links, and thus also their mass points in $0\xi_0\eta_0\zeta_0$. Hence we may replace link $i-1$ by an arbitrary link j, $j = 1,\ldots,n$, $j \neq i$, of this assembly, and consider it having not one, but a number of follower satellite links i from the assembly, $i = 1,\ldots,k \leq n$, $i \neq j$, each of these links i being rotated *and* translated respective to joint i at the end of link j or more precisely at the mass point m_j. The relative position of each link i with the end m_i, versus link j with the end m_j is the same as before—between link i with m_i and link $i-1$ with m_{i-1}. We measure this position by an angle θ_i between link j and link i, and with the distance r_i between the mass points m_j (=joint i) and m_i (see Section 1.1, Fig. 1.1). Note here that there may be, in general, k joints i and mass points m_j at the same geometric location in $0\xi_0\eta_0\zeta_0$. We can call these joints *colocated* or *k-multiple*.

The above leads all in all to the set of $\frac{1}{2}n(n-1)$ *optional* links i, each joining a corresponding pair of mass points m_i, m_j, $i, j = 1,\ldots,n$, $i \neq j$, in $0\xi_0\eta_0\zeta_0$, see the dashed and dotted lines in Fig. 1.2, with *another option* of either one of the two motions (rotation, translation) to be selected for each chosen link. With single motion for each link and provided the links are not flexible, the designer must choose less than $2n-3$ links, otherwise he makes the structure rigid (not moving). Also, without discussing the rational of admissible and useful types of such structure, we assume here that some of the joints must be fastened to the base and some must be free-ended, allowing for the gripper. Then it is natural to expect that the *designer will choose a collection of chains joining the base points with the free ends*; see the tree marked with dashed and full lines in Fig. 1.2, as an instance. A single chain adjoined to the base and open-ended is the option marked by a full line. The chains may have colocated joints, but are made kinematically independent. On the other hand, the external and structural forces acting upon the links would interact through the joints making the structure dynamically coupled.

The location of the mass center or the mass point m_i at the end of link i was convenient in Section 1.1 and is maintained here for clarity of exposition (radius vectors coinciding with joints). However, our discussion may be immediately adjusted to points m_i located elsewhere on link i, provided the position is known. The adjustment by adding or subtracting a constant, is obvious. See also the example in Section 1.6.

A specified decision on the linkage i, j gives a plan for connections within the system of our n masses, forming the mechanical network mentioned in Section 1.1 (see Skowronski [24]) and producing the desired Cartesian model. There are several codes for specifying the so-called *association table* in particular cases of the linkage, based on propositions by Khalil, Renauld and Zapala, etc. For a detailed description of the codes and the bibliography, the reader is directed to Coiffet [1].

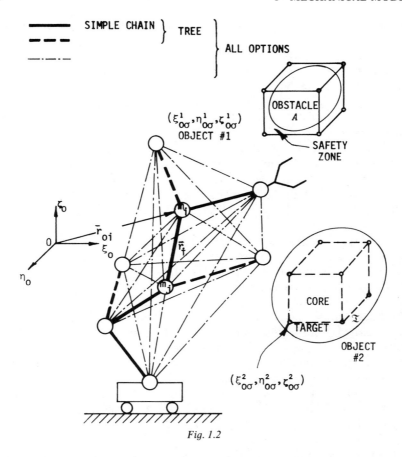

Fig. 1.2

As justified in Section 1.1 we include into the n masses the appendages attached to the arm. One (or several) of them may thus represent the gripper (or several grippers) as well as own link and joint. A few other masses may cover the vehicle. Almost every mass will include inertia of a sensor and an actuator, as well as its own link and joint.

Now, to produce the equations of our model, we again need to return to the all-options case, and start by looking at a single but arbitrary pair m_i, m_j from within the assembly. First, let us state the elementary kinematics involved.

Let the position of m_j be specified by the radius vector $\bar{r}_{0j}$ in $0\xi_0\eta_0\zeta_0$, and let $d_i = $ const be the length of the link i.

The position $\zeta_i = 0$ of m_i is fixed on the link i, that is, determined with respect to the link coordinates $0_i\xi_i\eta_i\zeta_i$, while at the same time this link i is rotated by the angles $\alpha(t)$, $\beta(t)$, $\gamma(t)$ and translated along the axis $0_j\zeta_j$ with respect to the joint i located at m_j. It is convenient to record the translation by

1.2 MULTIPLE MECHANICAL SYSTEM. CARTESIAN MODEL

its equivalent displacement $\bar{r}_i(t)$, $t \geq t_0$, $r_i^0 = |\bar{r}_i(t_0)| = d_i$, of m_i along $0_j\zeta_j$ with respect to 0_j at the joint i and mass point m_j. Figure 1.3 explains the situation. Note that if m_i is not located at the end of link i, $r_i^0 < d_i$.

The motion of m_i ($\xi_{0i}(t), \eta_{0i}(t)$) referred to $0\xi_0\eta_0\zeta_0$ is expressed in general by the vectors

$$\bar{r}_{0i}(t) = \bar{r}_{0j}(t) + \bar{r}_i(t), \tag{1.2.1}$$

$$\dot{\bar{r}}_{0i}(t) = \dot{\bar{r}}_{0j}(t) + \dot{\bar{r}}_i(t) + \bar{\omega}_i(t) \times \bar{r}_i(t), \tag{1.2.2}$$

where $\bar{r}_{0j}$ is the radius vector of mass m_j or the joint i from the origin of the world coordinates, and $\bar{\omega}_i$ is the angular velocity vector of rotation of link i. Note that $|\bar{r}_i| \triangleq r_i$ is the distance between m_i and m_j at any instant $t \geq t_0$ during the motion:

$$r_i(t) = [(\xi_{0i} - \xi_{0j})^2 + (\eta_{0i} - \eta_{0j})^2 + (\zeta_{0i} - \zeta_{0j})^2]^{1/2}. \tag{1.2.3}$$

The position and orientation of the link i is related to the previous adjacent link j by the homogeneous transform A_i known from Section 1.1. The matrix A_i is specified in terms of the angle between the links $\theta_i(t)$ and the distance $r_i(t)$ (1.1.1), (1.1.2) and represents what is called *forward kinematics*, as opposed to the *inverse kinematics* generated at each link i by an inverse transform to A_i (1.1.10). Obviously, the matrix A_i is either rotational (1.1.6) or translational (1.1.7), but to be consistent with our considering both motions, we should generally make it a product of two matrices, rotational and translational.

In turn the link j may rotate and translate about some joint j located at any other mass in the network, which serves as a reference to the motion of this

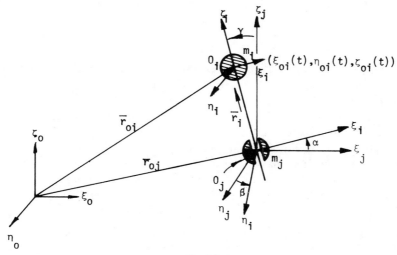

Fig. 1.3

link. We may then apply the above description successively within the particular structure of the network imposed by the designer. It should eventually refer to at least one joint fixed to the base through some regression of links. This means that, similarly as in Section 1.1, the relation of m_i to the base coordinates will be obtained by the transformation matrix $T_i = A_1 \cdot A_2 \cdot \ldots, \cdot A_i$ in terms of all the independent variables $\theta_1, \ldots, \theta_i, r_1, \ldots, r_i$ predecessing the joint i (1.1.2), within the structure concerned.

We have a very similar arrangement about velocities, the corresponding transformation for each mass m_i given by the Jacobian (1.1.5) in the forward kinematics and by the inverted Jacobian (1.1.14) in the inverse kinematics. In general, however, we have a 6×6 Jacobian matrix that relates the velocities (translational $\dot\xi_{0i}, \dot\eta_{0i}, \dot\zeta_{0i}$, and angular $\omega_{i\xi}, \omega_{i\eta}, \omega_{i\zeta}$) of the link i to the base through all the preceding independent velocity variables $\dot\theta_1, \ldots, \dot\theta_i, \dot r, \ldots, \dot r_i$:

$$(\dot{\bar r}_{0i}, \omega_i)^T = J \cdot (\dot\theta_1, \ldots, \dot\theta_i, \dot r_1, \ldots, \dot r_i). \qquad (1.2.4)$$

The inverse of the 6×6 Jacobian is more difficult to obtain as it may seem from the procedure leading to (1.1.14). There does not exist at this time an algorithm for obtaining closed-form solutions to such an inverse transform for arbitrary arm geometries (see Lenarcic [1]). However, at any instant of time, the arm is in a single configuration and all the elements of the Jacobian have unique numerical values. Thus, at a given instant, the Jacobian is an array of 36 numbers, and numerical analysis provides techniques for inverting such matrices. Obviously, if the analytic solution could be obtained, it is faster to calculate, and more agreeable physically.

Now, in order to investigate the dynamics, we return to our modular link i, and differentiate (1.2.2) with respect to time, to obtain the acceleration of the mass m_i:

$$\ddot{\bar r}_{0i} = \ddot{\bar r}_{0j} + \bar\omega \times (\bar\omega \times \bar r_i) + \dot{\bar\omega} \times \bar r_i + \ddot{\bar r}_i + 2\bar\omega \times \dot{\bar r}_i. \qquad (1.2.5)$$

The first and fourth terms on the right-hand side represent the translational acceleration, the second and third centrifugal, and the last term Cariolis acceleration.

If the designer chooses the case when m_i is not translated along $0_j\zeta_j$ but only rotating with the link coordinates $0_i\xi_i\eta_i\zeta_i$, we have $\dot{\bar r}_i = \ddot{\bar r}_i = 0$, and the last two terms of (1.2.5) vanish, while $\bar r_i = $ constant $= r_i^0$. On the other hand, when m_i is only translated we have $\ddot{\bar r}_{0i} = \ddot{\bar r}_{0j} + \ddot{\bar r}_i$.

Consider now forces or torques applied upon each point m_i. Without narrowing the applicability of our model, we can split them into the following classes:

Potential forces: $\mathscr{P}_{ji}(\bar r_i)$, energy conservative, representing gravity and restoring forces, either within link i or acting as outside compensation still through link i.

1.2 MULTIPLE MECHANICAL SYSTEM. CARTESIAN MODEL

Damping forces: $\mathscr{D}_{ji}(\bar{r}_i, \dot{\bar{r}}_i)$, energy dissipative or accumulative (negative damping), representing structural damping in link i, Coulomb or viscous damping in joint i, etc.

Input forces: $\mathscr{F}_i(\bar{r}_{0i}, \dot{\bar{r}}_{0i}, u_i)$, with u_i control variable produced by the actuator at joint i (force or torque).

Some of the masses include appendages attached to the arm, and thus must relate to *points of suspension* of these appendages (wheels of the vehicle, wires of sensors or actuators, etc.). These points may be fixed or move, depending upon constraints imposed on them. Depending on the type of constraints and type of suspension, we have the *reaction forces* $\mathscr{R}_i(\bar{r}_{0i}, \dot{\bar{r}}_{0i})$ acting upon the points m_i, to complement the previous three classes. We shall make the constraints stationary and holonomic, and the reaction ideal $\equiv$ producing no work done.

The reader is invited to look for details on particular shapes of the functions $\mathscr{P}_{ji}$, $\mathscr{D}_{ji}$, $\mathscr{F}_i$, R_i in Benedict and Tesar [1], Skowronski and Ziemba [1], and Skowronski [12], [24].

Summing up the forces upon m_i from various links, in view of (1.2.5), the Newton equation of motion for all options reads

$$m_i(\ddot{\bar{r}}_{0j} + \ddot{\bar{r}}_i) + m_i\bar{\omega} \times (\bar{\omega} \times \bar{r}_i) - m_i\dot{\bar{\omega}} \times \bar{r}_i \times 2m_i\bar{\omega} \times \dot{\bar{r}}_i$$
$$= \sum_j \mathscr{P}_{ji} + \sum_j \mathscr{D}_{ji} + \mathscr{F}_i + \mathscr{R}_i, \qquad (1.2.6)$$

where the first term represents the translative inertia force, the second and third terms the centrifugal forces, and the fourth the Coriolis force.

Due to their origin, we shall use for these forces the joint name *kinetic* (rather than "fictitious"), as opposed to the *applied* forces $\mathscr{P}, \mathscr{D}, \mathscr{F}$. In reference to the base, the equation (1.2.6) becomes

$$m_i\ddot{\bar{r}}_{0i} = \mathscr{P}_i(\bar{r}_{0i}) + \mathscr{D}_i(\bar{r}_{0i}, \dot{\bar{r}}_{0i}) + \mathscr{F}_i(\bar{r}_{0i}, \dot{\bar{r}}_{0i}, u_i), \qquad (1.2.7)$$

where

$$\mathscr{P}_i = \sum_j \mathscr{P}_{ji}, \qquad \mathscr{D}_i = \sum_j \mathscr{D}_{ji}.$$

The work done by the potential forces on virtual displacements within the whole network is

$$\delta W_{\mathscr{P}} = \sum_i \mathscr{P}_i(\bar{r}_{0i}) \, \delta \bar{r}_{0i}. \qquad (1.2.8)$$

Similarly for the damping and input:

$$\delta W_{\mathscr{D}} = \sum_i \mathscr{D}_i(\bar{r}_{0i}, \dot{\bar{r}}_{0i}) \, \delta \bar{r}_{0i}, \qquad (1.2.9)$$

$$\delta W_{\mathscr{F}} = \sum_i \mathscr{F}_i(\bar{r}_{0i}, \dot{\bar{r}}_{0i}, u_i) \, \delta \bar{r}_{0i}, \qquad (1.2.10)$$

respectively. Whence the work done by all the *applied forces* is

$$\delta W = \sum_i (\mathscr{P}_i + \mathscr{D}_i + \mathscr{F}_i) \delta \bar{r}_{0i}, \qquad (1.2.11)$$

while the ideal reactions produce no work.

From (1.2.8) we immediately have the *potential* of the conservative forces

$$\mathscr{U}_\mathscr{P} = \sum_i \oint \mathscr{P}_i(\bar{r}_{0i}) \, dr_{0i} \qquad (1.2.12)$$

and thus their components along the base axes ξ_0, η_0, ζ_0 are

$$\mathscr{P}_\xi = \partial \mathscr{U}_\mathscr{P}/\partial \xi_0, \qquad \mathscr{P}_\eta = \partial \mathscr{U}_\mathscr{P}/\partial \eta_0, \qquad \mathscr{P}_\zeta = \partial \mathscr{U}_\mathscr{P}/\partial \zeta_0, \qquad (1.2.13)$$

while the potential energy is $\mathscr{V} = -\mathscr{U}_\mathscr{P}$. Similarly (1.2.9) may produce the generalized *dissipation function*

$$\mathscr{U}_\mathscr{D} = \sum_i \oint \mathscr{D}_i(\bar{r}_{0i}, \dot{\bar{r}}_{0i}) \, dr_{0i}, \qquad (1.2.14)$$

with damping components along ξ_0, η_0, ζ_0,

$$\mathscr{D}_\xi = \partial \mathscr{U}_\mathscr{D}/\partial \xi_0, \qquad \mathscr{D}_\eta = \partial \mathscr{U}_\mathscr{D}/\partial \eta_0, \qquad \mathscr{D}_\zeta = \partial \mathscr{U}_\mathscr{D}/\partial \zeta_0. \qquad (1.2.15)$$

For linear damping, (1.2.14) obviously becomes the Rayleigh function.

Talking about Cartesian models, let us remark that our general structure just introduced is capable of covering the *arms with flexible links* (see Ardafio [1], Dubovsky and Gardner [1], and Gevarter [1]). All we would have to do in such a case is introduce in each structural link "dummy joints" which will make the link a subchain of arbitrary configuration and arbitrary type (open, constrained, locked, etc.) but with flexibly suspended mass points. Such suspension is then entering into the spring forces in the motion equation (1.2.7).

Having the means to describe the motion of the manipulator, the next most important operation is static or dynamic *positioning of the objects* (Cartesian targets, obstacles). Since the gripper eventually must use the coordinate system $0\xi_0\eta_0\zeta_0$, we refer to points in this system for our positioning. The objects will thus be given in terms of a *volume* in $0\xi_0\eta_0\zeta_0$ specified by a single (in case of lumped mass) or several points $(\xi^\kappa_{0\sigma}, \eta^\kappa_{0\sigma}, \zeta^\kappa_{0\sigma})$, $\sigma = 1, \ldots, s$, positioning the boundary of the volume (see Fig. 1.2). The superscript $\kappa = 1, \ldots, m$ refers to the object, without differentiating between the target and the obstacles, at this stage. Such a volume may either envelop the obstacle or be inscribed into a target (see Fig. 1.2) to make sure in the first case that the grown obstacle is avoided and in the second case that the target is reached. The volume enveloping an obstacle, or several obstacles, determines a *safety zone* about them (see Section 1.1). The volume inscribed in the target approximates the target from below and will be called its *core*. The volume may also be described by intervals along ξ_0, η_0, ζ_0 specified by the boundary points, cf. (1.1.6).

An object κ may either *rest*: $\xi_{0\sigma}^\kappa$, $\eta_{0\sigma}^\kappa$, $\zeta_{0\sigma}^\kappa$ = const, $\sigma = 1,\ldots,s$, or *move*: $\xi_{0\sigma}^\kappa(t)$, $\eta_{0\sigma}^\kappa(t)$, $\zeta_{0\sigma}^\kappa(t)$, $t \geq t_0$, generally with deformation. If deformation is excluded, then distances between the boundary points are preserved:

$$|\xi_{0\sigma}^\kappa - \xi_{0\mu}^\kappa|, |\eta_{0\sigma}^\kappa - \eta_{0\mu}^\kappa|, |\zeta_{0\sigma}^\kappa - \zeta_{0\mu}^\kappa| = \text{const}, \qquad \sigma, \mu = 1,\ldots,s. \quad (1.2.16)$$

1.3 GENERALIZED COORDINATES. LAGRANGIAN MODEL

At each joint i of our model in Section 1.2 we have allowed one motion only to be selected from two options; either rotation measured by the angle $\theta_i(t)$ between links, or translation measured by the displacement extension $r_i(t)$ of a link, much in the same fashion as we did in Section 1.1 for the RP-manipulator. The selected $\theta_i(t)$ or $r_i(t)$, but not both, is called a *joint variable* for the joint i and the link i, briefly *i-joint variable*. Let us recall that we have as many joints as links, no matter how the links are connected, that is, the k-multiple joint is viewed as k joints colocated at one point. Hence we have n joint variables, which represent a minimal number of independent variables that suffices to describe the configuration of the manipulator. They will thus be considered Lagrangian (generalized) coordinates $q_1,\ldots,q_n$ of the system. Together with n corresponding generalized velocities $\dot{q}_1,\ldots,\dot{q}_n$, they determine the motion of the manipulator. The vector $\bar{q}(t) = (q_1(t),\ldots,q_n(t))^T$ determines the configuration of the system in the configuration space R^n (see Section 1.1). Letting the velocities vector $\dot{\bar{q}}(t) = (\dot{q}_1(t),\ldots,\dot{q}_n(t))^T$ in the tangent space, the motion is then determined by the behavior of the *representing vector* $\bar{x}(t) = (\bar{q}(t), \dot{\bar{q}}(t))^T$ in the *phase space* R^{2n}, which is in our case the same as a *state vector* in the state space R^N, $N = 2n$.

In Section 1.2 we agreed that the designer will choose his options on linkage by selecting a family of chains that connect base points with some free ends for the gripper. Let us consider an open chain which does exactly that (see the full-line connection in Fig. 1.2). To *allow for* the DOF and thus the generalized coordinates assumed in the network, the chain would have to agree with what we assumed in Section 1.2, namely, the *kinematic independence* of other chains, in spite of the fact that some of its joints may be colocated with joints of other chains. The latter may generate dynamic force coupling. In additional recognition of the fact that the links are, in general, subject to dynamic loads as well as structural dynamic forces, we shall call such a chain *kinetic*. The independence mentioned will allow us to *investigate one kinetic chain in lieu of the family, provided its dynamical coupling is accounted for*, and provided its features are general enough to include all the properties to be discussed.

Since each q_i is either θ_i or r_i it can be found by inverting the transformation matrix A_i of Sections 1.1–1.2 [see also (1.1.10) and (1.2.3)]. The corresponding generalized velocities can be found by inverting (1.2.4), [see (1.1.13)]. The above is meant, however, with all the qualifications made in Section 1.2 regarding both the inverse of A_i and the Jacobian.

In the general case, the difficulty in calculating the inverses corresponding to (1.1.10) and (1.1.13) is quite obvious; hence the problems with attaining the fast on-line calculation if needed for kinematic control purposes (see Chapter 3). Fortunately, in our case, as seen from Sections 1.1–1.2, we need inverses mainly for the off-line (precalculation) job; hence the speed of calculation is not essential.

Now let us write the transformation $T_i = A_i \cdot, \ldots, \cdot A_i$ in the classical form referring to $q_1, \ldots, q_n$:

$$\bar{r}_{0i} = \bar{r}_{0i}(q_1, \ldots, q_n), \tag{1.3.1}$$

which is stationary due to the assumed constraints. Then the velocity is

$$\dot{\bar{r}}_{0i} = \frac{\partial \bar{r}_{0i}}{\partial q_1} \dot{q}_1 + \cdots + \frac{\partial \bar{r}_{0i}}{\partial q_n} \dot{q}_n, \tag{1.3.2}$$

where the partial derivatives generate the Jacobian. Consequently the kinetic energy is

$$T = \frac{1}{2} \sum_i m_i \left(\sum_j \frac{\partial \bar{r}_{0i}}{\partial q_j} \dot{q}_j \right)^2 \tag{1.3.3}$$

which after developing the square becomes the square form

$$T(q, \dot{q}) = \frac{1}{2} \dot{\bar{q}}^T A(\bar{q}) \dot{\bar{q}} = \frac{1}{2} \sum_i \sum_j a_{ij}(\bar{q}) \dot{q}_i \dot{q}_j \tag{1.3.4}$$

where $a_{ij}(\bar{q}) = a_{ji}(\bar{q})$ are coefficients of the corresponding positive definite *generalized inertia matrix* $A(q)$. Observe also that (1.3.1), (1.3.2) generate

$$\delta \bar{r}_{0i} = \frac{\partial \bar{r}_{0i}}{\partial q_1} \delta q_1 + \cdots + \frac{\partial \bar{r}_{0i}}{\delta q_n} \delta q_n. \tag{1.3.5}$$

To transform the kinetic forces from Cartesian to Lagrange coordinates, we calculate

$$\sum_i m_i \ddot{\bar{r}}_{0i} \delta \bar{r}_{0i} = \sum_i m_i \ddot{\bar{r}}_{0i} \sum_j \frac{\partial \bar{r}_{0i}}{\partial q_j} \delta q_j = \sum_j \sum_i m_i \ddot{\bar{r}}_{0i} \frac{\partial \bar{r}_{0i}}{\partial q_j} \delta q_j.$$

On the other hand,

$$\ddot{\bar{r}}_{0i} \frac{\partial \bar{r}_{0i}}{\partial q_j} = \frac{d}{dt} \left(\dot{\bar{r}}_{0i} \frac{\partial \bar{r}_{0i}}{\partial q_j} \right) - \dot{\bar{r}}_{0i} \frac{d}{dt} \left(\frac{\partial \bar{r}_{0i}}{\partial q_j} \right),$$

1.3 GENERALIZED COORDINATES. LAGRANGIAN MODEL

whence

$$\sum_i m_i \ddot{\bar{r}}_{0i} \delta \bar{r}_{0i} = \sum_j \left[\frac{d}{dt} \left(\sum_i m_i \dot{\bar{r}}_{0i} \frac{\partial \bar{r}_{0i}}{\partial q_j} \right) - \sum_i m_i \dot{\bar{r}}_{0i} \frac{d}{dt} \left(\frac{\partial \bar{r}_{0i}}{\partial q_j} \right) \right] \delta q_j.$$

Differentiating (1.3.3), we obtain that the first sum in the square bracket equals $\partial T/\partial \dot{q}_j$, while the second sum equals $\partial T/\partial q_j$, which makes

$$\sum_i m_i \ddot{\bar{r}}_{0i} \delta \bar{r}_{0i} = \sum_j \left[\frac{d}{dt} \left(\frac{\partial T}{\partial \dot{q}_j} \right) - \frac{\partial T}{\partial q_j} \right] \delta q_j. \tag{1.3.6}$$

The fact that the work done by the applied forces on virtual displacements in Cartesian coordinates equals that in generalized coordinates, that is,

$$\sum_i (\mathscr{P}_i + \mathscr{D}_i + \mathscr{F}_i) \delta \bar{r}_{0i} = \sum_j Q_j \delta q_j, \qquad i, j = 1, \ldots, n, \tag{1.3.7}$$

is used to transform the resultant of $\mathscr{P}_i, \mathscr{D}_i, \mathscr{F}_i$ into the generalized applied force $Q_j(\bar{q}, \dot{\bar{q}}, u_j)$. In view of (1.2.7) and (1.2.11), comparing (1.3.6) with (1.3.7), we have

$$\sum_j \left[\frac{d}{dt} \left(\frac{\partial T}{\partial \dot{q}_j} \right) - \frac{\partial T}{\partial q_j} \right] \delta q_j = \sum_j Q_j \delta q_j.$$

Since the above must hold for all arbitrary δq_j, $j = 1, \ldots, n$, we have

$$\frac{d}{dt} \left(\frac{\partial T}{\partial \dot{q}_j} \right) - \frac{\partial T}{\partial q_j} = Q_j, \qquad j = 1, \ldots, n, \tag{1.3.8}$$

which is the Lagrange motion equation of the second kind.

The transform (1.3.7) can be made explicit. From (1.2.11) and (1.3.5), we have

$$\partial W = \sum_i (\mathscr{P}_i + \mathscr{D}_i + \mathscr{F}_i) \sum \frac{\partial \bar{r}_{0i}}{\partial q_j} \delta q_j$$

$$= \sum_j \delta q_j \sum_i (\mathscr{P}_i + \mathscr{D}_i + \mathscr{F}_i) \frac{\partial \bar{r}_{0i}}{\partial q_j},$$

which, by (1.3.7), gives the generalized force defined as

$$Q_j(\bar{q}, \dot{\bar{q}}, u_j) = \sum_i (\mathscr{P}_i + \mathscr{D}_i + \mathscr{F}_i) \frac{\partial \bar{r}_{0i}}{\partial q_j}. \tag{1.3.9}$$

Since the force transforms (1.3.7) and (1.3.9) work for the particular forces $\mathscr{P}_i, \mathscr{D}_i, \mathscr{F}_i$ as well as for their resultant, we may write

$$Q_j(\bar{q}, \dot{\bar{q}}, u_j) = Q_j^P(\bar{q}) + Q_j^D(\bar{q}, \dot{\bar{q}}) + Q_j^F(\bar{q}, \dot{\bar{q}}, u_j), \tag{1.3.10}$$

representing successively the potential, damping and input force generalized counterparts of $\mathscr{P}_i, \mathscr{D}_i, \mathscr{F}_i$. Substituting (1.3.10) into (1.3.8), and *swapping the index j for i*, gives

$$\frac{d}{dt}\left(\frac{\partial T}{\partial \dot{q}_i}\right) - \frac{\partial T}{\partial q_i} = Q_i^P(\bar{q}) + Q_i^D(\bar{q},\dot{\bar{q}}) + Q_i^F(\bar{q},\dot{\bar{q}},u_i), \qquad i = 1,\ldots,n, \quad (1.3.11)$$

with the kinetic forces on the left and applied forces on the right hand side. Since by the defining property of potential

$$Q_i^P(\bar{q}) = \frac{\partial \mathcal{U}_P}{\partial q_i}, \qquad i = 1,\ldots,n, \quad (1.3.12)$$

[see (1.2.13)], introducing the Lagrangian function $L(\bar{q},\dot{\bar{q}}) = T(\bar{q},\dot{\bar{q}}) + \mathcal{U}_P(\bar{q})$, we have from (1.3.11) an equivalent Lagrange equation

$$\frac{d}{dt}\left(\frac{\partial L}{\partial \dot{q}_i}\right) - \frac{\partial L}{\partial q_i} = Q_i^D(\bar{q},\dot{\bar{q}}) + Q_i^F(\bar{q},\dot{\bar{q}},u_i), \quad (1.3.13)$$

which is the standard form to use, when one likes to describe the motion of the manipulator in the Lagrangian equation form.

The Lagrangian function is also called the *kinetic potential*. Acknowledging that the potential energy $\mathscr{V}(\bar{q}) = -\mathcal{U}_P(\bar{q})$, we may consider the Lagrangian to be the difference in energies

$$L(\bar{q},\dot{\bar{q}}) = T(\bar{q},\dot{\bar{q}}) - \mathscr{V}(\bar{q}). \quad (1.3.14)$$

Let us now substitute the general expression (1.3.4) for the kinetic energy in (1.3.11). First we denote

$$\frac{\partial T}{\partial \dot{q}_i} = \sum_j a_{ij}(\bar{q})\dot{q}_j \triangleq b_i(\bar{q},\dot{\bar{q}}),$$

$$\frac{\partial T}{\partial q_i} = \sum_j \sum_i \frac{\partial a_{j\sigma}}{q_i}\dot{q}_j\dot{q}_\sigma \triangleq c_i(\bar{q},\dot{\bar{q}}).$$

Then

$$\frac{d}{dt}\left(\frac{\partial T}{\partial \dot{q}_i}\right) - \frac{\partial T}{\partial q_i} = \sum_j a_{ij}\ddot{q}_j + \sum_j \frac{\partial b_i}{\partial q_j}\dot{q}_j - c_i(\bar{q},\dot{\bar{q}}).$$

The first term on the right-hand side represents translative inertia, the next two represent centrifugal and Coriolis forces (jj—centrifugal; $j\sigma$, $\sigma \neq j$—Coriolis). We aggregate them by denoting

$$K_i(\bar{q},\dot{\bar{q}}) = \sum_j \frac{\partial b_i}{\partial q_j}\dot{q}_j - c_i(\bar{q},\dot{\bar{q}}).$$

Then the equations (1.3.11) become the *explicit Lagrange equations* or the *Newtonian form* in generalized coordinates

$$\sum_j a_{ij}(\bar{q})\ddot{q}_j + K_i(\bar{q},\dot{\bar{q}}) - Q_i^D(\bar{q},\dot{\bar{q}}) - Q_i^P(\bar{q}) = Q_i^F(\bar{q},\dot{\bar{q}},u_i), \qquad i = 1,\ldots,n,$$

$$(1.3.15)$$

1.3 GENERALIZED COORDINATES. LAGRANGIAN MODEL

with an obvious meaning of particular terms. The centrifugal Coriolis term may be calculated explicitly as

$$K_i = \sum_j \frac{\partial b_i}{\partial q_j} \dot{q}_j - c_i(\bar{q},\dot{\bar{q}}) = \sum_j \sum_\sigma \left(\frac{\partial a_{ji}}{\partial q_\sigma} - \frac{1}{2}\frac{\partial a_{j\sigma}}{\partial q_i} \right) \dot{q}_j \dot{q}_\sigma$$

$$= \sum_j \sum_\sigma \left(\frac{\partial a_{ji}}{\partial q_\sigma} + \frac{\partial a_{\sigma i}}{\partial q_j} - \frac{\partial a_{j\sigma}}{\partial q_i} \right) \dot{q}_j \dot{q}_\sigma, \qquad i,j,\sigma = 1,\ldots,n.$$

The term in the brackets is customarily abbreviated as the Christoffel symbol $[j_i^\sigma]$ (see Whittaker [1]), hence

$$K_i(\bar{q},\dot{\bar{q}}) = \sum_j \sum_\sigma [j_i^\sigma] \dot{q}_j \dot{q}_\sigma. \qquad (1.3.16)$$

The equations (1.3.15) may be also rewritten in the vectorial form

$$A(q)\ddot{\bar{q}} + \bar{K}(\bar{q},\dot{\bar{q}}) - \bar{Q}^D(\bar{q},\dot{\bar{q}}) - \bar{Q}^P(\bar{q}) = \bar{Q}^F(\bar{q},\dot{\bar{q}},\bar{u}), \qquad (1.3.17)$$

where

$$\bar{K} = (K_1,\ldots,K_n)^T, \bar{Q}^{D,P,F} = (Q_1^{D,P,F},\ldots,Q_n^{D,P,F})^T.$$

The kinetic energy of our case is the square form (1.3.4), which is positive definite, that is, vanishes only when all the joints are at rest: $\dot{q}_i = 0, i = 1,\ldots,n$, while the corresponding generalized inertia matrix $A(\bar{q})$ has coefficients satisfying the Sylvester criteria. It means that its determinant and all the diagonal minors are *positive*. Hence, A is also nonsingular. Then, as (1.3.15) is linear in accelerations, we can solve it for $\ddot{q}_i$. The inverse matrix $A^{-1}(\bar{q})$ has coefficients $a_{ij}^{-1} = A_{ji}/\det A$, $j, i = 1,\ldots,n$, where A_{ji} is a cofactor to the element a_{ji} of the determinant of the matrix A. Compare here (1.1.19) and the substitutions that follow in Section 1.1. Then we multiply each of the n equations (1.3.15) by a_{ij}^{-1} and sum them up to obtain

$$\ddot{q}_i + \sum_j a_{ij}^{-1}(K_j - Q_j^D - Q_j^P) = \sum_j a_{ij}^{-1} Q_j^F, \qquad (1.3.18)$$

remembering that

$$\sum_i a_{ki} a_{ij}^{-1} = \frac{1}{\det A} \sum_i a_{ki} A_{ji} = \delta_{kj} = \begin{cases} 0, & k \neq j, \\ 1, & k = j. \end{cases}$$

We now introduce the following functions called the *force characteristics*:

centrifugal-Coriolis characteristics,

$$\Gamma_i(\bar{q},\dot{\bar{q}}) = \sum_j a_{ij}^{-1} K_j(\bar{q},\dot{\bar{q}}), \qquad (1.3.19)$$

damping characteristics,

$$D_i(\bar{q},\dot{\bar{q}}) = -\sum_j a_{ij}^{-1} Q_j^D(\bar{q},\dot{\bar{q}}), \qquad (1.3.20)$$

potential characteristics,
$$\Pi_i(\bar{q}) = -\sum_j a_{ij}^{-1} Q_j^P(\bar{q}), \tag{1.3.21}$$

and *input* characteristics,
$$F_i(\bar{q}, \dot{\bar{q}}, u_i) = \sum_j a_{ij}^{-1} Q_j^F(\bar{q}, \dot{\bar{q}}, u_j), \tag{1.3.22}$$

where $u = (u_1, \ldots, u_n)^T$.

Moreover, we may often need to separate gravity forces from elastic in Q_i^P, thus it is convenient to write
$$\Pi_i(\bar{q}) = G_i(\bar{q}) + \Psi_i(\bar{q}) \tag{1.3.23}$$

the first term being the *gravity characteristic*, the second the *restoring (spring) forces characteristic*. We may now rewrite (1.3.15) in the following Newtonian form
$$\ddot{q}_i + \Gamma_i(\bar{q}, \dot{\bar{q}}) + D_i(\bar{q}, \dot{\bar{q}}) + G_i(\bar{q}) + \Psi_i(\bar{q}) = F_i(\bar{q}, \dot{\bar{q}}, u_i), \tag{1.3.24}$$

[see (1.1.27)], or vectorially
$$\ddot{q} + \bar{\Gamma}(\bar{q}, \dot{\bar{q}}) + \bar{D}(\bar{q}, \dot{\bar{q}}) + \bar{G}(\bar{q}) + \bar{\Psi}(\bar{q}) = \bar{F}(\bar{q}, \dot{\bar{q}}, \bar{u}) \tag{1.3.25}$$

where $\bar{\Gamma} = (\Gamma_1, \ldots, \Gamma_n)^T$, $\bar{D} = (D_1, \ldots, D_n)^T$, $\bar{G} = (G_1, \ldots, G_n)^T$, $\bar{\Psi} = (\Psi_1, \ldots, \Psi_n)^T$, $\bar{F} = (F_1, \ldots, F_n)^T$, with the following meaning of these vectorial characteristics [cf. (1.3.17)]:

$$\begin{aligned}\bar{\Gamma}(\bar{q}, \dot{\bar{q}}) &= A^{-1} \bar{K}(\bar{q}, \dot{\bar{q}}), \\ \bar{D}(\bar{q}, \dot{\bar{q}}) &= -A^{-1} \bar{Q}^D(\bar{q}, \dot{\bar{q}}), \\ \bar{\Pi}(\bar{q}) &= \bar{G}(\bar{q}) + \bar{\Psi}(\bar{q}) = -A^{-1} \bar{Q}^P(\bar{q}), \\ \bar{F}(\bar{q}, \dot{\bar{q}}, u) &= A^{-1} \bar{Q}^F(\bar{q}, \dot{\bar{q}}, \bar{u}).\end{aligned} \tag{1.3.26}$$

Observe that, since $\det A$ is nonvanishing and positive, the characteristics (1.3.19)–(1.3.22) as well as their vectorial counterparts (1.3.26), have the same domain of definition as the corresponding forces that generate them. Moreover, they have *at least* the same zero points and the points of changing their sign, obviously with the sign itself dependent on $a_{ij}^{-1}(\bar{q})$ as well.

Whittaker [1] gives the condition by which we may check whether the equations (1.3.24) may be equivalently written in the inertially coupled form (1.3.11). Whittaker proves that if the system
$$\ddot{q}_i + \Phi_i(\bar{q}, \dot{\bar{q}}) + \Pi_i(\bar{q}) = 0, \qquad i = 1, \ldots, n, \tag{1.3.27}$$

with Φ_i homogeneous of degree two in velocities, is reducible to the form

1.3 GENERALIZED COORDINATES. LAGRANGIAN MODEL

(1.3.8), then T must be an integral of the system

$$\frac{1}{2}\sum_j \frac{\partial(\Phi_i + \Pi_i)}{\partial \dot{q}_i}\frac{\partial T}{\partial \dot{q}_j} + \frac{\partial T}{\partial q_i} = 0, \qquad i = 1,\ldots,n, \tag{1.3.28}$$

where, as denoted,

$$\Phi_i(\bar{q},\dot{\bar{q}}) \triangleq \Gamma_i(\bar{q},\dot{\bar{q}}) + D_i(\bar{q},\dot{\bar{q}}). \tag{1.3.29}$$

We may check the consistency of our Newtonian form by substituting Γ_i, D_i, Π_i, and T into (1.3.28) (see Section 1.6).

The best method to obtain the Lagrange equations in particular cases is that which is commonly employed in manipulator dynamics and explained in detail by Paul [1]. We use the homogeneous transformations (see Section 1.1) $\bar{r}_{0i} = T_i \bar{r}_{0i-1}$, $T_i = A \cdots A_i$, and their derivatives to calculate both kinetic and potential energies. Then, substituting them into the Lagrange equations, we find the kinetic forces which must be equal to the applied forces. We illustrate the calculations in Section 1.6.

For simulation purposes, the applied forces are known. For the open loop control, all the forces are calculated from assumed desired $\bar{q}(t)$, $\dot{\bar{q}}(t)$ wherefrom the controller is found by making it equal to the sum of such forces (see Section 3.1); for more advanced types of control (feedback, adaptive, optimal) and for more complicated models (including damping, elastic forces, etc.), one needs the Lagrangian equations as in the simulation case: all the applied forces physically determined while the open loop (forward) part of the controller is obtained by making the resultant equal to the kinetic forces (see Section 3.1 and further text).

We have derived the manipulator equations in a manner that is convenient for applying the control theory. It is also natural to use the Lagrangian model with joint coordinates, which can be directly measured by the manipulator itself and then also directly controlled. However, there obviously are other methods like the popular Newton–Euler method (see for comparison an excellent review by Silver [1]), the Kane's method (see Kane and Levinson [1] and Kane, Likins, and Levinson [1]), and the computer aided modeling developed by the Vukobratovic group (see Vukobratovic and Potkonjak [1]), etc.

Exercises and Comments

1.3.1 Consider a 3R manipulator shown in Fig. 1.4 with the masses m_1, m_2, m_3 and the moment of inertia J_1 located as marked. It is assumed to be subject to gravity forces and linear damping in joints only. The case could be the study of the first three joints of a PUMA robot in which $d_0 = 66$ cm, $d_1 = d_2 = 43$ cm, $r_1 = r_2 = 21.6$ cm, $m_1 = 77.3$ kg, $m_2 = 36.3$ kg, $m_3 = 12.3$ kg. Introducing the radius vectors $\bar{r}_{0i} = \bar{R}_0 + \bar{R}_i$, $i = 1, 2, 3$, calculate the potential

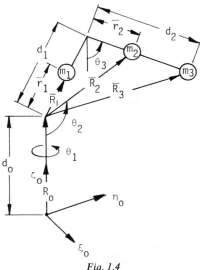

Fig. 1.4

and kinetic energies and derive the Lagrange equations (1.3.13) as well as the Newtonian forms (1.3.15), (1.3.24). The vectors $\bar{R}_j$, $j = 0, 1, 2, 3$, are available from the geometry, and $\bar{r}_{0i}$, $i = 1, 2, 3$, are calculated in terms of the A_i matrices.

1.3.2 From basic mechanics (see, for instance, Meirovitch [1]), derive the extended least action principle

$$\int_{t_0}^{t_f} (\delta T + \delta W)\, dt = 0,$$

and using (1.3.9) show its equivalent for Cartesian and joint coordinate representation of the manipulator dynamics. Illustrate the case on the system in Exercise 1.3.1. The principle gives the formal basis for the transformation (1.3.9).

1.4 WORK REGIONS. TARGET AND OBSTACLES

Formally, for suitable smooth functions Γ_i, D_i, G_i, Ψ_i, and F_i, the equations (1.3.24) have uniquely determined solutions through each point in some bounded *admissible* subset of the space R^N. This way they determine the motion of the representative point $(\bar{q}(t), \dot{\bar{q}}(t))$, $t \geq t_0$, and hence also the motion of our manipulator. The conditions allowing the above will be discussed in Section 1.5. At the moment we assume that the admissible subset is at least as

1.4 WORK REGIONS. TARGET AND OBSTACLES

large as the region in R^N where the motion is physically possible, and make an attempt to estimate this region.

As indicated in Section 1.1, the reaching ability of the joints, particularly of the gripper, expressed in terms of stop-limits on $\bar{q}, \dot{\bar{q}}$ is an essential problem in planning the work, and thus in designing the manipulator. The so-called *work space* in $0\xi_0\eta_0\zeta_0$ is the volume of space comprising all points that a *reference point* attached to the free end of a manipulator can reach (see Fig. 1.5). This point is posed at the center of the gripper, the tip of it or a so-called wrist point of the manipulator. We identify it with the position m_i at the end of some (or

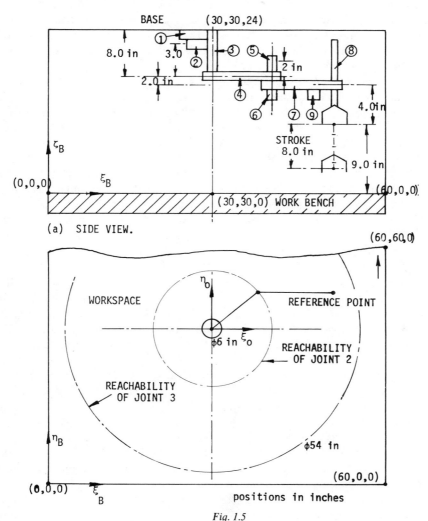

Fig. 1.5

several) free link. It is convenient if the work space is simply connected so that a target can be approached by the reference point from an arbitrary direction. If this is not possible, we at least attempt to have a nonvoid portion of the work space having the above property, usually called the *primary* or dextrous work space. The remainder is then classified as a *secondary* work space, with limited orientations of the gripper available at its points.

The primary work space is illustrated in Fig. 1.5 on a 3-joint ceiling-based RRP-manipulator working on a planar bench. It is of the type often met in the electronic assembly workshops. The inner boundary radius of the work space is 3.0 in., and the outer boundary radius is 27.0 in. The height stroke of the gripper is 8.0 in. The suspension base 0 in $0\xi_0, \eta_0, \zeta_0$ is translated from the work space coordinates $0_B \xi_B \eta_B \zeta_B$ by a constant vector $\bar{r}_{OB} = \text{const}$, which does not introduce any changes in our model since there is no motion between the base and the workbench.

While the reference point reaches, the whole arm is confronted with obstacles which may disallow the reaching. Thus the work space must be specified in terms of the all-joint *configuration representing points* $\bar{q}$ in R^n that are reachable. This is how we obtain the bounded work region Δ_q (see Section 1.1). To determine the extent of Δ_q, we may use static investigation. Under an action of a static force, the manipulator reaches a stable equilibrium configuration in which it assumes maximum extension with respect to the base, just like in Section 1.1 it was the r_{max} position, in the direction of the force. In this configuration, the reference point lies on the bounding surface of Δ_q. Since the moment of the force about each revolute joint axis must be zero in the static equilibrium configuration, the line of action of the force must intersect all the revolute joint axes. This idea underlies the algorithm for determining Δ_q (see Kumar and Waldron [1] and Desa and Roth [1]).

An alternative method for determining Δ_q directly in joint coordinates follows from the fact that the manipulator Jacobian is singular at the boundary of the work space. Placing the reference point at the two extremes of the work space, we may easily show that a pair of rows or columns of the Jacobian become linearly dependent. Figure 1.6 shows the extremal positions (inner and outer) for the manipulator of Fig. 1.5. Specifying the joint variables there and substituting into the Jacobian of rank 4, we reduce the rank. The condition is necessary only, thus it works when there are no other singularities.

The extension of Δ_q to the *state work region* $\Delta = \Delta_q \times \Delta_{\dot{q}}$ in R^N, i.e., limiting velocities within the bounded set $\Delta_{\dot{q}}$ in the tangent space, is specified by the amount of power (force × velocity) which the actuators can supply, or by an attempt to avoid "velocity obstacles" over some configurations in Δ_q (see Section 1.1) like stopping before some antitargets, having a limited speed within the target, etc. Apart from the method mentioned above for determining Δ_q, there obviously are other methods for finding both Δ_q and the corresponding $\Delta_{\dot{q}}$. We refer the reader to the literature on manipulator design

1.4 WORK REGIONS. TARGET AND OBSTACLES

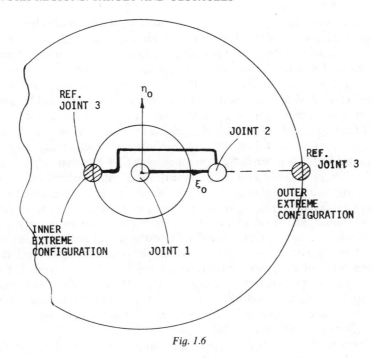

Fig. 1.6

and kinematic control quoted already, since the topic is beyond our interest in this book.

Correspondingly to the cases in $0\xi_0\eta_0\zeta_0$ we also have a simply connected *primary work region* Δ^p in Δ, and a secondary work region $\Delta - \Delta^p$ in the state space. These sets will be classified closer in Chapter 3 together with controllability and maneuverability regions. It is up to the designer to produce Δ that encloses a target, possibly placing it in Δ^p, and to exclude the obstacles. If the latter is not convenient, for instance in the case of the planned motions surrounding the obstacle and thus causing the exclusion to make all of Δ secondary, we leave the obstacle in Δ and use control forces or torque to generate its avoidance. In any case we call the complement of the union of obstacles to Δ the *free-to-work* region and aim at making it primary whenever possible.

It appears easier to maneuver a point $(\bar{q}(t), \dot{\bar{q}}(t))$ through Δ towards the transformed target, than to do the same maneuvering with the whole three dimensional chain of n links in $0\xi_0\eta_0\zeta_0$. For instance using configuration space methods, Lozano and Perez [1] and Brooks [1] have demonstrated important automatic path-finding algorithms for determining the optimal, primarily translational path of a polygonal object through a set of polygonal obstacles. It seems that these methods may also, with further development, be an important precursor to automated techniques that will allow robots, through artificial intelligence, to perceive roadways within suitable internally generated and stored configuration or phase-space subregions. Graphical simulations (see Red and Truong [1]), of configuration space maps provide the interactive flexibility necessary to plan manipulator tasks. The graphic cursor may be then used to select configuration and phase-space subregions suitable for work, and to "see" the required collision-free paths (see Red [1]).

Now the environmental objects: Cartesian targets and obstacles must be repositioned to R^N. As seen in Section 1.1 reaching a Cartesian target requires a certain configuration or a set of configurations of joint variables of all joints which is allowed by obstacles. Such configurations $\bar{q}$ are expressed by the configuration target $\mathfrak{T}_q$ in Δ_q which is then augmented by a set of desired velocities that avoid the velocity obstacles (restricted actuator power) to form the state space target $\mathfrak{T}$. Consequently repositioning of the Cartesian target produces only a part of the conditions which determine the state space target. It is, however, an integrated part through the coupling between the joint coordinates, as was seen in Section 1.1 (1.1.11). Thus the procedure of repositioning as formulated above requires the inverse kinematics (1.1.10), which is very inconvenient, particularly when the Cartesian target moves and we must do the calculations on-line.

Incidentally, there is an alternative procedure which transforms only the joint variables of the gripper into the world space $0\xi_0\eta_0\zeta_0$ of the Cartesian target—the so-called *task-space* method. This approach is particularly useful when the extra 3 DOF of the gripper itself must be considered. On the other hand, when the work space is not free and we must aim at certain configurations of all joints (the state-space target) and moreover when dynamical feedback (state related) control is involved, we still finish with transforming the motion from the task space to the state space by on-line inverse transformation anyway. So the saving of effort through the task space may not be that significant.

Let us consider the repositioning of a target from Cartesian coordinates to the state space. When the Cartesian *target is at rest*, either a point or a body of any shape, we can follow the technique described in Section 1.2 and transform off-line, before the motion begins, the boundary points of the target core from $0\xi_0\eta_0\zeta_0$ to R^n using the inverse transformation (1.1.10), (1.1.11). Then we complement it by extra demand on velocities, possibly directly in terms of $\dot{\bar{q}}$ or using the inverted Jacobian (1.1.14), (1.1.15) (see Dailly [1]).

The same technique is applied to the so-called *path planning*, that is, when the Cartesian target is given in the form of a *stipulated* Cartesian path. The problem has a vast literature on transforming such a path into joint variables (see the review by Luh [1,2]). Basically the used methods apply inverse kinematic transformation on- or off-line to time-cut points on the Cartesian

1.4 WORK REGIONS. TARGET AND OBSTACLES

path to get the configuration and thus the state-space target path. Again we should remember here that the latter differs from the Cartesian path in that it involves all the joints.

The problem is different when the motion of the Cartesian target is not stipulated for some reason. For instance, because it is uncertain, as it may be the case when the target is a flexible body with certain volume. We may then use the *product-space method* discussed in Sections 4.6, and 4.7. In certain circumstances, it disposes of the inverse transformations altogether, even if the Cartesian target is given in $0\xi_0\eta_0\zeta_0$.

Suppose a Cartesian object k is represented by several moving mass points (at variable distances if it is flexible). The variable coordinates may be transformed or *directly taken* as Lagrangian coordinates $q_1^k(t), \ldots, q_r^k(t)$, usually $r \leq n$, which have so far nothing to do with the joint variables of the manipulator. They form the object's configuration vector

$$\bar{q}^k(t) \triangleq (q_1^k(t), \ldots, q_r^k(t))^T. \tag{1.4.1}$$

In case the object is a target, we may also impose some requirements on the velocity over it, this way introducing the corresponding desired $\dot{q}_1^k, \ldots, \dot{q}_r^k$. They again may come through the inverted kinematics from the Cartesian velocities, or be imposed independently. They form the generalized velocity vector $\dot{\bar{q}}^k(t) = (\dot{q}_1^k, \ldots, \dot{q}_r^k)$ of the object which, together with (1.4.1), generates the target's state $\bar{x}^k(t) = (\bar{q}^k(t), \dot{\bar{q}}^k(t))$ representing the motion of the object (target) in the object's state space R^{2r}. Here again $\bar{x}^k(t)$ has nothing to do with the manipulator state vector $\bar{x}(t)$ introduced before. Observe however that if one forms the product vector $\bar{X}^k(t) = (\bar{x}(t), \bar{x}^k(t))^T$ in the product space $R^{2(n+r)}$ and introduces a "diagonal" set

$$M^k = \{\bar{X}^k(t) \,|\, \bar{x}(t) = \bar{x}^k(t)\}, \tag{1.4.2}$$

then $\bar{X}^k(t)$ approaching M^k secures the fact that the manipulator approaches the target. When the target is a rigid body, it will have generally 6 DOF or less, hence $r \leq 6$, but apart from that, the procedure is the same. The latter case is described in Section 4.6. The flexible case is a separate study not covered by this book (see Flashner and Skowronski [1], and Skowronski [30]).

Finally, the target may be given in terms of a solution to the motion equation of some *dynamical model* which is *to be tracked*. The product-space method is applicable to this case as well, with the solution considered $x^k(t)$, $t \geq t_0$. Such targets occur as objectives in adaptive control, and we shall return to this topic in Sections 4.6 and 4.7.

When the obstacles are objects at rest, their transformation to the joint-coordinate frame is very much the same as for the target, except instead of dealing with boundary points of the core, we deal with those of the envelope. Using inverse kinematics for the velocities to be avoided as well, we obtain the avoidance set in R^N, which we try to push out of Δ. If this is not feasible, then avoidance controllers must be designed to avoid a specified *avoidance subset* $\mathscr{A}$ of the work region Δ.

For moving obstacles, the product-space method applies even better than to moving targets. Instead of the product vector $\bar{X}^k(t)$ chasing M^k (k for given obstacle), it must avoid this "diagonal". Note here that the method takes care of the motion of all joints of the manipulator, which is especially important for avoidance. We return to this topic in Chapter 5.

Exercises and Comments

1.4.1 Consider the manipulator of Figs. 1.5 and 1.6 with the data given in the adjacent text. Calculate the forward and inverse kinematics including the Jacobian. Demonstrate that the Jacobian becomes singular at the boundary of the work space.

1.4.2 For the same manipulator, derive the Lagrange equations of motion, assuming only gravity forces acting.

1.4.3 Suppose a Cartesian path is given in terms of a curve. To produce a target path in Δ:

(a) choose the total work time and subdivide it into equal intervals $\delta t = 5\text{ms}$;

(b) for ends of each interval, find the points $(\xi_{0\sigma}, \eta_{0\sigma}, \zeta_{0\sigma})$, $\sigma = 1, \ldots, n$, along the Cartesian path;

(c) transfer the points into the values $\bar{q}^\sigma$, $\sigma = 1, \ldots, n$;

(d) specify velocities at each $\bar{q}^\sigma$;

(e) assume the target path as the polynomial

$$q_{mi}(t) = a_{4i}t^4 + a_{3i}t^3 + a_{2i}t^2 + a_{1i}t + a_{0i}, \qquad i = 1, \ldots, n,$$

and differentiate it with respect to time;

(f) substituting the values $\bar{q}^\sigma$, $\dot{\bar{q}}^\sigma$ determine the coefficients.

1.5 STATE-SPACE REPRESENTATION

The standard form of the equations of motion used in control theory to govern the state vector $\bar{x}(t) = (x_1(t), \ldots, x_N(t))^T$ in the state space R^N is in the so-called *normal form* of first-order differential equations

$$\dot{x}_i = f_i(\bar{x}, \bar{u}), \qquad i = 1, \ldots, N, \qquad (1.5.1)$$

where the functions $f_i(\cdot)$, $i = 1, \ldots, N$, must be at least such as to produce solutions to (1.5.1) on Δ. It will be discussed later in this section. The state vector $\bar{x}(t)$ has been already introduced. We also have the *control vector* $\bar{u}(t) = (u_1(t), \ldots, u_n(t))^T$, $t \geq t_0$, with values in a given *closed* and *bounded* ($\equiv$ compact) set $U \subset R^n$ of *control constraints*. The control vector is to

1.5 STATE-SPACE REPRESENTATION

be determined by a feedback and adaptive signal related *control program* $\bar{P}(\bar{x}, \bar{\lambda})$ chosen to secure a required control objective for the manipulator. The program is thus to be defined by

$$u_i(t) \triangleq P_i(\bar{x}(t), \bar{\lambda}(t)), \qquad \forall t \geq t_0 \tag{1.5.2}$$

specified at each joint i, $\bar{P} \triangleq (P_1, \ldots, P_n)^T$. The vector $\bar{\lambda}(t) = (\lambda_1(t), \ldots, \lambda_l(t))^T \subset \Lambda \subset R^l$ of *adaptive parameters* bounded in a band Λ, will be defined later. In this and the next chapter, we deal with modelling (simulation) rather than control or adaptive control, thus the *controller remains passive* and we may as well *assume* P_i, λ_i *given*, or if convenient $u_i(t) \equiv 0$, $i = 1, \ldots, n$, $\lambda_k(t) \equiv 0$, $k = 1, \ldots, l$.

The normal form (1.5.1) may be vectorially written as

$$\dot{\bar{x}} = \bar{f}(\bar{x}, \bar{u}), \tag{1.5.3}$$

where $\bar{f} \triangleq (f_1, \ldots, f_N)^T$. To make most of the results in control theory applicable, we should write the manipulator equations in the above normal form. This has been the reason for our attempt to write (1.3.15) as the inertially decoupled (1.3.24). For, if this step is possible, we may *choose the state vector* $\bar{x}(t)$ by assuming $x_i \triangleq q_i$, $x_{n+i} \triangleq \dot{q}_i$, $i = 1, \ldots, n = \frac{1}{2}N$, or $\bar{x}(t) = (\bar{q}(t), \dot{\bar{q}}(t))^T$. Then (1.3.24) becomes

$$\dot{x}_i = x_{n+i},$$
$$\dot{x}_{n+i} = -\Gamma_i(\bar{x}) - D_i(\bar{x}) - G_i(x_1, \ldots, x_n)$$
$$\qquad - \psi_i(x_1, \ldots, x_n) + F_i(\bar{x}, u_i) \tag{1.5.4}$$

which is of the form (1.5.1), with $f_i = \dot{q}_i$, $f_{n+i} = -\Gamma_i - D_i - G_i - \psi_i + F_i$, $i = 1, \ldots, n = \frac{1}{2}N$. Such a subcase of (1.5.1) is called *symplectic*. Note that some of the characteristics *will include the parameters* $\lambda_1, \ldots, \lambda_l$, which thus may appear in (1.5.1), (1.5.3) as well, but we prefer to introduce them later, when needed. The equation (1.5.4) may be also written as

$$\dot{x}_i = \gamma_i(\bar{x}) + d_i(\bar{x}) + g_i(\bar{x}) + \psi_i(\bar{x}) + f_i^F(\bar{x}, u_i), \qquad i = 1, \ldots, N, \tag{1.5.5}$$

or, vectorially, as

$$\dot{\bar{x}} = \bar{\gamma}(\bar{x}) + \bar{d}(\bar{x}) + \bar{g}(\bar{x}) + \bar{\psi}(\bar{x}) + \bar{f}^F(\bar{x}, \bar{u}), \tag{1.5.6}$$

where

$$\bar{\gamma} = (\gamma_1, \ldots, \gamma_N)^T \triangleq (0, \ldots, 0, -\Gamma_1, \ldots, -\Gamma_n)^T;$$
$$\bar{d} = (d_1, \ldots, d_N)^T \triangleq (0, \ldots, 0, -D_1, \ldots, -D_n)^T;$$
$$\bar{g} = (g_1, \ldots, g_N)^T \triangleq (0, \ldots, 0, -G_1, \ldots, -G_n)^T,$$
$$\bar{\psi} = (\psi_1, \ldots, \psi_N) \triangleq (\dot{q}_1, \ldots, \dot{q}_n, -\psi_1, \ldots, -\psi_n)^T,$$
$$\bar{f}^F = (f_1^F, \ldots, f_N^F)^T \triangleq (0, \ldots, 0, F_1, \ldots, F_n)^T.$$

The above establishes our notation for particular *state-force* representations $\gamma_i(\bar{x}), d_i(\bar{x}), g_i(\bar{x}), \psi_i(\bar{x})$ and $f_i^F(\bar{x}, \bar{u})$, as well as for their functional shapes in terms of the above vectors, with the meaning of $\Gamma_i, \ldots, F_i$ determined by (1.3.19)–(1.3.23).

Observe that eliminating dt from the equations (1.5.1) and considering the points of R^N, where $f_i(\bar{x}, \bar{u}) \neq 0$, we have

$$\frac{dx_1}{f_1(\bar{x}, \bar{u})} = \cdots = \frac{dx_N}{f_N(\bar{x}, \bar{u})}, \tag{1.5.7}$$

which is called the *symmetric form* and determines the field of directions in R^N. Substituting the characteristics, Eqs. (1.5.7) become

$$\frac{dq_1}{\dot{q}_1} = \cdots = \frac{dq_n}{\dot{q}_n} = \frac{d\dot{q}_1}{-\Gamma_1 - D_1 - G_1 - \psi_1 + F_1} = \cdots$$
$$= \frac{d\dot{q}_n}{-\Gamma_n - D_n - G_n - \psi_n + F_n},$$

wherefrom we have immediately

$$\frac{d\dot{q}_i}{dq_i} = \frac{-\Gamma_i(\bar{q}, \dot{\bar{q}}) - D_i(\bar{q}, \dot{\bar{q}}) - G_i(\bar{q}) - \psi_i(\bar{q}) + F_i(\bar{q}, \dot{\bar{q}}, u_i)}{\dot{q}_i}, \quad i = 1, \ldots, n, \tag{1.5.8}$$

which is called the *phase-space* (trajectory) *equation*. It follows that the motion in R^N, $N = 2n$, is representable simultaneously (couplings!) on n projection planes $0 q_i \dot{q}_i$ selected in R^N.

In case the kinetic energy T is not the square form (1.3.4), which happens for instance if the suspension constraints of the manipulator network are nonstationary, that is, dependent explicitly on t, we will not be able to decouple (1.3.15) with respect to the inertia and write (1.3.24) leading to the form (1.5.1). In such cases, or in general when such decoupling presents difficulty perhaps only due to the complicated form of characteristics, we may use an *alternative direct* approach to writing state equations. With moving suspension constraints of the network, the transformation (1.3.1) depends on t explicitly and thus so does the kinetic energy $T(\bar{q}, \dot{\bar{q}}, t)$ which makes the Lagrangian also explicitly time dependent $L(\bar{q}, \dot{\bar{q}}, t)$. Recall (1.3.13) and introduce a new variable called *generalized momentum*

$$p_i(t) \triangleq \frac{\partial L(q, \dot{q}, t)}{\partial \dot{q}_i} = \frac{\partial T(\bar{q}, \dot{q}, t)}{\partial \dot{q}_i} = \sum_j a_{ij}(\bar{q}) \dot{q}_i, \tag{1.5.9}$$

and form the function

$$H(\bar{q}, \bar{p}, t) = \sum_i p_i \dot{q}_i - L(\bar{q}, \dot{\bar{q}}, t) \tag{1.5.10}$$

1.5 STATE-SPACE REPRESENTATION

called *Hamiltonian*, where $\bar{p} = (p_1, \ldots, p_n)^T$. The transformation using (1.5.9) is called *Legendre's transformation* in mechanics. From (1.5.10),

$$dH = \sum_i \dot{q}_i \, dp_i + \sum_i p_i \, d\dot{q}_i - dL. \tag{1.5.11}$$

But

$$dL = \sum_i \frac{\partial L}{\partial q_i} dq_i + \sum_i \frac{\partial L}{\partial \dot{q}_i} d\dot{q}_i + \frac{\partial L}{\partial t} dt$$

$$= \sum_i \frac{\partial L}{\partial q_i} dq_i + \sum_i p_i \, d\dot{q}_i + \frac{\partial L}{\partial t} dt,$$

which substituted into (1.5.11) yields

$$dH = -\sum_i \frac{\partial L}{\partial q_i} dq_i + \sum_i \dot{q}_i \, dp_i - \frac{\partial L}{\partial t} dt. \tag{1.5.12}$$

On the other hand, purely formally,

$$dH = \sum \frac{\partial H}{\partial q_i} dq_i + \sum \frac{\partial H}{\partial p_i} dp_i + \frac{\partial H}{\partial t} dt. \tag{1.5.13}$$

Comparing coefficients of (1.5.12) and (1.5.13) gives

$$\frac{\partial H}{\partial q_i} = -\frac{\partial L}{\partial q_i}, \qquad \frac{\partial H}{\partial p_i} = \dot{q}_i, \qquad \frac{\partial H}{\partial t} = -\frac{\partial L}{\partial t}. \tag{1.5.14}$$

The relation (1.5.9) implies

$$\frac{d}{dt}\left(\frac{\partial L}{\partial \dot{q}_i}\right) = \frac{dp_i}{dt} = \dot{p}_i.$$

If so, then from (1.3.13), $\dot{p}_i = Q_i^D + Q_i^F + \partial L/\partial q_i$ and in view of (1.5.14),

$$\dot{p}_i = Q_i^D + Q_i^F - \partial H/\partial q_i.$$

The latter, together with (1.5.14), produces *Hamilton's equations of motion*

$$\dot{q}_i = \frac{\partial H(\bar{q}, \bar{p}, t)}{\partial p_i},$$

$$\dot{p}_i = -\frac{\partial H(\bar{q}, \bar{p}, t)}{\partial q_i} + Q_i^D(\bar{q}, \dot{\bar{q}}) + Q_i^F(\bar{q}, \dot{\bar{q}}, u_i), \qquad i = 1, \ldots, n, \tag{1.5.15}$$

$$\frac{\partial L}{\partial t} = -\frac{\partial H}{\partial t}.$$

As mentioned, the Hamiltonian form may be used to derive the state motion equations, when T is still the square form (1.3.1), but the inertial decoupling of

(1.3.15) into (1.3.24) proves difficult. Then T is not an explicit function of t and so neither are L and H. The third equation in (1.5.15) becomes the tautology $0 = 0$ and we have

$$\dot{q}_i = \frac{\partial H(\bar{q}, \bar{p})}{\partial p_i},$$
$$\dot{p}_i = -\frac{\partial H(\bar{q}, \bar{p})}{\partial q_i} + Q_i^D(\bar{q}, \dot{\bar{q}}) + Q_i^F(\bar{q}, \dot{\bar{q}}, u_i), \quad (1.5.16)$$

which may be used to produce (1.5.1). Technically, we proceed as follows. By (1.3.4) and (1.5.9),

$$\sum_i \dot{q}_i \frac{\partial T(\bar{q}, \dot{\bar{q}})}{\partial \dot{q}_i} = \sum_i p_i \dot{q}_i = \sum_i \sum_j a_{ij} \dot{q}_i \dot{q}_j = 2T, \quad (1.5.17)$$

and the Hamiltonian function (1.5.10) becomes

$$H(\bar{q}, \bar{p}) = \sum_i p_i q_i - L(\bar{q}, \dot{\bar{q}}) = T(\bar{q}, \dot{\bar{q}}) + \mathscr{V}(\bar{q}),$$

which is the *total energy* of the system. Then again from (1.5.9),

$$\bar{p} = A(\bar{q})\dot{\bar{q}}, \quad (1.5.18)$$

and

$$\dot{\bar{q}} = A^{-1}(\bar{q})\bar{p}. \quad (1.5.19)$$

Substituting (1.5.19) into $T(\bar{q}, \dot{\bar{q}})$, we obtain $T(\bar{q}, \bar{p})$. Indeed, by (1.5.17) and (1.5.18),

$$2T = \sum_i \sum_j a_{ij} \dot{q}_i \dot{q}_j = \sum_i p_i \dot{q}_i = \sum_i \sum_j a_{ij}^{-1} p_i p_j, \quad (1.5.20)$$

which yields

$$T(\bar{q}, \bar{p}) = \frac{1}{2} \sum_i \sum_j a_{ij}^{-1} p_i p_j. \quad (1.5.21)$$

Then we immediately have $H(\bar{q}, \bar{p}) = T(\bar{q}, \bar{p}) + \mathscr{V}(\bar{q})$. Similarly, substituting (1.5.19) into $Q_i^{D,F}$, we obtain $Q_i^D(\bar{q}, \bar{p})$, $Q_i^F(\bar{q}, \bar{p}, u_i)$.

With the above, Eqs. (1.5.16) become

$$\dot{q}_i = \frac{\partial T(\bar{q}, \bar{p})}{\partial p_i} = \sum_j a_{ij}^{-1}(\bar{q}) p_j,$$
$$\dot{p}_i = -\frac{\partial T(\bar{q}, \bar{p})}{\partial q_i} + Q_i^P(\bar{q}) + Q_i^D(\bar{q}, \bar{p}) + Q_i^F(\bar{q}, \bar{p}, u_i), \quad i = 1, \ldots, n. \quad (1.5.16')$$

Now choosing the state variables $x_i = q_i$, $x_{n+i} = p_i$, $i = 1, \ldots, n$, produces the state equations (1.5.1). However, they are not in the symplectic form (1.5.4),

1.5 STATE-SPACE REPRESENTATION

since the first n equations each relate to the vector $\bar{p}$ rather than the corresponding component p_i. The detailed calculation leading to (1.5.16′) is shown on the example of a unit manipulator in the next section (Section 1.6).

If we want to write the motion equations in the control form (1.5.1), the type (1.5.16′) differs from (1.5.4) in that the penalizing multiplication by $A^{-1}(\bar{q})$ is shifted from the forces Q_i, K_i to the velocity equations. Note that the nonpotential force vectors $\bar{Q}^D$, $\bar{Q}^F$ still have to be multiplied by A^{-1} because of the transformation (1.5.19) used to obtain $\bar{Q}^{D,F}(\bar{q}, \bar{p})$.

We can write (1.5.16′) in the form (1.5.5), (1.5.6) obviously adjusting the state-force representation vectors $\bar{\gamma}$, $\bar{d}$, $\bar{g}$, $\bar{\psi}$, and $\bar{f}^F$. Should we now wish to return to (1.5.4), we may resubstitute (1.5.18). Again the calculation is shown in Section 1.6.

Having dealt with the shapes of (1.5.1), let us suppose now that we wish to solve it. To investigate the case, we need the following notions.

We say that a function $\bar{f}(\cdot)$ satisfies *locally* the *Lipschitz condition* with respect to $\bar{x}$ in Δ, if and only if for any closed and bounded subset Δ^L of Δ there is a constant $K(\Delta^L)$ such that $|\bar{f}(\bar{x}) - \bar{f}(\bar{y})| \leq K|\bar{x} - \bar{y}|$ for all $\bar{x}, \bar{y} \in \Delta$. The latter is always satisfied for $f(\cdot)$ continuously differentiable.

Let us now assume that a given control program (1.5.2) has been substituted to the state equations (1.5.1). No matter where the equations (1.5.1) came from, either the Lagrange or the Hamiltonian equations, the functions $f_i(\cdot)$ together with their additive terms $\gamma_i(\cdot), \ldots, f_i^F(\cdot)$ must be at least locally Lipschitz continuous on the work region $\Delta \subset R^N$ in order to secure at each initial state $\bar{x}^0 = \bar{x}(t_0)$, t_0 initial instant, of this region, given $\bar{u}(\cdot)$, a unique solution $k(\bar{u}, \bar{x}^0, \cdot): R \to \Delta$ to (1.5.3), identified at $k(\bar{u}, \bar{x}^0, t_0) = \bar{x}^0$, $\forall t_0 \in R$. The control program (1.5.2) which generates such $\bar{u}(\cdot)$ is called *admissible*. We make the blanket assumption that *all control programs discussed in this text are admissible*, unless otherwise stated.

Note that the solutions do not depend on the initial instant t_0 explicitly, or as we say later, are t_0-*uniform*: if $k(\bar{u}, \bar{x}^0, t)$ is a solution of (1.5.3), then for any $t_0 \in R$, $k(\bar{u}, \bar{x}^0, t - t_0)$ is also a solution, since (1.5.3) remains unchanged under the translation [check by substituting both into (1.5.3)]. Hence it may be convenient to take $t_0 = 0$. We shall indicate our intention of doing so by showing $\bar{x}^0 = \bar{x}(0)$. The above property of t_0-independence of the solutions made some control theorists call the system "time invariant".

The solution curve in Δ denoted $k(\bar{u}, \bar{x}^0, R)$ is called a *trajectory* in this set. Observe that the $\bar{x}^0$ family of trajectories over Δ gives a first integral of (1.5.3). In topological dynamics, it is obtained as a homeomorphic map (one-to-one, onto, and continuous) of R^N into itself and called a dynamical system. Its basic feature is the so-called *group property*: $k(\bar{u}, k(\bar{u}, \bar{x}^0, t_1), t_2) = k(\bar{u}, \bar{x}^0, t_1 + t_2)$ for all $t_1, t_2 \in R$, which makes each trajectory retraceable for $-t$. Another feature is that some subsets $\tilde{M}$ of Δ may be *invariant* under the map described: $\bar{x}^0 \in \tilde{M} \Rightarrow k(\bar{u}, \bar{x}^0, R) \subset \tilde{M}$ and one of the minimal invariant sets is the *singular trajectory* specified by

$$k(\bar{u}, \bar{x}^e, R) = \bar{x}^e, \quad (1.5.22)$$

where $\bar{x}^e$ is the obvious *rest position* of the system. Since trajectories are unique (not crossing) in Δ, any other trajectory may approach $\bar{x}^e$ only asymptotically.

We observe from (1.5.3) that the vector $\bar{f}$ is tangent to the trajectory at each position of the representing point $\bar{x}(t) = k(\bar{u}, \bar{x}^0, t)$. This vector slides along the trajectory forming a vector field that coincides with the direction field specified by (1.5.7). The vector field is defined everywhere in Δ except at the points where

$$\bar{f}(\bar{x}, \bar{u}) = 0, \quad (1.5.23)$$

which are termed *singular points*. The points that are not singular are called *regular*. If point $\bar{x}^e$ satisfies (1.5.23) with $\bar{u} \equiv 0$, then by (1.5.3), $k(\bar{u}, \bar{x}^e, t) = \text{const} = \bar{x}^e$, $\forall t \in R$, that is, (1.5.22). Conversely, (1.5.22) substituted into (1.5.1) makes (1.5.23). We conclude that (1.5.22) and (1.5.23) are equivalent. Since (1.5.23) defines the singular points of (1.5.1), such points coincide with the rest positions of the system (note that (1.5.23) yields $\dot{q}_i = 0$, $i = 1, \ldots, n$) producing the *equilibria*. In particular, if $\bar{x}^e = 0$, the corresponding equilibrium is called zero equilibrium and the trajectory a *zero trajectory* or *trivial solution* of (1.5.3).

Since $\dot{q}_i = 0, i = 1,\ldots,n$, generates $\ddot{q}_i = 0$ and $K_i(\bar{q}, \dot{\bar{q}}) = 0$, the equilibria in terms of (1.3.15), (1.3.24) are respectively defined by

$$\begin{aligned}\dot{q}_i &= 0, \quad Q_i^P(\bar{q}, \dot{\bar{q}}) + Q_i^P(\bar{q}) + Q_i^F(\bar{q}, \dot{\bar{q}}, u_i) = 0 \\ \dot{q}_i &= 0, \quad D_i(\bar{q}, \dot{\bar{q}}) + G_i(\bar{q}) + \Psi_i(\bar{q}) - F_i(\bar{q}, \dot{\bar{q}}, u_i) = 0, \quad i = 1,\ldots,n.\end{aligned} \quad (1.5.24)$$

Similarly in terms of (1.5.16'), the equilibria are defined by

$$p_i = 0, \quad Q_i^P(\bar{q}) + Q_i^D(\bar{q}, \bar{p}) + Q_i^F(\bar{q}, \bar{p}, u_i) = 0. \quad (1.5.25)$$

The above equations will still be simplified considerably due to assumptions made upon the applied forces and characteristics in Sections 1.7 and 1.8.

The region in Δ which forms a neighborhood of a single isolated equilibrium is called *local*, whatever its size. In particular, when it covers Δ, it will be called globally local. Any region in Δ which is not local (includes several equilibria) is called global.

Let us now observe that, with the control program (1.5.2) substituted, our system (1.5.1) is *autonomous*. Formally, it means that the functions $f_i(\cdot)$, together with their additive terms $\gamma_1(\cdot), \ldots, f_i^F(\cdot)$, are not explicitly dependent on the time t. Physically speaking, it is traditional to call the system autonomous if it includes its source of power (see Kononenko [1]). In practice, however, the measurement of the supply is possible at an output of a programming device, in our case it is the supporting minicomputer. Thus one

1.5 STATE-SPACE REPRESENTATION

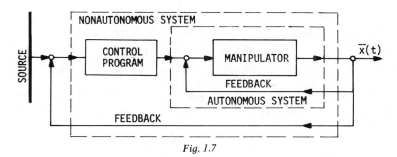

Fig. 1.7

maintains the sense of the above calling the system autonomous (see Fig. 1.7), if the control program P_i and other applied forces $Q_i^S, Q_i^G, Q_i^D, Q_i^F$ are subject to the environmental influence only through the system variables $\bar{q}, \dot{\bar{q}}$ or system parameters, so that the *system programs its own supply of power* (see Skowronski [24] and Kononenko [1]).

Such is, however, quite often not the case in real work conditions of a manipulator, particularly when working in a difficult, often unpredictable environment. We may have forces which act from outside without any influence from the system or with only partial influence. Such a system is then *nonautonomous* or *open*. Formally, it means that explicit functions of time will appear in addition or within the applied forces in (1.2.7) and thus in the generalized force $Q_j(\bar{q}, \dot{\bar{q}}, t)$ in (1.3.8) and all subsequent motion equations. For most practical purposes in the case of our manipulator model, it will be in the form of an additive *perturbation force* $Q_i^R(\bar{q}, \dot{\bar{q}}, t)$, partially feedback influenced, appearing as an extra term:

$$\sum_j a_{ij}(\bar{q})\ddot{q}_j + K_i(\bar{q}, \dot{\bar{q}}) - Q_i^D(\bar{q}, \dot{\bar{q}}) - Q_i^P(\bar{q}) = Q_i^F(\bar{q}, \dot{\bar{q}}, u_i) + Q_i^R(\bar{q}, \dot{\bar{q}}, t), \quad (1.3.15')$$

or as a *perturbation characteristic*:

$$R_i(\bar{q}, \dot{\bar{q}}, t) = \sum_j a_{ij}^{-1} Q_j^R(\bar{q}, \dot{\bar{q}}, t), \quad (1.5.26)$$

additive in (1.3.24):

$$\ddot{q}_i + \Gamma_i(\bar{q}, \dot{\bar{q}}) + D_i(\bar{q}, \dot{\bar{q}}) + G_i(\bar{q}) + \Psi_i(q) = F_i(\bar{q}, \dot{\bar{q}}, u_i) + R_i(\bar{q}, \dot{\bar{q}}, t). \quad (1.3.24')$$

Obviously we also have vectors $\bar{Q}^R = (Q_1^R, \ldots, Q_n^R)^T$ and $\bar{R} = (R_1, \ldots, R_n)^T$ additive in the right hand sides of the vectorial equations (1.3.17) and (1.3.25) respectively. Observe that adding the external perturbation does not affect the derivation of our models, either Cartesian or Lagrangian except for the changes just imposed. The comment refers also to the case when the *perturbation force is not additive* as we have assumed, but becomes an integrated part of any of the applied forces making their functions depending explicitly upon t—the phenomenon called in dynamics *parametric excitation*.

In our state equations (1.5.5) we add the perturbation by forming the state characteristic vector $\bar{f}^R \triangleq (f_1^R, \ldots, f_N^R) = (0, \ldots, 0, R_1, \ldots, R_n)^T$ which makes (1.5.6) into

$$\dot{\bar{x}} = \bar{\gamma}(\bar{x}) + \bar{d}(\bar{x}) + \bar{g}(\bar{x}) + \bar{\psi}(\bar{x}) + \bar{f}^F(\bar{x}, \bar{u}) + \bar{f}^R(\bar{x}, t). \tag{1.5.6'}$$

Finally, in the Hamiltonian equations (1.5.16), the perturbation appears as the generalized force $Q_i^R(\bar{q}, \bar{p}, t)$:

$$\dot{q}_i = \frac{\partial T(\bar{q}, \bar{p})}{\partial p_i}$$

$$\dot{p}_i = -\frac{\partial H}{\partial q_i} + Q_i^D(\bar{q}, \bar{p}) + Q_i^F(\bar{q}, \bar{p}, u_i) + Q_i^R(\bar{q}, \bar{p}, t). \tag{1.5.16''}$$

This leads to the general normal form for the nonautonomous case written as

$$\dot{x}_i = f_i(\bar{x}, \bar{u}, t), \qquad i = 1, \ldots, N, \tag{1.5.1'}$$

or, vectorially, as

$$\dot{\bar{x}} = \bar{f}(\bar{x}, \bar{u}, t). \tag{1.5.3'}$$

We use the same notation for $f_i, \bar{f}$ as in (1.5.1), (1.5.3) in the belief that changing it would cause more confusion than leaving it alone. We assume that the display of arguments of the function in each case may make the difference visible. The parameters $\bar{\lambda}$ are hidden: at present $\bar{\lambda}(t) \equiv 0$.

The solutions to (1.5.3') will not be unique in Δ, they may cross each other. To secure their uniqueness, we must *geometrize* the parametric representation $\bar{x}(t)$ in Δ by looking at points $(\bar{x}, t) = (x_1, \ldots, x_N, t)$ called *events* in $\Omega \triangleq \Delta \times R$, in the *space of events* R^{N+1}. Granted that the control program (stationary or not) is substituted, the functions $f_i(\bar{x}, \bar{u}, t)$ together with their additive terms $\gamma_i(\cdot), \ldots, f_i^R(\cdot)$ must be at least locally Lipschitz continuous in $\bar{x}$ and measurable in t on Ω in order to produce at each $(\bar{x}^0, t_0) \in \Omega$ a unique solution $k(\bar{u}, \bar{x}^0, t_0, \cdot): R_0^+ \to \Omega$ to (1.5.3'). Here we denote $R_0^+ \triangleq [t_0, \infty) \subset R$. Later we may also need $R_0^f \triangleq [t_0, t_f]$, where $t_f < \infty$ is a terminal instant. Now the solutions do depend upon the initial instant t_0, and the corresponding curves in Ω are called *motions*. Again the $\bar{x}^0, t_0$ family of motions in Ω forms a first integral to (1.5.3') and even a dynamical system in Ω, but when projected down to Δ exhibits none of the nice properties of the autonomous system. The motions are not unique there; that is, they may intersect. However, the vector $\bar{f}(\bar{x}, \bar{u}, t)$ is obviously still tangent to the motion and forms a vector field with singularities. We shall assume that

$$\bar{f}(\bar{x}^e, 0, t) = 0 \qquad \forall t \geq t_0, \tag{1.5.27}$$

specifying equilibria, means that $\bar{x} = \bar{x}^e = $ const is a solution. It is a half-line in Ω and a point in Δ.

Exercises and Comments

1.5.1 Equation (1.5.8) represents the instantaneous slope of the projection of the trajectory on the $0q_i\dot{q}_i$ plane. Find the slope at regular points where this projection intersects the $\dot{q}_i$ axis.

1.5.2 Show that $\bar{f}(\bar{x},t)$ is Lipschitz continuous on some Δ if it is continuous in $(\bar{x},t)$ and linear in $\bar{x}$ on this set. Give examples of continuous functions which are not Lipschitz continuous. Give examples of Lipschitz functions for which "universal" (in contrast to local) Lipschitz constants exist.

1.5.3 If $0 < n < 1$, show that the normal form $\dot{x} = |x|^n$, $x(0) = 0$, has infinitely many solutions. In fact, show that for any $c > 0$ there exists a solution satisfying $x = 0$ for $t \in [0,c]$ and $x \neq 0$ for $c < t$. Note that the right-hand side is continuous but not Lipschitz continuous near $x(0) = 0$.

Generalize the above to $\dot{x} = f(x)$, $x(0) = 0$, with f continuous and positive for $x \neq 0$ and $f(0) = 0$.

1.5.4 Describe the difference in $\Delta \times R \subset R^{N+1}$ between the motions of $\dot{\bar{x}} = \bar{f}(\bar{x},t)$ and $\dot{\bar{x}} = \bar{f}(\bar{x})$. Can unique motions of $\dot{\bar{x}} = \bar{f}(\bar{x},t)$ be mapped from R^{N+1} into unique trajectories in R^N?

1.5.5 Consider the system $\ddot{q} + \omega^2 \sin q = 0$ and find all the equilibria and singular trajectories. Sketch the state-space pattern.

1.5.6 Is the system $\ddot{q} + \Phi(q,\dot{q},t) + \Psi(q,t) = 0$ autonomous?

1.6 THE MODULAR MANIPULATOR REVISITED

We illustrate our dynamic simulation on a unit manipulator slightly modified with respect to that of Section 1.1. It is the PR-type (instead of RP as in Section 1.1), ceiling suspended on a spring, with planar operation and with the gripper working along a horizontal bench. The mass centers of amalgamated links and joints, that is, the positions of m_i are calculated and positioned at a fixed distance on the link axis. We also recognize the contribution to inertia by actuators separately. The structural details, dimensions and the kinematic arrangement, including the coordinate frames, is shown in Fig. 1.8. Following it, we have $\bar{r}_{01} = x_1 \bar{i}$, $\bar{r}_{02} = (x_1 + l_1 + l_2 \cos\theta)\bar{i} + (l_2 \sin\theta)\bar{j}$, and thus $\dot{\bar{r}}_{01} = \dot{x}_1 \bar{i}$, $\dot{\bar{r}}_{02} = (\dot{x}_1 - l_2 \sin\theta \cdot \dot{\theta})\bar{i} + (l_2 \cos\theta \cdot \dot{\theta})\bar{j}$.

We calculate the kinetic energy for a particular m_i:

$$T_1 = \tfrac{1}{2} m_1 \dot{\bar{r}}_{01} \cdot \dot{\bar{r}}_{01} = \tfrac{1}{2} m_1 \dot{q}_1^2,$$

$$T_2 = \tfrac{1}{2} m_2 \dot{\bar{r}}_{02} \cdot \dot{\bar{r}}_{02} = \tfrac{1}{2} m_2 [(\dot{q}_1 - l_2 \dot{q}_2 \sin q_2)^2 + (l_2 \dot{q}_2 \cos q_2)^2]$$

$$= \tfrac{1}{2} m_2 (\dot{q}_1^2 - 2 l_2 \dot{q}_1 \dot{q}_2 \sin q_2 + l_2^2 \dot{q}_2^2),$$

and then for the actuators

$$T_{1a} = \tfrac{1}{2} m_{a1} \dot{q}_1^2, \qquad T_{2a} = \tfrac{1}{2} I_{a2} \dot{q}_2^2,$$

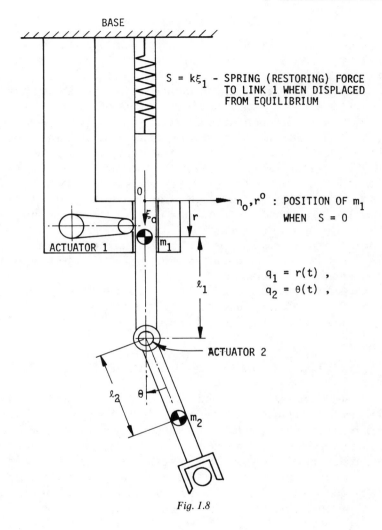

Fig. 1.8

where I_{a2} is the rotary actuator inertia coefficient. Hence, the total kinetic energy for the manipulator is

$$T = \tfrac{1}{2}(m_1 + m_{a1} + m_2)\dot{q}_1^2 - m_2 l_2 \dot{q}_1 \dot{q}_2 \sin q_2 + \tfrac{1}{2}(m_2 l_2^2 + I_{a2})\dot{q}_2^2, \quad (1.6.1)$$

or in matrix-vector form $T = \tfrac{1}{2}\dot{\bar{q}}^T A(q)\dot{\bar{q}}$, where

$$A(\bar{q}) = \begin{pmatrix} m_1 + m_{a1} + m_2 & -m_2 l_2 \sin q_2 \\ -m_2 l_2 \sin q_2 & m_2 l_2^2 + I_{a2} \end{pmatrix} \quad (1.6.2)$$

1.6 THE MODULAR MANIPULATOR REVISITED

is the inertia matrix of our case. Similarly, we calculate the potential energy to obtain

$$\mathscr{V} = \tfrac{1}{2}kq_1^2 - 9.81(m_1 + m_2)q_1 - 9.81m_2(l_2\cos q_2 + l_1). \tag{1.6.3}$$

Hence the Lagrangian $L = T - \mathscr{V}$ is

$$\begin{aligned}L = \tfrac{1}{2}a_{11}\dot{q}_1^2 + a_{12}(q_2)\dot{q}_1\dot{q}_2 + \tfrac{1}{2}a_{22}\dot{q}_2^2 + \tfrac{1}{2}kq_1^2 - 9.81(m_1 + m_2) \\ - 9.81m_2(l_2\cos q_2 + l_1)\end{aligned} \tag{1.6.4}$$

where a_{ij} are the coefficients of the matrix $A(\bar{q})$ of (1.6.2).

Introducing the damping forces in each joint as

$$Q_i^D = \lambda_i |\dot{q}_2| \dot{q}_i, \qquad \dot{q}_i \neq 0, \quad i = 1, 2,$$

and substituting all the above into (1.3.13) we obtain the motion equations

$$\begin{aligned}(m_1 + m_{a1} + m_2)\ddot{q}_1 - m_2 l_2 \ddot{q}_2 \sin q_2 - m_2 l_2 \dot{q}_2^2 \cos q_2 + kq_1 \\ - 9.81(m_1 + m_2) + \lambda_1 |\dot{q}_1|\dot{q}_1 = Q_1^F \\ (m_2 l_2^2 + I_{a2})\ddot{q}_2 - m_2 l_2 \ddot{q}_1 \sin q_2 - 9.81 m_2 l_2 \sin q_2 \\ + \lambda_2 \dot{q}_2 |\dot{q}_2| = Q_2^F\end{aligned} \tag{1.6.5}$$

or in vector form (1.3.17)

$$\begin{pmatrix} m_1 + m_{a1} + m_2 & -m_2 l_2 \sin q_2 \\ -m_2 l_2 \sin q_2 & m_2 l_2^2 + I_{a2} \end{pmatrix} \begin{pmatrix} \ddot{q}_1 \\ \ddot{q}_2 \end{pmatrix} + \begin{pmatrix} \lambda_1 |\dot{q}_1|\dot{q}_1 \\ \lambda_2 |\dot{q}_2|\dot{q}_2 \end{pmatrix} \\ + \begin{pmatrix} kq_1 - 9.81(m_1 + m_2) \\ -9.81 m_2 l_2 \sin q_2 \end{pmatrix} = \begin{pmatrix} Q_1^F \\ Q_2^F \end{pmatrix}, \tag{1.6.6}$$

which is far more difficult to decouple inertially than the equations in Section 1.1. On the other hand, such decoupling is needed to write (1.6.6) in the state form (1.5.3). To do so we have the alternative of using the Hamiltonian form (1.5.16). From (1.5.9) and (1.6.4), we have [cf. (1.5.18)]

$$\begin{aligned}p_1 = \frac{\partial L}{\partial \dot{q}_1} = a_{11}\dot{q}_1 + a_{12}(q_2)\dot{q}_2, \\ p_2 = \frac{\partial L}{\partial \dot{q}_2} = a_{12}(q_2)\dot{q}_1 + a_{22}\dot{q}_2,\end{aligned} \tag{1.6.7}$$

with a_{ij} specified by (1.6.2). Then also [cf. (1.5.19)]

$$\dot{q}_1 = \frac{a_{22}p_1 - a_{12}(q_2)p_2}{\det A}, \qquad \dot{q}_2 = \frac{-a_{12}(q_2)p_1 + a_{11}p_2}{\det A}. \tag{1.6.8}$$

Substituting (1.6.8) into (1.6.1), we have

$$T = \frac{1}{2(\det A)^2}\left\{a_{11}\left[\det\begin{pmatrix}p_1 a_{12}\\p_2 a_{22}\end{pmatrix}\right]^2 + 2a_{12}(q_2)\det\begin{pmatrix}p_1 a_{12}\\p_2 a_{22}\end{pmatrix}\det\begin{pmatrix}a_{11} p_1\\a_{12} p_2\end{pmatrix}\right. \\ \left. + a_{22}\left[\det\begin{pmatrix}a_{12} p_1\\a_{12} p_2\end{pmatrix}\right]^2\right\},$$

which gives [cf. (1.5.21)]

$$T(\bar{q},\bar{p}) = \tfrac{1}{2}a_{11}^{-1}p_1^2 + a_{12}^{-1}p_1 p_2 + \tfrac{1}{2}a_{22}^{-1}p_2^2, \qquad (1.6.9)$$

where

$$\begin{aligned}a_{11}^{-1} &= a_{22}/\det A,\\ a_{12}^{-1} &= -a_{12}/\det A, \\ a_{22}^{-1} &= a_{11}/\det A.\end{aligned} \qquad (1.6.10)$$

Then, recalling that $H(\bar{q},\bar{p}) = T(\bar{q},\bar{p}) + \mathscr{V}(\bar{q})$, we have

$$\frac{\partial H}{\partial p_1} = a_{11}^{-1}p_1 + a_{12}^{-1}p_2, \qquad \frac{\partial H}{\partial p_2} = a_{12}^{-1}p_1 + a_{22}^{-1}p_2, \qquad (1.6.11)$$

$$\frac{\partial H}{\partial q_1} = \frac{\partial \mathscr{V}}{\partial q_1} = kq_1 - 9.81(m_1 + m_2),$$

$$\frac{\partial H}{\partial q_2} = \frac{\partial T}{\partial q_2} + \frac{\partial \mathscr{V}}{\partial q_2} = m_2 l_2(9.81\sin q_2 - \dot{q}_1\dot{q}_2\cos q_2),$$

and also for our specified damping forces

$$Q_1^D(\bar{q},\bar{p}) = \lambda_1[a_{22}p_1 - a_{12}(q_2)p_2]\cdot|a_{22}p_1 - a_{12}(q_2)p_2|, \qquad (1.6.12)$$
$$Q_2^D(\bar{q},\bar{p}) = \lambda_2[a_{12}(q_2)p_1 - a_{11}p_2]\cdot|a_{12}(q_2)p_1 - a_{11}p_2|. \qquad (1.6.13)$$

Given $Q_i^F(\bar{q},\dot{\bar{q}})$, $i = 1,2$, we can transform them exactly in the same way. The above gives the Hamilton canonical equations (1.5.16') for our case as

$$\begin{aligned}\dot{q}_1 &= a_{11}^{-1}(q_2)p_1 + a_{12}^{-1}(q_2)p_2,\\ \dot{q}_2 &= a_{12}^{-1}(q_2)p_1 + a_{22}^{-1}(q_2)p_2,\\ \dot{p}_1 &= kq_1 - 9.81(m_1 + m_2) - Q_1^D(\bar{q},\bar{p}) + Q_1^F,\\ \dot{p}_2 &= m_2 l_2(9.81\sin q_2 - \dot{q}_1\dot{q}_2\cos q_2) - Q_2^D(\bar{q},\bar{p}) + Q_2^F,\end{aligned} \qquad (1.6.14)$$

where $Q_i^D(\bar{q},\bar{p})$, $i = 1,2$, are specified by (1.6.12) and (1.6.13), respectively. We may now let $x_1 \triangleq q_1$, $x_2 \triangleq q_2$, $x_3 \triangleq p_1$, $x_4 \triangleq p_2$ which makes (1.6.14) into the

1.6 THE MODULAR MANIPULATOR REVISITED

state equations

$$\begin{aligned}
\dot{x}_1 &= a_{11}^{-1}(x_2)x_3 + a_{12}^{-1}(x_2)x_4, \\
\dot{x}_2 &= a_{12}^{-1}(x_2)x_3 + a_{22}^{-1}(x_2)x_4, \\
\dot{x}_3 &= -kx_1 + 9.81(m_1 + m_2) - Q_1^D(x_1,\ldots,x_4) + Q_1^F, \\
\dot{x}_4 &= -9.81 m_2 l_2 \sin x_2 - Q_2^D(x_1,\ldots,x_4) + Q_2^F,
\end{aligned} \quad (1.6.15)$$

with the coefficients $a_{11}^{-1}, a_{12}^{-1}, a_{22}^{-1}$ given by (1.6.10) and the functions $Q_i^D(\bar{x})$ obtained from (1.6.12) and (1.6.13) by substituting the transformation $\bar{q}, \bar{p} \to \bar{x}$.

Note that (1.6.15) has the general form of (1.5.1) but not the same state vector and not the same specified meaning of the functions f_i as in the symplectic form (1.5.4). In order to make (1.6.14) into the shape of (1.5.4), we would have to substitute (1.6.7) back into it.

Chapter 2

Force Characteristics, Energy, and Power

2.1 POTENTIAL FORCES. COUPLING CHARACTERISTICS

Since the potential energy is the negative of potential, the generalized potential forces are defined by the partial derivatives (1.3.12)

$$Q_i^P(\bar{q}) = -\frac{\partial \mathscr{V}(\bar{q})}{\partial q_i}, \qquad i = 1,\ldots,n, \qquad (2.1.1)$$

whence the potential energy is specified by

$$\mathscr{V}(\bar{q}) = \mathscr{V}(\bar{q}^0) - \sum_i \int_{q_i^0}^{q_i} Q_i^P(\bar{q}) dq_i. \qquad (2.1.2)$$

The function $\mathscr{V}(\cdot)$ is assumed to be defined at least on Δ_q, single valued, and smooth to the degree required by (2.1.2) and whatever smoothness is needed for $Q_i^P(\cdot)$. Thus, in most cases $\mathscr{V}(\cdot)$ is taken as analytic.

By (2.1.1) and (1.3.21) the potential characteristics become

$$\Pi_i(\bar{q}) = \sum_j a_{ij}^{-1}(\bar{q}) \frac{\partial \mathscr{V}(\bar{q})}{\partial q_j}, \qquad (2.1.3)$$

and thus also

$$\Pi_i(\bar{q}) = -\sum_j a_{ij}^{-1}(\bar{q}) \frac{\partial \mathscr{U}_p}{\partial q_j}. \qquad (2.1.4)$$

Observe moreover that

$$\Pi_i(\bar{q}) = -\sum_j a_{ij}^{-1}(\bar{q}) \cdot Q_j^P(\bar{q}) = \frac{-1}{\det A} \sum_j A_{ji}(\bar{q}) \cdot Q_j^P(\bar{q}),$$

which shows that physically Π_i represents the negative of the ratio of potential forces per inertia, or shall we say, *force per mass*, measured by acceleration of a conservative system.

Alternatively to the latter observation, we have

$$\Pi_i(\bar{q}) = -\bar{A}^i(\bar{q}) \cdot \bar{Q}^P(\bar{q}), \qquad (2.1.5)$$

2.1 POTENTIAL FORCES. COUPLING CHARACTERISTICS

where

$$\bar{A}^i = (a_{i1}^{-1}, \ldots, a_{in}^{-1}) = (1/\det A)(A_{1i}, \ldots, A_{ni}),$$

and $\bar{Q}^P = (Q_1^P, \ldots, Q_n^P)^T$ is the vector introduced in Section 1.3. Cramer's rule also gives

$$\Pi_i(\bar{q}) = -\det Q_p^i/\det A, \qquad (2.1.6)$$

where Q_p^i is a matrix which agrees with A except for the ith column replaced by the vector $\bar{Q}^P$.

The force Q_i^P is composed of gravity and spring forces

$$Q_i^P(\bar{q}) \triangleq Q_i^G(\bar{q}) + Q_i^S(\bar{q}), \qquad (2.1.7)$$

with formulas (1.3.21), (1.3.25), (2.1.5), and (2.1.6) holding for each of the two component terms in (2.1.7) separately:

$$G_i(\bar{q}) = -\sum_j a_{ij}^{-1} Q_j^G(\bar{q}), \qquad (2.1.8)$$

$$\Psi_i(\bar{q}) = -\sum_j a_{ij}^{-1} Q_j^S(\bar{q}), \qquad (2.1.9)$$

and similarly regarding (1.3.25), (2.1.5), and (2.1.6).

In turn, the functions $Q_i^G(\bar{q})$, $Q_i^S(\bar{q})$ have *special shapes due to coupling* of the particular spring forces that may act on m_i. We had a look at that in Section 1.1 [see (1.1.19), (1.1.20), (1.1.21)].

The gravity forces are in general trigonometric functions (sine or cosine) in rotary joint coordinates and linear functions in prismatic joint coordinates, coupled by the joint variables added together. The trigonometric part is obviously power series developable with some cutoff approximation (see Section 1.1).

The spring forces are quite often analytic functions as well and thus are also power series developable. Frequently, they may be represented as the sum of terms with direct action between m_i and the base $Q_{0i}^S(q_i)$ and those between links representing spring couplings $Q_{ij}^S(q_i - q_j)$, $i \neq j$. Obviously, if the neighboring joints are different $(R - P)$ one of the variables is constant (see Section 1.1).

In general, however,

$$Q_i^S(\bar{q}) = Q_{0i}^S(q_i) + \sum_j Q_{ij}^S(q_i - q_j), \qquad (2.1.10)$$

or, vectorially,

$$\bar{Q}^S = \bar{Q}^{S0}(\bar{q}) + \sum_j \bar{Q}^{Sj}(\bar{q}), \qquad (2.1.11)$$

where

$$\bar{Q}^{S0} \triangleq (Q_{01}^S, \ldots, Q_{0n}^S)^T, \qquad \bar{Q}^{Sj} \triangleq (Q_{1j}^S, \ldots, Q_{nj}^S)^T.$$

For the characteristics we have correspondingly

$$\bar{\Psi}(\bar{q}) = \bar{\Psi}^0(\bar{q}) + \sum_j \bar{\Psi}^j(\bar{q}), \qquad (2.1.12)$$

where

$$\bar{\Psi}^0(\bar{q}) = -A^{-1} \cdot \bar{Q}^{S0}(\bar{q}) \qquad (2.1.13)$$

and

$$\bar{\Psi}^j(\bar{q}) = -A^{-1} \cdot \bar{Q}^{Sj}(\bar{q}). \qquad (2.1.14)$$

Writing (2.1.12) in scalar terms, that is, as $\bar{\Psi} = (\Psi_1, \ldots, \Psi_n)$, we have

$$\Psi_i(\bar{q}) = \Psi_{0i}(\bar{q}) + \sum_j \Psi_{ij}(\bar{q}), \qquad i,j = 1,\ldots,n, \quad i \neq j, \quad (2.1.15)$$

where

$$\Psi_{0i}(\bar{q}) = -\sum_\sigma a_{i\sigma}^{-1} Q_{0\sigma}^S(q_\sigma), \qquad (2.1.16)$$

$$\Psi_{ij}(\bar{q}) = -\sum_\sigma a_{i\sigma}^{-1} Q_{i\sigma}^S(q_i - q_\sigma). \qquad (2.1.17)$$

Alternatively, using the (2.1.5)-type representation, we have

$$\Psi_{0i}(\bar{q}) = -\bar{A}^i(\bar{q}) \cdot \bar{Q}^{S0}(\bar{q}), \qquad (2.1.18)$$

$$\Psi_{ij}(\bar{q}) = -\bar{A}^i(\bar{q}) \cdot \bar{Q}^{Sj}(\bar{q}), \qquad (2.1.19)$$

where $\bar{A}^i = (a_{i1}^{-1}, \ldots, a_{in}^{-1})$. Furthermore, according to the (2.1.6)-type of representation, we have

$$\Psi_{0i}(\bar{q}) = (1/\det A) \det Q_{S0}^i, \qquad (2.1.20)$$

$$\Psi_{ij}(\bar{q}) = (-1/\det A) \det Q_{Sj}^i, \qquad (2.1.21)$$

where Q_{S0}^i, Q_{Sj}^i are matrices that agree with the matrix $A(\bar{q})$ but for the ith column which is replaced by the vectors $\bar{Q}^{S0}$ and $\bar{Q}^{Sj}$, respectively.

The example below illustrates the above notation on the case of 3 DOF system which is subject to the spring forces only in order to concentrate on our present problem.

Example 2.1.1. Consider the system

$$\begin{aligned} a_{11}\ddot{q}_1 + a_{12}\ddot{q}_2 + a_{13}\ddot{q}_3 - Q_{01}^S(q_1) - Q_{12}^S(q_1-q_2) - Q_{13}^S(q_1-q_3) &= 0 \\ a_{21}\ddot{q}_1 + a_{22}\ddot{q}_2 + a_{23}\ddot{q} - Q_{02}^S(q_2) - Q_{21}^S(q_2-q_1) - Q_{23}^S(q_2-q_3) &= 0 \quad (2.1.22) \\ a_{31}\ddot{q}_1 + a_{33}\ddot{q}_2 + a_{33}\ddot{q}_3 - Q_{03}^S(q_3) - Q_{31}^S(q_3-q_1) - Q_{32}^S(q_3-q_2) &= 0 \end{aligned}$$

2.1 POTENTIAL FORCES. COUPLING CHARACTERISTICS

which we rewrite as

$$a_{11}\ddot{q}_1+a_{12}\ddot{q}_2+a_{13}\ddot{q}_3-Q^S_{01}(q_1)-Q^S_{11}(q_1-q_1)-Q^S_{12}(q_1-q_2)-Q^S_{13}(q_1-q_3)=0$$
$$a_{21}\ddot{q}_1+a_{22}\ddot{q}_2+a_{23}\ddot{q}_3-Q^S_{02}(q_2)-Q^S_{21}(q_2-q_1)-Q^S_{22}(q_2-q_2)-Q^S_{23}(q_2-q_3)=0$$
$$a_{31}\ddot{q}_1+a_{33}\ddot{q}_2+a_{33}\ddot{q}_3-Q^S_{03}(q_3)-Q^S_{31}(q_3-q_1)-Q^S_{32}(q_3-q_2)-Q^S_{33}(q_3-q_3)=0$$
(2.1.23)

assuming $Q^S_{ii}(q_i - q_i) \equiv 0$. Then the above takes the vector form

$$A\ddot{\bar{q}} - \bar{Q}^{S0}(\bar{q}) - \sum_j \bar{Q}^{Sj}(q_1 - q_j, q_2 - q_j, q_3 - q_j) = 0, \qquad (2.1.24)$$

where

$$\bar{Q}^{S0} = (Q^S_{01}(q_1), Q^S_{02}(q_2), Q^S_{03}(q_3))^T,$$
$$\bar{Q}^{Sj} = (Q^S_{1j}(q_1 - q_j), Q^S_{2j}(q_2 - q_j), Q^S_{3j}(q_3 - q_j))^T, \qquad j = 1, 2, 3.$$

We can now write (2.1.23) in inertially decoupled shapes successively corresponding to (2.1.16)–(2.1.21). From (2.1.24),

$$\ddot{\bar{q}} - A^{-1} \cdot \bar{Q}^{S0} - A^{-1} \cdot \bar{Q}^{S1} - A^{-1} \cdot \bar{Q}^{S2} - A^{-1} \cdot \bar{Q}^{S3} = 0. \qquad (2.1.25)$$

Then, if

$$-\bar{\Psi}^0 \triangleq A^{-1}\bar{Q}^{S0} = \begin{pmatrix} a^{-1}_{11}Q^S_{01}+a^{-1}_{12}Q^S_{02}+a^{-1}_{13}Q^S_{03} \\ a^{-1}_{21}Q^S_{01}+a^{-1}_{22}Q^S_{02}+a^{-1}_{32}Q^S_{03} \\ a^{-1}_{31}Q^S_{01}+a^{-1}_{32}Q^S_{02}+a^{-1}_{33}Q^S_{03} \end{pmatrix} = \begin{pmatrix} \sum_\sigma a^{-1}_{1\sigma}Q^S_{0\sigma} \\ \sum_\sigma a^{-1}_{2\sigma}Q^S_{0\sigma} \\ \sum_\sigma a^{-1}_{3\sigma}Q^S_{0\sigma} \end{pmatrix} = \begin{pmatrix} \Psi_{01} \\ \Psi_{02} \\ \Psi_{03} \end{pmatrix},$$
(2.1.26)

$$-\bar{\Psi}^j \triangleq A^{-1}\bar{Q}^{Sj} = \begin{pmatrix} 0 & +a^{-1}_{12}Q^S_{12}+a^{-1}_{13}Q^S_{12} \\ a^{-1}_{12}Q^S_{21}+ & 0 & +a^{-1}_{23}Q^S_{23} \\ a^{-1}_{31}Q^S_{31}+a^{-1}_{32}Q^S_{32}+ & 0 \end{pmatrix} = \begin{pmatrix} \sum_\sigma a^{-1}_{1\sigma}Q^S_{1\sigma} \\ \sum_\sigma a^{-1}_{2\sigma}Q^S_{2\sigma} \\ \sum_\sigma a^{-1}_{3\sigma}Q^S_{3\sigma} \end{pmatrix} = \begin{pmatrix} \Psi_{1j} \\ \Psi_{2j} \\ \Psi_{3j} \end{pmatrix},$$
(2.1.27)

we can write (2.1.25) as

$$\ddot{\bar{q}} + \bar{\Psi}^0(\bar{q}) + \sum_j \bar{\Psi}^j(\bar{q}) = 0, \qquad (2.1.28)$$

or in scalar form of its components as

$$\ddot{q}_i + \Psi_{0i}(\bar{q}) + \sum_j \Psi_{ij}(q_i - q_j) = 0, \qquad i = 1, \ldots, n.$$

that is,

$$\ddot{q}_i - \sum_\sigma a_{i\sigma}^{-1} Q_{0\sigma}^S - \sum_j \sum_\sigma a_{i\sigma}^{-1} Q_{i\sigma}^S = 0, \qquad i = 1, \ldots, n. \qquad (2.1.29)$$

From (2.1.26), (2.1.27) we also have two other forms of the motion equations corresponding to (2.1.18), (2.1.19):

$$\ddot{q}_i - (a_{i1}^{-1}, a_{i2}^{-1}, a_{i3}^{-1})\bar{Q}^{S0} - \sum_j (a_{i1}^{-1}, a_{i2}^{-1}, a_{i3}^{-1})\bar{Q}^{Sj} = 0, \qquad (2.1.30)$$

and corresponding to (2.1.20), (2.1.21):

$$\ddot{q}_1 - \frac{1}{\det A}\left[\det\begin{pmatrix} Q_{01}^S & a_{12} & a_{13} \\ Q_{02}^S & a_{22} & a_{23} \\ Q_{03}^S & a_{32} & a_{33} \end{pmatrix} + \det\begin{pmatrix} 0 & a_{12} & a_{13} \\ Q_{12}^S & a_{22} & a_{23} \\ Q_{13}^S & a_{32} & a_{33} \end{pmatrix}\right] = 0, \qquad (2.1.31)$$

and analogously for $i = 2, 3$. ∎

The reader may have noticed already that $\bar{\Psi}$ of (2.1.12), due to its structure, may also be written in the matrix form $\Psi(\bar{q}) = (\Psi_{ij}(\bar{q}))$, $i, j = 0, 1, \ldots, n$, which results in summing up two matrices obtained by the procedure that clearly follows from the display in (2.1.24) and (2.1.25). The role of such a matrix Ψ is that of the *elasticity matrix* of the system and we shall call it so.

2.2 CONSERVATIVE REFERENCE FRAME. ENERGY SURFACE

Consider now the system (1.3.11) with the potential forces represented by (2.1.1). We have

$$\frac{d}{dt}\frac{\partial T}{\partial \dot{q}_i} - \frac{\partial T}{\partial q_i} = -\frac{\partial \mathscr{V}}{\partial q_i} - Q_i^P(\bar{q}, \dot{\bar{q}}) + Q_i^F(\bar{q}, \dot{\bar{q}}, \bar{u}), \qquad i = 1, \ldots, n, \qquad (2.2.1)$$

Multiplying it by $\dot{q}_i$ and summing up over i,

$$\sum_i \left(\frac{d}{dt}\frac{\partial T}{\partial \dot{q}_i}\right)\dot{q}_i - \sum_i \frac{\partial T}{\partial q_i}\dot{q}_i + \sum_i \frac{\partial \mathscr{V}}{\partial q_i}\dot{q}_i = \sum_i (Q_i^F + Q_i^P)\dot{q}_i \qquad (2.2.2)$$

or

$$\frac{d}{dt}\left[\sum_i \dot{q}_i \frac{\partial T}{\partial \dot{q}_i} - T + \mathscr{V}\right] = \sum_i (Q_i^F + Q_i^P)\dot{q}_i. \qquad (2.2.3)$$

Since

$$\sum_i \dot{q}_i \frac{\partial T}{\partial \dot{q}_i} = \sum_i \dot{q}_i \sum_j a_{ij}(\bar{q})\dot{q}_j = \sum_i \sum_j a_{ij}(\bar{q})\dot{q}_i\dot{q}_j = 2T,$$

2.2 CONSERVATIVE REFERENCE FRAME. ENERGY SURFACE

and

$$\frac{d}{dt}[T(\bar{q},\dot{\bar{q}}) + \mathscr{V}(\bar{q})] = \dot{H}(\bar{q},\dot{\bar{q}}),$$

(2.2.3) becomes

$$\dot{H}(\bar{q},\dot{\bar{q}}) = \sum_i [Q_i^F(\bar{q},\dot{\bar{q}},\bar{u}) + Q_i^P(\bar{q},\dot{\bar{q}})]\dot{q}_i, \qquad (2.2.4)$$

representing the instantaneous change of total energy.

Let us now assume that only the potential forces are active: $Q_i^F \equiv 0, Q_i^D \equiv 0$. Then (2.2.4) yields $\dot{H}(\bar{q},\dot{\bar{q}}) = 0$ or $H(\bar{q},\dot{\bar{q}}) = $ const, which means that the system described by the equations

$$\frac{d}{dt}\frac{\partial T}{\partial \dot{q}_i} - \frac{\partial T}{\partial q_i} = Q_i^P(\bar{q}), \qquad i = 1,\ldots,n,$$

or, equivalently,

$$\sum_j a_{ij}(\bar{q})\ddot{q}_i + K_i(\bar{q},\dot{\bar{q}}) - Q_i^P(\bar{q}) = 0, \qquad (2.2.5)$$

is an energy preserving or *conservative* system. Whatever of the amount $H(\bar{q},\dot{\bar{q}})$ is used by the kinetic energy in motion, it is balanced by $\mathscr{V}(q)$ restituted by $\bar{Q}^P$. Consequently Q_i^P are called *restitutive* (*restoring*) forces as an alternative name to *potential* or *conservative forces*. The latter as opposed to Q_i^D, Q_i^F which by (2.2.4) alter the power balance. Note that the centrifugal Coriolis forces hide in $\dot{H}(\bar{q},\dot{\bar{q}})$ in the passage from (2.2.2) to (2.2.4) and do not alter the power balance either—they are then called *energy neutral*.

Observe that an estimate of the change of energy corresponding to (2.2.4) can be derived from (1.5.16′), or more conveniently from its equivalent (1.5.16). To do so, consider $H(\bar{q},\bar{p})$ of (1.5.16) and calculate

$$\dot{H}(\bar{q},\bar{p}) = \sum_i \frac{\partial H}{\partial q_i}\dot{q}_i + \sum_i \frac{\partial H}{\partial p_i}\dot{p}_i.$$

Substituting $\dot{q}_i, \dot{p}_i$ obtained from (1.5.16), we have

$$\dot{H}(\bar{q},\bar{p}) = \sum_i \left(\frac{\partial H}{\partial q_i}\frac{\partial H}{\partial p_i} - \frac{\partial H}{\partial p_i}\frac{\partial H}{\partial q_i}\right) + \sum_i \frac{\partial H}{\partial p_i}(Q_i^P + Q_i^F), \qquad (2.2.4')$$

wherefrom (2.2.4) follows immediately.

Hence, the conservative system (2.2.5) has its Hamiltonian counterpart in the form obtained from (1.5.16′) by making Q_i^D, Q_i^F to vanish identically

$$\dot{q}_i = \frac{\partial T(\bar{q},\bar{p})}{\partial p_i} = \sum_j a_{ij}^{-1}(\bar{q})p_j, \qquad \dot{p}_i = -\frac{\partial T(\bar{q},\bar{p})}{\partial q_i} + Q_i^P(\bar{q}). \qquad (2.2.6)$$

The assumed single valuedness of $\mathscr{V}$ means that the restitution tracks back precisely the graph of the usage. It also means that Q_i^P, Π_i are single valued, which, particularly with spring forces, corresponds to elastic behavior, the matrix Ψ representing the mechanical stiffness of the system. Such modeling is not entirely realistic. There will be hysteresis in links, a nonadiabatic compression in pneumatic or hydraullic compensation, etc., so we should apply a multivalued function Q_i^S. Then the system would cease to be conservative. However, one may maintain the restitution property and thus the single valuedness of Q_i^S and $\mathscr{V}$ by shifting the nonelastic phenomena to the damping characteristics which complement Q_i^S in our overall model (see Skowronski [24]). Without going into details, let us say that this procedure is physically justified and often used (see Christensen [1]).

Substituing $Q_i^D \equiv 0$, $Q_i^F \equiv 0$ into (1.5.4), (1.5.16'), we obtain the state representations of the conservative system. Substituting the same into the equilibria equations (1.5.24), (1.5.25), we see that the equilibria of the conservative system are defined by

$$\dot{q}_i = 0, \qquad Q_i^P(\bar{q}) = 0, \qquad i = 1,\ldots,n, \tag{2.2.7}$$

that is, by the vanishing of the potential forces. The above implies immediately

$$\dot{q}_i = 0, \qquad \Pi_i(\bar{q}) = 0, \qquad i = 1,\ldots,n. \tag{2.2.7'}$$

Conversely, (2.2.7') implies zero acceleration, that is, motion with constant velocity. But *isolated roots* of (2.2.7) have zero velocity, that is, they are rest positions or *equilibria*. Consequently, we consider (2.2.7), (2.2.7') to be equivalent.

Observe that (2.1.1) and (2.2.7) together imply that the equilibria lie at the extremal points of the potential energy surface $\mathscr{B}: v = \mathscr{V}(\bar{q})$ in the space $v \times R^n$ over the set Δ_q.

A little geometry will be instructive. The surface $\mathscr{B}$ is singlesheeted and smooth, since $\mathscr{V}$ was assumed to be single valued and continuously differentiable. There is little chance that any element of the manipulator may become an energy sink with negative storage. Thus there is no narrowing in assumption that if $\mathscr{V}$ admits negative values at all, they are bounded. With the latter and with the origin placed at an absolute minimum of $\mathscr{V}$ over $\bar{\Delta}_q$, we can always adjust the free constant $\mathscr{V}(\bar{q}^0)$ of (2.1.2) so as to obtain $\mathscr{V}$ positive semidefinite, if the minimum is not single, or positive definite, if it is.

Since the equilibria coincide with the extrema of $\mathscr{B}$, they are isolated (see our comment in Section 1.5) and there is a finite sequence of them in bounded Δ. See the two-dimensional illustration in Fig. 2.1.

As stated by Lagrange and proved by Dirichlet, the isolated minima of $\mathscr{B}$ are the stable equilibria. We call them *Dirichlet stable* and use this term in the plain meaning of elementary physics in high school. The reader may as well

2.2 CONSERVATIVE REFERENCE FRAME. ENERGY SURFACE

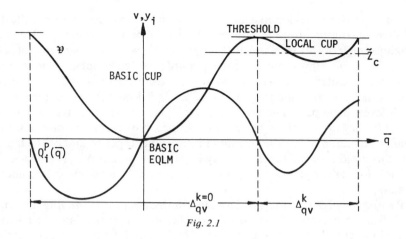

Fig. 2.1

bring back memories of a ball returning to a stable position at the bottom of a convex two-dimensional $\mathscr{V}$.

The matrix $[-\partial Q_i^P/\partial q_j]$ is called the *functional coefficient of restitution* of the system. The restitution will be said to be positive if the matrix is positive definite, that is, all its principal minors are positive, which suffices for a local minimum. Thus we have

Property 2.2.1. An equilibrium is Dirichlet stable if the restitution in its neighborhood is positive.

Note that the matrix $[-\partial Q_i^P/\partial q_j]$ may be immediately replaced by $[\partial \Pi_i/\partial q_j]$, without a change of conditions.

The sections $\mathscr{V}(\bar{q}) = \text{const} = v_c$ of $\mathscr{V}$ are the potential energy levels $\tilde{Z}_c$, with $v_c \in [0, v_\Delta)$, $v_\Delta = \sup \mathscr{V}(\bar{q}) | \bar{q} \in \Delta$. We shall assume that each $\tilde{Z}_c$ separates a region of $\mathscr{V}$ from its complement. Note that $\tilde{Z}_c$ may be disjoint so that the region separated may be disjoint as well.

The level corresponding to the absolute minimum of $\mathscr{V}$ on $\bar{\Delta}_q$ is called *basic* and the corresponding equilibria (one or more) the *basic equilibria*. We have placed the origin of R^n at a basic equilibrium and arranged for the constant $\mathscr{V}(\bar{q}^0)$ of (2.1.2) to be such that $\mathscr{V}(0) = 0$. This means also

$$Q_i^P(0) = 0, \quad i = 1,\ldots,n, \qquad (2.2.8)$$

and, consequently,

$$\Pi_i(0) = 0, \quad i = 1,\ldots,n, \qquad (2.2.9)$$

in agreement with the zero equilibrium of Section 1.5. Then there is a neighborhood $\Delta_{qv} \subset \Delta_q$ of the origin on which $\mathscr{V}(\bar{q})$ increases in $|q_1|,\ldots,|q_n|$.

The pattern continues for all q_i up to the lowest level corresponding to the first relative maximum or inflection of $\mathcal{V}$ with respect to some q_i. The part neighboring 0 of this level is called the *potential energy threshold*, denoted by Z_{cv}, $v_c = v_{cv}$. The threshold separates a simply connected region from the rest of $\mathcal{V}$, and is called the *potential energy cup* (basic), denoted by Z_v.

Consider the continuous v_c family of levels below Z_{cv}, that is, for $v \le v_{cv}$. Such levels have parts enclosed in Z_v (see Fig. 2.1); these parts *form a nest* about 0; they do not cross one another and each of them separates a simply connected region of $\mathcal{V}$. The threshold itself is the upper bound of the family. The threshold projected orthogonally into Δ_q determines Δ_{qv}. The projection of the said family of levels behaves identically in Δ_{qv}, and the map is isometric (distance preserving).

By Property 2.2.1 we may pick up other stable equilibria. Moving the origin of R^n by a simple transformation of variables, we may repeat our construction of the local cup about any other Dirichlet stable equilibrium. We then obtain a sequence of Z_v^k, $k = 0, 1, \ldots, m < \infty$, finite since Δ is bounded, with $v = 0$ referring to the basic cup. Correspondingly there is a sequence of Δ_{qv}^k. It is shown, Shestakov [1], that the stable equilibria and thus Δ_{qv}^k are separated by the Dirichlet unstable equilibria (maxima).

Can we estimate the size of Δ_{qv}^k? Evidently, there is the possibility of a single equilibrium occurring in Δ_q when the functions Q_i^P or Π_i are "hard enough", that is, with sufficiently large positive partial derivatives yielding a steep monotone increase. Then no threshold ever appears and Z_v^k stretches indefinitely yielding $\Delta_{qv} = \Delta_q$. In the remaining cases one encounters at least two equilibria, then there is at least one that neighbors the basic equilibrium. Thus the threshold must appear, if not sooner, then at this neighboring equilibrium, the threshold yielded by (2.2.7). Since the positive coefficient of restitution suffices to the minimum at any equilibrium, thus conversely such a minimum is necessary for positive restitution. We said (see Shestakov [1]), that the neighboring equilibrium cannot be a minimal point again, so the positive restitution is contradicted there. Then, provided the threshold crosses over the said equilibrium (Dirichlet unstable), the positive coefficient of restitution defines Δ_{qv}. The above proviso must hold for $N = 2$, but quite often holds also for $N > 2$. Since $\mathcal{V}(q)$ is symmetric by definition, the basic cup is always symmetric and our proviso holds for the basic cup for $N > 2$. We thus have the following conclusion:

Property 2.2.2. The region Δ_{qv} is defined by the positive coefficient of restitution with accuracy to within the distance between the threshold and the nearest unstable equilibrium.

We have, however, an alternative physically motivated criterion for Δ_{qv}. By the very nature of equilibrium, when a motion goes outwards from a Dirichlet

2.2 CONSERVATIVE REFERENCE FRAME. ENERGY SURFACE

stable equilibrium, the restoring force must tend toward this equilibrium and vice versa; the sense of motion and restoring force vectors are opposite,

$$Q_i^P(\bar{q}) \cdot q_i < 0, \qquad q_i \neq 0, \quad i = 1, \ldots, n, \qquad (2.2.10)$$

yielding in turn the so-called *restitution law*

$$\sum_i Q_i^P(\bar{q})q_i = \bar{Q}^P(\bar{q}) \cdot \bar{q} < 0, \qquad q \neq 0. \qquad (2.2.11)$$

Correspondingly, in terms of characteristics, the (2.2.11) becomes

$$\sum_i \Pi_i(\bar{q})q_i = \bar{\Pi}(\bar{q})\bar{q} > 0, \qquad \bar{q} \neq 0. \qquad (2.2.12)$$

Since (2.2.10)–(2.2.12) are implied by the stable equilibrium, they are necessary conditions for the minimum. Consider (2.1.10). At the neighboring unstable equilibrium, all Q_i^P cross zero and since they are smooth functions, they must gain the opposite sign, which contradicts (2.2.10). Contradicting the necessary conditions produces the opposite property, hence the *boundary of the set defined by (2.2.12) determines the threshold.*

Let us return now to the sequence of cups Z_v^k. Since the cups and their corresponding thresholds are all defined, so is the highest threshold defined as the maximal over bounded Δ_q. Let $\tilde{Z}_L$ be the infimum of the levels $\tilde{Z}_c$ above the highest threshold. Then $\tilde{Z}_L$ separates a region enclosing all the cups Z_v^k from the rest of $\mathfrak{P}$. This region projected into Δ_q defines the set Δ_{qL} which encloses the union of Δ_{qv}^k, $k = 0, 1, \ldots, m$. Since there are no thresholds in $\Delta_q - \Delta_{qL}$, this region forms a structure that resembles a local cup about ∞ and will thus be called the potential energy cup in the large.

Example 2.2.1. Let $Q^P(q) = aq + bq^3 + cq^5$, $q(t) \in R$, $a, c > 0$, with certain limitations imposed upon the "softness" coefficient b. We have $\mathscr{V}(q) = \frac{1}{2}aq^2 + \frac{1}{4}bq^4 + \frac{1}{6}cq^6$. The equilibria are obtained from $q(a + bq^2 + cq^4) = 0$, which yields $q = 0$ and

$$q = \pm\left(\frac{-b \pm \sqrt{b^2 - 4ac}}{2c}\right)^{1/2}.$$

For $b \geq 0$, the spring is "hard": monotone increasing with q, and there is only one Dirichlet stable equilibrium $q^e = 0$. The same occurs for $b < 0$, but small: $b^2 < 4ac$. On the other hand, for $b < 0$, $b^2 = 4ac$ we have three, and for $b^2 > 4ac$ all five equilibria. Now letting $a = c = 1$, $b = -2.1$, we have the shape of $\mathfrak{P}$ specified quite clearly by Table 2.2.1.

Let us now introduce the space $h \times R^N$, $N = 2n$, and consider the surface $\mathscr{H}$ in $h \times R^N$ generated by the functions, either $h = H(\bar{q}, \dot{\bar{q}})$ according to (1.5.4) or $h = H(\bar{q}, \bar{p})$ according to (1.5.16') over $\Delta \subset R^N$, that is, confined to the set $Z \triangleq [0, h_\Delta) \times \Delta$, $h_\Delta = \sup H(\bar{x}) | \bar{x} \in \Delta$. For concreteness we shall use $H(\bar{q}, \dot{\bar{q}})$,

Table 2.2.1

q		-1.14		-0.9		0		0.9		1.14
$\mathscr{V}(q)$		0.16		0.31		0		0.31		0.16
	↘	min	↗	max	↘	min	↗	max	↘	min
$\mathscr{V}' = -Q^S$	⊖	0	⊕	0	⊖	0	⊕	0	⊖	0
$-\partial Q^S/\partial q$		1.2		-0.84		1.0		-0.84		1.2
$Q^S \cdot q$		⊕		Threshold		⊖		Threshold		⊕

but there is no difference to our further discussion whether we follow $H(\bar{q}, \dot{\bar{q}})$ of (1.5.4) or $H(\bar{q}, \bar{p})$ of (1.5.16) [see (1.5.20)]. The set Z is bounded as Δ is bounded (see Fig. 2.2). Next, we introduce a level surface of H, $Z_C \triangleq \{(h, \bar{q}, \dot{\bar{q}}) \in Z \mid H(\bar{x}) = \text{const} = h_C\}$. We say h_C is a *regular value* if $(h, \bar{q}, \dot{\bar{q}}) \in Z_C$ implies $\nabla H(\bar{q}, \dot{\bar{q}}) \neq 0$. It follows from the assumption on equilibria (isolated) that almost all values of h are regular. It follows from the derivation of (2.2.4) and (2.2.5) or (2.2.6) that the trajectories of the conservative (2.2.5) or (2.2.6) form an h_C family of curves on the surface $\mathscr{H}$. Thus, the latter represents the first integral of (2.2.5) or (2.2.6). To every N-tuple $\bar{x}^0 = (x_1^0, \ldots, x_N^0)$ of the constants of integration of (2.2.5), (2.2.6) in Δ, there corresponds a value $h_C \in [0, h_\Delta)$ which defines some Z_C, called alternatively the integral level. The inverse map is obviously set valued. Nevertheless, exhausting all $\bar{x}^0 \in \Delta$ we obtain the surface $\mathscr{H}$.

Let us project orthogonally the h_C family of levels onto Δ. For each level we obtain a set H_C in Δ. It follows from the implicit function theorem of the calculus that, given regular h_C, the corresponding H_C is an $(N-1)$-dimensional hypersurface in R^N, with the degree of smoothness determined by the smoothness of $\mathscr{H}$. Such surface is called *topographic* by analogy to the

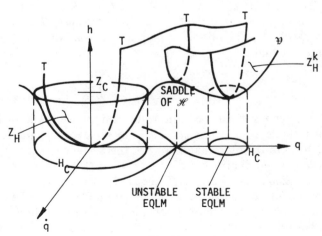

Fig. 2.2

2.2 CONSERVATIVE REFERENCE FRAME. ENERGY SURFACE

geographic card images of the height levels in a terrain. According to the above, the first integral of the conservative system (2.2.5) is accommodated by the family of H_C's in Δ, and H_C representing a trajectory picked up by specifying the initial state $\bar{x}^0 = (\bar{q}^0, \dot{\bar{q}}^0)$. The trajectories do not cross, neither do the levels, each defined by $H(\bar{q}, \dot{\bar{q}}) = \text{const} = h_C$.

The shape of $\mathcal{H}$ is obviously determined by the fact that its generator $h = H(\bar{q}, \dot{\bar{q}}) = T(\bar{q}, \dot{\bar{q}}) + \mathscr{V}(\bar{q})$. By (1.3.4) or (1.5.21), T is a single-valued, smooth, positive-definite function, even and symmetric with respect to the hyperplane $\dot{\bar{q}} = 0$ in Z. For all practical purposes, we can envisage it as an $(N-1)$-dimensional extension of a parabola. Since the surface $\mathcal{H}$ is obtained by superposition of $T(\bar{q}, \dot{\bar{q}})$ and $\mathscr{V}(\bar{q})$, the latter analyzed in terms of the surface $\mathfrak{V}$, it is the superposition of the parabolic T over $\mathfrak{V}$ along a bottom line of the latter (see Fig. 2.2). There are no changes in the notions introduced for $\mathfrak{V}$ except for expanding them along the additional set of velocity axes $\dot{q}_1, \ldots, \dot{q}_n$ or momentum axes $p_1, \ldots, p_n$ and thus adopting slightly different notation.

The equilibria have been shown to coincide with the extremal arguments of $\mathscr{V}$. Since $T(\bar{q}, 0) \equiv 0$, they also coincide with the extremal arguments of $H(\bar{q}, \dot{\bar{q}})$, thus underlying the extrema of $\mathcal{H}$.

By the shape of $T(\bar{q}, \dot{\bar{q}})$ the Dirichlet stable equilibria of $\mathfrak{V}$ define minima of $\mathcal{H}$ (double with respect to both $\mathscr{V}$ and T) and the Dirichlet unstable equilibria define the saddles of $\mathcal{H}$ (maxima for $\mathscr{V}$ and minima for T). The isoenergy levels Z_C are defined by the cuts $T(\bar{q}, \dot{\bar{q}}) + \mathscr{V}(\bar{q}) = \text{const} = h_C$ of $\mathcal{H}$ and the corresponding topographic surfaces H_C accommodate trajectories of (2.2.5) starting there and only these trajectories.

The conservative (2.2.5) serves thus as the conservative reference system. Since both T and $\mathscr{V}$ are positive-definite functions, H is positive-definite and again we let the origin of Z at the basic equilibrium, corresponding to the absolute minimum of $\mathcal{H}$ and to the basic total energy level, briefly *basic energy level* $h = 0$.

Then, since the basic level is a double minimum of both $\mathscr{V}$ and T, there is a neighborhood $\Delta_H \subset \Delta$ of the basic equilibrium on which H increases with $|q_1|, \ldots, |q_n|, |\dot{q}_1|, \ldots, |\dot{q}_n|$. Since T has no extremal values except 0, it is only the potential energy threshold which may break down the increase. To any potential energy level $\mathscr{V}(q) = \text{const}$, there corresponds some Z_C in $\mathcal{H}$. Thus the potential energy threshold becomes the *energy threshold* $Z_{CH} : H(\bar{q}, \dot{\bar{q}}) = h_{CH}$.

Correspondingly, the basic potential energy cup Z_v becomes the *basic energy cup* Z_H. It corresponds to Δ_H by the fact that the boundary $\partial \Delta_H = H_{CH}$ of Δ_H is an isometric image of Z_{CH}. Since the thresholds of $\mathfrak{V}$ and $\mathcal{H}$ coincide, H_{CH} is that H_C which passes through the boundary of Δ_{qv} and since both H_C and Δ_{qv} are defined, the H_{CH} is defined as well, and so is Δ_H. The set H_{CH} is called the *separating set* (generalized separatrix) of Δ_H. Hence, if the potential threshold does not appear: $\Delta_{qv} = \Delta$ (linear or hard Q_i^p), the energy threshold Z_{CH} never appears either and $\Delta_H = \Delta$.

It is now easy to describe the interior of Z_H. Similar to its potential counterpart Z_v, it consists in the continuous h family, $h \in [0, h_{CH}]$, of components of Z_C nested about the basic equilibrium, each separating a simply connected region from the rest of $\mathscr{H}$. The surfaces H_C in Δ_H behave correspondingly. The threshold Z_{CH} is the least upper bound of this family of Z_C and separates Z_H. So does H_{CH} for Δ_H. Since $\mathfrak{V}$ is symmetric on Δ_H, the threshold Z_{CH} coincides with the first maxima with respect to all q_i, $i = 1, \ldots, n$, at the two neighboring Dirichlet unstable equilibria.

It is easy to see that the global shape of $\mathscr{H}$ depends on $\mathfrak{V}$ ($\mathscr{V}$ is called the *driving function*). Once $\bar{Q}^P$ and thus $\mathscr{V}$ are given ($\mathscr{V}$ with accuracy to the constant $\mathscr{V}(t_0)$), (2.2.6) picks up the equilibria, (2.2.11) selects the Dirichlet stable ones. Similarly as for $\mathfrak{V}$, we may repeat our construction about other Dirichlet stable equilibria transforming the coordinates suitably. We obtain the local cups Z_H^k, $k = 0, 1, \ldots, m$, with underlying Δ_H^k, $k = 0, 1, \ldots, m$. With all the cups defined, the highest threshold is also defined by

$$h_{CH}^{\max} = \max(h_{CH}^1, \ldots, h_{CH}^m).$$

Then we again let

$$Z_L \mid h_L = \inf h_C > h_{CH}^{\max}.$$

Here Z_L separates the region $Z_{HL} \subset \mathscr{H}$ that includes all the local cups Z_H^k and defines the corresponding Δ_L in turn, including $\bigcup_k \Delta_H^k$. The shape of $\mathfrak{V}$ in the large extends to that of $\mathscr{H}$ because of the shape of T. On the complement of Δ_L to Δ, denoted by $C_\Delta \Delta_L$, there are no critical points and in $C_{\mathscr{H}} Z_{HL}$ there are no extrema of $\mathscr{H}$. Thus the structure of $C_{\mathscr{H}} Z_{HL}$ resembles that of a local cup with the nest centered about infinity. Therefore $C_{\mathscr{H}} Z_{HL}$ is called the *energy cup in the large*. The consequences for its image in Δ are obvious.

Consider the Z_C over $\mathscr{H}$ and denote by Δ_C the regions separated by the corresponding $H_C = \partial \Delta_C$ in Δ. It is instructive to envisage $\mathscr{H}$ in three dimensions as a well with the bottom line $\mathfrak{V}$ and walls $T(\bar{q}, \dot{\bar{q}})$ thus consisting of a number of valleys (cups) separated by saddles. We let the well be exposed to an intensive rain scattered evenly all over the place, thus the well being successively filled up with water. The rising water levels illustrate the varying Z_C, while the area of the free water surface at each level illustrates the corresponding sets Δ_C.

The reader will conclude readily that in general the levels Z_C produce the sets Δ_C either simply connected with simply connected complements $C_\Delta \Delta_C = \Delta - \Delta_C$ or disjoint with the complements multiply connected. For example, in Fig. 2.3, the set Δ_{C_3} is simply connected and H_{C_3} separates it from the simply connected $C_\Delta \Delta_{C_3}$. On the other hand, the set Δ_{C_2} is disjoint and H_{C_2} separates this set from $C_\Delta \Delta_{C_2}$ which is multiply connected. The levels Z_C below all the local minima produce simply connected Δ_C's (see Δ_{C_1} in Fig. 2.3), and so do the levels Z_C, $h_C \geq h_L$ (see Δ_{C_5} in Fig. 2.3).

2.2 CONSERVATIVE REFERENCE FRAME. ENERGY SURFACE

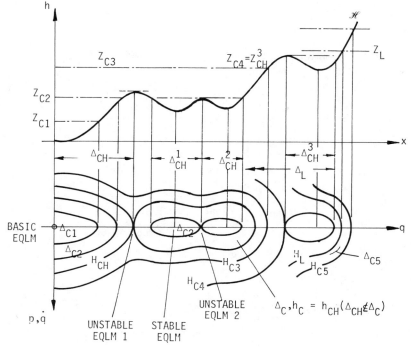

Fig. 2.3

Since there are no openings between the cups except over the thresholds, the rain fills up the cups quite independently of each other, yielding small disjoint lakes until the free water surface rises above the thresholds or separation sets. The study within cups is local (see Section 2.1). With rising water, more and more lakes become connected, requiring a global study. Finally at the level Z_L we obtain a large single lake surrounded by the remainder of $\mathscr{H}$ without thresholds, that is, the *cup in the large*. The investigation exercised on this remaining part of $\mathscr{H}$, that is, over $C_\Delta \Delta_L$ is called the *study in the large*.

Example 2.2.2. We continue the first Example 2.2.1, extending it now to the surface $\mathscr{H}$, the latter defined by

$$H(q, \dot{q}) = \frac{1}{2}\dot{q}^2 + \int Q^P(q)\,dq = \frac{1}{2}\dot{q}^2 + \frac{1}{2}aq^2 + \frac{1}{4}bq^4 + \frac{1}{6}cq^6.$$

The extremal points are defined by $\partial H/\partial q = aq + bq^3 + cq^5 = 0$, $\partial H/\partial \dot{q} = \dot{q} = 0$, thus coinciding with the equilibria $q^{(0)}, \ldots, q^{(4)}$. Using the same data, namely, $a = c = 1, b = -2.1$, we obtain the threshold $Z_{CH}: H(x) = h_{CH}$ where

$$h_{CH} = H(q^{(1)}, 0) = H(q^{(2)}, 0) = 0.31$$

(see Table 2.2.1). The H_{CH} passing through $q^{(1)}$, $q^{(2)}$ is

$$0.5\,\dot{q}^2 + 0.5\,q^2 - 0.53\,q^4 + 0.17\,q^6 = 0.31, \tag{2.2.13}$$

and it encloses not only Δ_H but also three disjoint Δ_H^k, the basic Δ_H inclusive. From Table 2.2.1, the threshold $h_{CH} = 0.31$ corresponding to it is the highest of the system on Δ so $H_{CH} = \partial \Delta_L$, and the complement of sets enclosed by this curve is the region in the large $C_\Delta \Delta_L$ that underlies the cup in the large. The reader may check, as an exercise, that (2.2.13) is an integral of

$$\ddot{q} + q - 2.1\,q^3 + q^5 = 0.$$

We shall turn now to a further extension, namely for the case $N = 2n = 4$. We see the reference system now as

$$\ddot{q}_i + \Psi_i(q_1, q_2) = 0, \qquad i = 1, 2 \tag{2.2.14}$$

with

$$\Psi_1(q_1, q_2) = \Psi_{11}(q_1) + \Psi_{12}(q_1 - q_2),$$
$$\Psi_2(q_1, q_2) = \Psi_{22}(q_2) + \Psi_{21}(q_1 - 1_2).$$

A large class of physical models allows symmetry in coupling, namely,

$$\Psi_{12}(q_1 - q_2) = -\Psi_{21}(q_1 - q_2) \tag{2.2.15}$$

which leads to significant consequences.

First, the restitution law becomes simplified and takes the form

$$\Psi_{11}(q_1)q_1 + \Psi_{22}(q_2)q_2 + \Psi_{12}(q_1 - q_2) \cdot (q_1 - q_2) \geq 0, \tag{2.2.16}$$

indicating that the restitutive forces have been directed opposite to the corresponding displacements which is a simple physical fact and must include the coupling force with respect to the relative displacement $(q_1 - q_2)$. Obviously (2.2.16) does not imply this fact. It is only the necessary condition for the minimum, thus also that for stability of equilibrium at the basic level.

Second, we obtain for $i = 1, 2$,

$$\Psi_i(-q_1, q_2) = -\Psi_i(q_1, -q_2),$$
$$\Psi_i(q_1, -q_2) = -\Psi_i(-q_1, q_2). \tag{2.2.17}$$

Let us now specify

$$\Psi_1(q_1, q_2) = a_1 q_1 + b_1 q_1^3 + c_1 q_1^5 + a_{12}(q_1 - q_2),$$
$$\Psi_2(q_1, q_2) = a_2 q_2 + b_2 q_2^3 + c_2 q_2^5 + a_{12}(q_1 - q_2),$$

that is, leave the restitutive coupling linear.

2.2 CONSERVATIVE REFERENCE FRAME. ENERGY SURFACE

We let $a_1, a_2, a_{12} > 0$, $c_1, c_2 > 0$, and $b_1, b_2 < 0$ with the same type of restrictions upon b_1, b_2 as made previously upon b. The total energy is

$$H(\bar{q}, \dot{\bar{q}}) = \tfrac{1}{2}(\dot{q}_1^2 + \dot{q}_2^2) + \tfrac{1}{2}(a_1 q_1^2 + a_2 q_2^2) + \tfrac{1}{4}(b_1 q_1^4 + b_2 q_2^4) + \tfrac{1}{6}(c_1 q_1^6 + c_2 q_2^6).$$

The equilibria are defined by the equations

$$a_1 q_1 + b_1 q_1^3 + c_1 q_1^5 + a_{12} q_1 + a_{12} q_2 = 0,$$

$$a_2 q_2 + b_2 q_2^3 + c_2 q_2^5 - a_{12} q_2 + a_{12} q_1 = 0.$$

The first obvious root is the basic equilibrium $q_1 = q_2 = 0$. Further, we add the equations to obtain

$$a_1 q_1 + b_1 q_1^3 + c_1 q_1^5 + a_2 q_2 + b_2 q_2^3 + c_2 q_2^5 = 0,$$

which for $q_1 \neq 0$ and $q_2 \neq 0$ is true if simultaneously

$$a_i + b_i q_i^2 + c_i q_i^4 = 0, \quad i = 1, 2,$$

that is, if

$$q_i = \pm \left(\frac{-b_i \pm \sqrt{b_i^2 - 4 a_i c_i}}{2 c_i} \right)^{1/2}.$$

Assuming $b_i = -2(a_i c_i)^{1/2}$, we shall indicate how to attempt the problem algebraically. The roots obtained are $q_1 = \pm(a_1/c_1)^{1/4}$ and $q_2 = \pm(a_2/c_2)^{1/4}$. On substitution in the equations for equilibria, one obtains

$$q_2^{(1)}, q_2^{(2)} = \pm \left(\frac{a_1}{c_1} \right)^{1/4} \left(1 + a_1 + \frac{a_1}{a_{12}} + \sqrt{a_1 b_1} \right),$$

$$q_1^{(1)}, q_1^{(2)} = \pm \left(\frac{a_2}{c_2} \right)^{1/4} \left(1 + a_2 + \frac{a_2}{a_{12}} + \sqrt{a_2 b_2} \right)$$

as the coordinates of the two equilibria that are other than stable. The threshold is obtained by substituting the values for $q_1^{(1)}, q_2^{(1)}$ into the energy. We have

$$h_{CH} = H(q_1^{(1)}, q_2^{(1)}, 0, 0)$$

$$= \frac{a_1}{2} \left(\frac{a_2}{c_2} \right)^{1/2} \left(1 + a_2 + \frac{a_2}{a_{12}} + \sqrt{a_2 c_2} \right)^2 + \frac{a_2}{2} \left(\frac{a_1}{c_1} \right)^{1/2} \left(1 + a_1 + \frac{a_2}{a_{12}} + \sqrt{a_1 c_1} \right)^2$$

$$- \frac{a_2 \sqrt{a_1 c_1}}{2 c_2} \left(1 + a_2 + \frac{a_2}{a_{12}} + \sqrt{a_2 c_2} \right)^4 - \frac{a_1 \sqrt{a_2 c_2}}{2 c_1} \left(1 + a_1 + \frac{a_1}{a_{12}} + \sqrt{a_1 c_1} \right)^4$$

$$+ \frac{c_1}{6} \left(\frac{a_2}{c_2} \right)^3 \left(1 + a_2 + \frac{a_2}{a_{12}} + \sqrt{a_2 c_2} \right)^6 + \frac{c_2}{6} \left(\frac{a_1}{c_1} \right)^3 \left(1 + a_1 + \frac{a_1}{a_{12}} + \sqrt{a_1 c_1} \right)^6.$$

Then H_{CH} is the three-dimensional closed surface in R^N

$$\dot{q}_1^2 + \dot{q}_2^2 + a_1 q_1^2 + a_2 q_2^2 - (\sqrt{a_1 c_1}\, q_1^4 + \sqrt{a_2 c_2}\, q_2^4) + \tfrac{1}{3}(c_1 q_1^6 + c_2 q_2^6) = h_{CH}$$

enclosing the region Δ_H. ∎

Exercises and Comments

2.2.1 The same formula (2.2.4) may be obtained from either the Hamiltonian form (1.5.16) or the Newtonian form (1.3.24), both inertially decoupled. To show the first, the reader is asked to write the full expression for $\dot{H}(\bar{q}, \bar{p})$ a total derivative and substitute (1.5.16). For the second, each of the equations (1.3.24) should be multiplied by $dq_i = \dot{q}_i dt$ and the results added and rearranged to obtain $dT + d\mathscr{V}$ on the left-hand side of the equality. Then the right-hand side will become the right-hand side of (2.2.4).

2.2.2 Consider a RPRP-manipulator and form the potential characteristics (gravity plus spring) for the coupled chain using the general formula (2.1.10) and making the slack joint variables constant. Determine the potential energy and equilibria; find the cups.

2.3 NONPOTENTIAL FORCES. POWER BALANCE. ENERGY FLOW

Let us now recall the relation (2.2.4), which represents the instantaneous rate of change of $H(\bar{q}, \dot{\bar{q}})$ or $H(\bar{q}, \bar{p})$ under the nonconservative forces. The rate is evaluated by the inner products

$$\bar{Q}^F(\bar{q}, \dot{\bar{q}}, \bar{u})\dot{\bar{q}} = \sum_i Q_i^F(\bar{q}, \dot{\bar{q}}, \bar{u})\dot{q}_i$$

and

$$\bar{Q}^D(\bar{q}, \dot{\bar{q}})\dot{\bar{q}} = \sum_i Q_i^D(\bar{q}, \dot{\bar{q}})\dot{q}_i$$

recognized as the *input* and *damping powers*, respectively. When a power is positive for $\dot{\bar{q}} \neq 0$, it accumulates the energy and the corresponding force is called *accumulative*. Nonpositive power dissipates or preserves the energy and the corresponding force is called *dissipative*. In particular, if the force is not potential but still produces a zero power over some time interval, it may be called *neutral*. Neutral forces are also known as nonenergetic or traditors (see Duinker [1]).

In the majority of cases, the damping forces are dissipative: dry friction or viscous damping at joints, caused by slips and other boundary sheer effects at mating surfaces; oil, water, or air resistance in the manipulator environment and/or its hydraulic or pneumatic power supply systems; and structural

2.3 NONPOTENTIAL FORCES. POWER BALANCE. ENERGY FLOW

damping in links caused by microscopic interface effects like internal friction or sliding. We shall specify such forces as *positive damping*, denoted by $Q_i^{DD}(\bar{q},\dot{\bar{q}})$ with corresponding characteristics $D_i^D(\bar{q},\dot{\bar{q}})$. Using the same transformation (1.3.7) as for the potential forces from the Cartesian to generalized damping, we may form the *dissipation function*

$$\mathscr{U}_D = \sum_i \oint Q_i^{DD}(\bar{q},\dot{\bar{q}})\,dq_i, \qquad (2.3.1)$$

which produces negative work [see also (1.2.14)]. The power of such positive damping forces is negative:

$$\sum_i Q_i^{DD}(\bar{q},\dot{\bar{q}})\,\dot{q}_i = \bar{Q}^{DD}(\bar{q},\dot{\bar{q}})\,\dot{q} < 0, \qquad \dot{\bar{q}} \neq 0, \qquad (2.3.2)$$

or, alternatively,

$$\sum_i D_i^D(\bar{q},\dot{\bar{q}})\dot{q}_i = \bar{D}^D(\bar{q},\dot{\bar{q}})\dot{\bar{q}} > 0, \qquad \dot{\bar{q}} \neq 0. \qquad (2.3.3)$$

On the other hand, we may reasonably assume that there is no damping at rest:

$$Q_i^{DD}(\bar{q}^e,0) = 0, \qquad D_i^D(\bar{q}^e,0) = 0, \qquad i = 1,\ldots,n. \qquad (2.3.4)$$

Note that the above assumptions make the positive damping an odd function with respect to the velocities $\dot{q}_i$. By the character of the applied damping forces, either viscous or Coulomb dry friction, we may also assert that they are at least nonincreasing:

$$-\partial Q_i^{DD}(\bar{q},\dot{\bar{q}})/\partial \dot{q}_j \geq 0, \qquad \dot{q}_i \neq 0, \qquad (2.3.5)$$

with respect to velocities. In most cases, there will also be a saturation of positive damping leading to an asymptotic leveling-off for the damping curves (see Fig. 2.4).

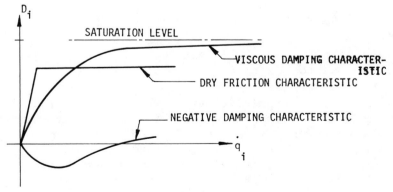

Fig. 2.4

Similar physical arguments lead to assuming that Q_i^{DD} depends upon $|\bar{q}|$ rather than upon an arbitrary $\bar{q}$ and that it monotone decreases with q_i:

$$-\frac{\partial Q_i^{DD}(|\bar{q}|,\dot{\bar{q}})}{\partial q_i} > 0, \qquad i = 1,\ldots,n, \tag{2.3.6}$$

for all $q_i \neq q_i^e$.

Whether or not the manipulator, working in certain conditions, exhibits auto-oscillations (or more generally any self-sustained motion) remains a matter for discussion, but the possibility is there. Then the implemented power is positive, at least for some values of the velocities $\dot{q}_i$, and we have the so-called *negative damping* forces. Such forces are obviously energy accumulative and will be denoted by $Q_i^{DA}(\bar{q},\dot{\bar{q}})$, with the corresponding characteristics $D_i^A(\bar{q},\dot{\bar{q}})$. Hence

$$\sum_i Q_i^{DA}(\bar{q},\dot{\bar{q}})\dot{q}_i = \bar{Q}^{DA}(\bar{q},\dot{\bar{q}})\dot{\bar{q}} > 0, \qquad \dot{\bar{q}} \neq 0, \tag{2.3.7}$$

and

$$\sum_i D_i^A(\bar{q},\dot{\bar{q}})\dot{q}_i = \bar{D}^A(\bar{q},\dot{\bar{q}})\dot{\bar{q}} < 0, \qquad \dot{\bar{q}} \neq 0. \tag{2.3.8}$$

Self-excitation is known as a so-called "velocity motor," and there is no self-excitation at rest; whence

$$Q_i^{DA}(\bar{q}^e,0) = 0, \qquad D_i^A(\bar{q}^e,0) = 0, \qquad i = 1,\ldots,n. \tag{2.3.9}$$

Once a self-sustained oscillation appears, it is reasonable to expect that there will be some neighborhood of $\dot{\bar{q}} = 0$ where the negative damping appears (see Fig. 2.2), and obviously it may appear as well at other intervals of velocity.

The functional shape of the positive damping Q_i^{DD} follows the coupling pattern described for the spring forces Q_i^S in Section 2.1. We have

$$\bar{Q}^{DD}(|\bar{q}|,\dot{\bar{q}}) \triangleq \bar{Q}^{DD0}(|\bar{q}|,\dot{\bar{q}}) + \sum_j \bar{Q}^{DDj}(|\bar{q}|,\dot{\bar{q}}), \tag{2.3.10}$$

and

$$\bar{D}^D(|\bar{q}|,\dot{\bar{q}}) \triangleq \bar{D}^{D0}(|\bar{q}|,\dot{\bar{q}}) + \sum_j D^{Dj}(|\bar{q}|,\dot{\bar{q}}) \tag{2.3.11}$$

decomposed exactly along the lines shown in the formulas between (2.1.13) and (2.1.21). The substitution is left to the reader. Assuming this done, we shall refer to the notions of $\bar{D}^{D0}(|\bar{q}|,\dot{\bar{q}})$, $\bar{D}^{Dj}(|\bar{q}|,\dot{\bar{q}})$, $D_i^D(|\bar{q}|,\dot{\bar{q}})$, $D_{0i}^D(|\bar{q}|,\dot{\bar{q}})$, $D_{ij}^D(|\bar{q}|,\dot{\bar{q}})$, $Q_{0i}^{DD}(|\bar{q}|,\dot{\bar{q}})$, $Q_{ij}^{DD}(|\bar{q}|,\dot{\bar{q}})$ as already introduced and familiar.

We also leave to the reader the decomposition of the assumptions (2.3.1)–(2.3.6) imposed on Q_i^{DD} and D_i^D in terms of (2.3.10) and (2.3.11), respectively.

2.3 NONPOTENTIAL FORCES. POWER BALANCE. ENERGY FLOW

With all the above, the damping forces result in both types of damping and we must let

$$Q_i^D(\bar{q},\dot{\bar{q}}) \triangleq Q_i^{DA}(\bar{q},\dot{\bar{q}}) + Q_i^{DD}(|\bar{q}|,\dot{\bar{q}}) \qquad (2.3.12)$$

and

$$D_i(\bar{q},\dot{\bar{q}}) \triangleq D_i^A(\bar{q},\dot{\bar{q}}) + D_i^P(|\bar{q}|,\dot{\bar{q}}). \qquad (2.3.13)$$

There cannot be any blanket assumptions upon the sign of the input power $\bar{Q}^F \cdot \dot{\bar{q}}$ since the controller may be used both ways, either to supply or withdraw the energy from the manipulator. The controller will also work at equilibria, thus implying there possibly nonzero values of Q_i^F, (see Benedict and Tesar [1]).

On the other hand, since the power from all energy sources (even nuclear) is always eventually limited, there may be no objection to our *axiom of bounded accumulation*, by which we mean the following. For any point $(\bar{q}^0, \dot{\bar{q}}^0) \in \Delta$ there is a number $N > 0$ large enough to secure

$$|\bar{Q}^{DA}(\bar{q},\dot{\bar{q}})\dot{\bar{q}} + \bar{Q}^F(\bar{q},\dot{\bar{q}},\bar{u})\dot{\bar{q}}| \le N < \infty \qquad (2.3.14)$$

for all $\bar{q}(t), \dot{\bar{q}}(t) \in \Delta$ along a trajectory from $(\bar{q}^0, \dot{\bar{q}}^0)$.

The axiom means that the external input, whatever its source, decreases whenever the velocity of motion increases and vice versa. The *regulating* role of this axiom is obvious. It reflects an optimistic belief in the overall stability of this world and may be considered our own derivative of the so-called and well-known *le Chatelier principle* that every action produces in a system (body) changes which tend to neutralize that action building up resistance to it (see Kononenko [2]).

With all the above discussion in mind, we may see that (2.2.4), which may be now written as

$$\dot{H}(\bar{q},\dot{\bar{q}}) = \bar{Q}^{DA}(\bar{q},\dot{\bar{q}})\dot{\bar{q}} + \bar{Q}^F(\bar{q},\dot{\bar{q}},\bar{u})\dot{\bar{q}} + \bar{Q}^{DD}(\bar{q},\dot{\bar{q}})\dot{\bar{q}}, \qquad (2.3.15)$$

gives the *power balance* along the trajectories of (1.5.4). Obviously, the same discussion refers to the forces of (1.5.16') with $H(\bar{q},\bar{p})$ and with $\bar{p}$ replacing $\dot{\bar{q}}$ in all the assumptions.

We return now to the surface $\mathscr{H}: h = H(\bar{x})$ in Z of the previous Section 2.2. Injecting (lifting) the trajectories $k(\bar{x}^0, R)$ of either (1.5.4) or (1.5.16') from Δ into Z, we obtain what may be called *energo-state trajectories* $h(\bar{u}, \bar{x}^0, h^0, R)$, where $h^0 = H(\bar{x}^0)$. Introduce now the scalar function $f_0(\bar{x},\bar{u}) \triangleq \dot{H}(\bar{x}) = \nabla H(\bar{x}) \cdot f(\bar{x},\bar{u})$, called the *power characteristic*, determined by (2.3.15). It follows that the energo-state trajectories are defined by the equations

$$\dot{h} = f_0(\bar{x},\bar{u}), \qquad \dot{\bar{x}} = \bar{f}(\bar{x},\bar{u}), \qquad (2.3.16)$$

with $\bar{f}$ coming from either (1.5.4) or (1.5.16').

The first scalar equation will be called the *energy equation* describing the balance of power or change of H subject to constraints specified by the second equation.

Mapping an energy-state trajectory $h(x^0, h^0, R)$ on the h axis of Z, we obtain the one-dimensional set of points $h(h^0, R)$ named the *autonomous energy flow* related continuously to the parameter $h^0 \in [0, h_\Delta)$, describing the initial amount of the total energy contained in the system before motion. Note that the above flow is t_0 independent since the trajectories are independent of t_0.

The values $h(h^0, t) = h_C \in [0, h^f) \subset [0, h_\Delta)$ define the amount of the energy at t, obtained as the result of the change of energy or *flux* during $[t_0, t)$, either *influx* (increase) or *outflux* (decrease). The latter two are also called the *energy input* and the *work done* by the energy changing actions, respectively. Zero flux might be incorporated in either the influx or the outflux. Integrating the energy equation, given $t < \infty$, we obtain

$$h(h^0, t) = h^0 + \int_{t_0}^{t} f_0(\bar{x}, \bar{u})\, ds, \tag{2.3.17}$$

where the integral represents the flux with respect to the initial flow value h^0 contained in the system at t_0. Note that the flow is additive on R:

$$h(h^0, t_1) + h(h(h^0, t_1), t_2) = h(h^0, t_2) \tag{2.3.18}$$

for any $t_1, t_2 \in R$, $t_2 > t_1$, and that $f_0(\bar{x}, \bar{u})$ is bounded [see (2.3.14)]. We shall also justify below that, for intervals with negative values of f_0, *the flux may not exceed h^0 so that there is no chance of $h(h^0, t)$ becoming negative.*

The existence and uniqueness of the energy-state trajectories follow from the properties of the state trajectories and the shape of $\mathcal{H}$.

Although $\mathcal{H}$ is N-dimensional like R^N, unlike R^N, it is a nonlinear creature (for illustration, the distance between two points in R^N will not be proportional to its image in $\mathcal{H}$). However, $\mathcal{H}$ is locally diffeomorphic to R^N, that is, a neighborhood of a point in $\mathcal{H}$ is the image of R^N under a one-to-one and onto map which is differentiable together with its inverse. Then it is possible to define a field of directions and thus the integral curves on every such neighborhood and hence on the whole of $\mathcal{H}$. Further, since $h = H(\bar{x})$ is single valued on Δ, to every point $\bar{x} \in \Delta$ there corresponds a unique point $(h, \bar{x}) \in \mathcal{H} \cap Z_\Delta$ and the energy-state trajectories are unique.

This is not the case with the energy flow. Although to each point $\bar{x}$ of Δ there corresponds a point $h(\bar{x}^0, h^0, t)$ on the h-axis, $h_C = h(h^0, t)$ of the energy flow, however, the inverse map is clearly set-valued, that is, the same h_C yields a continuum of points in $\mathcal{H}$, namely, the level Z_C and thus the topographic H_C in Δ. Thus a trajectory and/or an energy-state trajectory generate the corresponding energy flow, but the latter implements only the change of energy levels along the trajectory, and *we need to refer to known levels to be able to obtain even a qualitative behavior of the trajectory.*

2.3 NONPOTENTIAL FORCES. POWER BALANCE. ENERGY FLOW 65

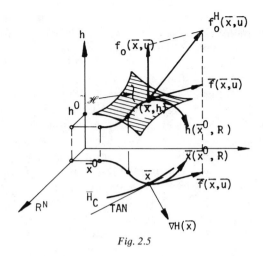

Fig. 2.5

From (2.3.16) it follows that f_0 is the h axis component of the vector $\bar{f}^H \triangleq (f_0, f_1, \ldots, f_N)^T$ (see Fig. 2.5). So the rise or fall of the energy levels along the trajectory depends upon the sign of f_0. Consider an arbitrary trajectory $k(\bar{x}^0, R)$, $\bar{x}^0 \in \Delta$ that is in general nonconservative, and given t, let $\bar{x}(t)$ be a regular point on this trajectory. Let H_C be the corresponding topographical (isoenergy) surface through that point (see Fig. 2.5), with the projection into Δ being shown in Fig. 2.6. Here H_C subdivides Δ into two regions. We call the region into which the positive gradient $\nabla H(\bar{x})$ is directed the *exterior*, and the other the *interior* of the surface H_C. If, given $\bar{u}$, we have $\bar{x}(t)$ such that $f_0(\bar{x}, \bar{u}) > 0$, then there is an energy influx at $\bar{x}(t)$ and the corresponding energy flow $h(h^0, R)$ is directed *upward* away from $h = 0$ on the h axis. This means that the other component of f^H, namely, the tangent vector $\bar{f}(\bar{x}, \bar{u})$ to

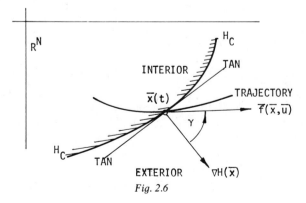

Fig. 2.6

the trajectory is directed toward the exterior of H_C, forming a sharp angle γ with the gradient (see Fig. 2.6). If $f_0(\bar{x}, \bar{u}) \geq 0$, then the thrust of the energy flow is the same, but the trajectory may have *contact* with H_C for longer than the instant t, that is, it may slide on H_C with $\bar{f}(\bar{x}, \bar{u})$ parallel to the tangent for some short interval of time. The latter happens at any point where $f_0(\bar{x}, \bar{u}) = 0$. On the other hand, when $f_0(\bar{x}, \bar{u}) < 0$, the opposite takes place. The sense of the energy flow is reversed, it is directed *downward* toward $h = 0$ on the h axis. This forces the tangent vector $\bar{f}(\bar{x}, \bar{u})$ to be directed toward the interior of H_C, forming an obtuse angle γ with $\nabla H(\bar{x})$.

Note that $f_0(\bar{x}, \bar{u}) = \nabla H(\bar{x}) \cdot \bar{f}(\bar{x}, \bar{u})$ is a point function on Δ and because of its continuity, the points where f_0 has a definite sign are not isolated. Hence, we can specify regions in Δ covered with the *fields* or *zones* $\mathcal{H}^+ : f_0(\bar{x}, \bar{u}) > 0$ (accumulative), $\mathcal{H}^0 : f_0(\bar{x}, \bar{u}) = 0$ (neutral), and $\mathcal{H}^- : f_0(\bar{x}, \bar{u}) < 0$ (strictly dissipative), on which the energy balance (2.3.15) is respectively positive, zero, or negative, and on which the trajectories leave, slide, or enter the interior of each current H_C in Δ that they cross. At the same time, the constant sign of f_0 decides the direction of the energy flux. The flux on some interval $[t_0, t) \in R$ may be positive, negative, or zero. No matter what happens at a particular $\tau \in [t_0, t)$, if, given t_0, we have

$$h(h^0, t) \leq h(h^0, t_0), \qquad (2.3.19)$$

then an energy flow of either (1.5.4) or (1.5.16′) is called *dissipative on* $[t_0, t)$; otherwise it is *accumulative on this interval*. Strong inequality in (2.3.19) produces *strict dissipativeness*. The relation (2.3.19) corresponds to what is sometimes called in the literature the *dissipation inequality*. The system (forces) may be called dissipative (accumulative) if the produced flows are such.

The same flow is called *monotone dissipative* (*monotone strictly dissipative*, *monotone accumulative*) on $[t_0, t)$ if for any $\tau \in [t_0, t)$ we have $f_0(\bar{x}(\tau)) \leq 0$ ($f_0(\bar{x}, (\tau)) < 0$, $f_0(\bar{x}(\tau)) > 0$). We conclude that the monotone dissipative arcs $[t_0, t)$ of trajectories are embedded in the dissipative field $\mathcal{H}^{0-}$, while the monotone accumulative arcs are in the accumulative field $\mathcal{H}^+$.

Monotone dissipativeness includes the *conservative* subcase: $f_0(x(\tau)) \equiv 0$. This is an often convenient convention, although the conservative systems will usually be specified separately. In such systems there is no energy flux; thus $h(h_C, R) = h_C$ for any h_C or $h(h^0, R) = h^0$ along any flow. Note that the above refers to the total energy. It is perhaps worth mentioning that if for some $t' \in R$ we obtain $h(h^0, t') = h^0$, it has nothing to do with the conservative system. For instance, it is not difficult to envisage some $t' \in [t_0, t_1)$ such that the flow dissipative on $[t_0, t')$ and accumulative on $[t', t_1)$ produces the flux

$$\int_{t_0}^{t_1} f_0(x) \, dt = 0. \qquad (2.3.20)$$

By (2.3.19) the flow is then dissipative on $[t_0, t_1)$, but it is not monotone

2.3 NONPOTENTIAL FORCES. POWER BALANCE. ENERGY FLOW

dissipative, neither is it conservative, although the flux (2.3.20) is the same for the conservative flow on $[t_0, t_1)$, since the latter has zero flux for any interval in R.

Consider the flow dissipative on $[t_0, t)$. By (2.3.19),

$$h(h^0, t_0) = h^0 \geq h(h^0, t).$$

The system acts during $[t_0, t)$ as an energy source for its environment (for instance, mechanical energy transferred into heat via dry friction). To do so, the system requires some supply of energy h^0, accumulated prior to the instant t_0, say during $[\tau_0, t_0)$, with $f_0(\bar{x}, \bar{u}) > 0$, $\tau \in [\tau_0, t_0)$, where $\bar{f}_0(\bar{x}, \bar{u})$ and $|t_0 - \tau_0|$ are large enough to provide recycling during dissipation on $[t_0, t)$. We assume this amount to be the minimum energy necessary and, by (2.3.14), finite:

$$H_r(x^0) = \inf \int_{\tau_0}^{t_0} f_0(\bar{x}, \bar{u}) \, d\tau > 0, \qquad k(\bar{x}(\tau_0), \tau) \to \bar{x}^0, \qquad (2.3.21)$$

as the *required energy*, as named by Willems [1]. Granted the required energy, we define the maximal amount of it which may, at t, have been extracted from the system ($f_0(\bar{x}, \bar{u}) \leq 0$) into the environment:

$$H_a(x) = \sup_u \int_{t_0}^t -f_0(\bar{x}, \bar{u}) \, ds, \qquad \bar{x} = k_u(x^0, s), \quad s \to t, \qquad (2.3.22)$$

and call it the *recoverable work* or the *available storage* (see Willems [2]). By dissipativeness, $0 \leq H_a(x) < \infty$. Obviously, the system cannot dissipate more energy than has been supplied to it, thus

$$h^0 = H_r(x^0) \geq h(h^0, t)$$

for any given $t \in R_0^+$.

On the other hand, $H_a(x(t))$ has been the maximum energy extracted, thus

$$0 \leq H_a(x) \leq h(h^0, t) \leq H_r(x^0) < \infty \qquad (2.3.23)$$

for any given $t \in R_0^+$ as long as the flow is dissipative. Hence the energy outflux is estimated by $H_r(x^0) - H_a(x) = h^0 - H_a(x)$, which agrees with (2.3.17), with nonpositive $f_0(x)$; check

$$\int_{t_0}^t f_0(x(s)) \, ds \leq H_r(x^0) - H_a(x). \qquad (2.3.24)$$

Since the flow is additive, we may introduce as well the maximum outflux as the *cycle energy* $H_k(x) = H_r(x^0) - H_a(x)$. Note that the cycle energy is in general different from the stored energy in the system at t, which is the flow value $h(h^0, t)$ defined by (2.3.17). The cycle energy serves as an estimate for the energy flow.

Recall the derivation of (2.3.15), and justify (2.3.4) and (2.3.9). We conclude that the autonomous uncontrolled system may not start from the basic equilibrium: $H_r(x^0) = 0$. This confirms our discussion in Section 1.5 [see (1.5.22)].

When the cycle energy vanishes, that is, when $H_r(x^0) = H_a(x)$, the dissipative flow is said to be *reversible*. The flow is *irreversible* if $H_k(x)$ does not vanish except at the equilibrium $\bar{x} = 0$.

Now let us suppose that the flow is accumulative on a given $[t_0, t)$. Our reasoning of all the above may be inverted. The system consumes the energy from the environment on the interval, similarly as the dissipative system had to do prior to the instant t_0 [see the derivation of (2.3.21)]. The available storage of the system at $t < \infty$ becomes the *produced storage*

$$H_p(\bar{x}) = \sup_{t_0 < t} \int_{t_0}^{t} f_0(\bar{x}, \bar{u}) \, dt > 0, \tag{2.3.25}$$

with $\tilde{f}_0(\bar{x}, \bar{u}) > 0$, obtained similarly to H_r for the dissipative system. This storage is fixed relatively to some N, introduced in (2.3.14), making $f_0(\bar{x}, \bar{u})$ bounded [see (2.3.17)]. Since no autonomous system may start from the basic equilibrium, we now estimate that

$$0 < h(h^0, t) \le H_p(\bar{x}) \tag{2.3.26}$$

for any given $t \in R$ as long as the flows are accumulative. We immediately have the influx estimated by $H_p(\bar{x})$ which by (2.3.26) is also the cycle energy.

Now suppose that the flow is dissipative on $[t_0, t_1]$ and accumulative on $[t_1, t_2]$. If the dissipative flow has been reversible, that is, $H_r(\bar{x}^0) = H_a(\bar{x}(t_1))$, we may envisage the situation that

$$H_a(\bar{x}(t_1)) = H_p(\bar{x}(t_2)) = H_r(\bar{x}^0),$$

that is, the energy flow is *recycled* to its value h^0, that is, $h(h^0, t_0) = h(h^0, t_2)$. When the above occurs for $[t_0, t_n)$, $t_n = nt_2$, with n a positive integer, the energy flow on this interval may be n times recycled, and we shall say that the flow is *multiply reversible*. In particular, if this multiple reversibility is such as to allow for a constant $L > 0$, yielding

$$h(h^0, t) = h(h^0, t + L), \qquad \forall t \in R, \tag{2.3.27}$$

that is, the function h is periodic with the period L, we shall say that the dissipative flow of $[t_0, t_1)$ is *L periodically reversible*. The flux on the recycled flow on $[t_0, t_2)$ vanishes [see (2.3.20)], and it does so for the entire interval of definition of the multiply reversible flow.

When the dissipative flow on $[t_0, t_1)$ is reversible, but $H_a(\bar{x}(t_1)) \neq H_p(\bar{x}(t_2))$, the flow is not recycled on $[t_0, t_2)$; but, in particular, one may still meet with circumstances in which for any $\varepsilon > 0$ there may be found a $\delta > 0$ sufficiently large for $t_2 - t_1 < \delta$ to yield

$$|H_a(\bar{x}(t_1)) - H_p(\bar{x}(t_2))| < \varepsilon. \tag{2.3.28}$$

2.3 NONPOTENTIAL FORCES. POWER BALANCE. ENERGY FLOW

One may say then that the energy flow on $[t_0, t_2)$ has been *almost recycled*. The reader may be able to form the concept of the *almost-periodic reversibility* of the flow on $[t_0, t_1)$.

The above discussion obviously applies to dissipative (accumulative) flows on suitable intervals. In most cases, the intervals of dissipativeness and accumulativeness may be partitioned further into intervals with monotone properties. Thus in a *time global sense, we have a collection of monotone dissipative and/or accumulative time intervals* to consider.

When a perturbation force appears in the resultant of the applied forces [see (1.3.15′), (1.3.24′), (1.5.16″)], the derivation of the power balance (2.3.15) remains the same, except that one more term of the perturbation power,

$$\sum_i Q_i^R(\bar{q}, \dot{\bar{q}}, t) \dot{q}_i \neq 0, \qquad \dot{q}_i \neq 0, \qquad (2.3.29)$$

is added to its right-hand side, yielding

$$\dot{H}(\bar{q}, \dot{\bar{q}}, t) = \bar{Q}^{DA}(\bar{q}, \dot{\bar{q}}) \dot{\bar{q}} + \bar{Q}^F(\bar{q}, \dot{\bar{q}}, \bar{u}) \dot{\bar{q}} + \bar{Q}^R(\bar{q}, \dot{\bar{q}}, t) \dot{\bar{q}} + \bar{Q}^{DD}(\bar{q}, \dot{\bar{q}}) \dot{\bar{q}} \quad (2.3.15')$$

Moreover, the axiom of bounded accumulation becomes

$$|\bar{Q}^{DA}(\bar{q}, \dot{\bar{q}}) \dot{\bar{q}} + \bar{Q}^F(\bar{q}, \dot{\bar{q}}, \bar{u}) \dot{\bar{q}} + \bar{Q}^R(\dot{\bar{q}}, \dot{\bar{q}}, t) \dot{\bar{q}}| \leq N < \infty, \qquad (2.3.14')$$

for all $\bar{q}(t), \dot{\bar{q}}(t) \in \Delta$ along a motion from $(\bar{q}^0, \dot{\bar{q}}^0, t_0)$.

The system under consideration is the nonautonomous (1.5.3′) with solutions motions $k(\bar{u}, \bar{x}^0, t_0, R_0^+)$ in $\Delta \times R \subset R^{N+1}$. Consequently, the conservative frame of reference $\mathcal{H}$ has to be augmented into $\mathcal{H} \times R$, together with all the related motions. The augmentation is cylindrical since $H(\bar{x})$ is not dependent explicitly on t. Hence the levels $H_C \times R$ are isometric to previous H_C in Δ. Upon $\mathcal{H} \times R$ we now have *energy-state motions* $h(\bar{x}^0, t_0, h^0, R)$ with the *energy flow* $h(h^0, t_0, \cdot): R_0^+ \to R$ being now t_0 dependent and with the power characteristic $f_0(\bar{x}, \bar{u}, t)$ being t explicit dependent and determined by (2.3.15′). Introducing the augmentation, we still have the same dissipation inequality and all its consequences are maintained.

Chapter 3

Controllability

3.1 FROM CLASSICAL TO RECENT PROGRAMS AND OBJECTIVES

Our previous two chapters dealt with modeling (simulation) only, so the control was kept dormant in its passive role of a given variable, in particular just zero. From now on, it becomes active, pursuing objectives of steering and design, although it may not always be the sole agent attempting such tasks. Also its role in these attempts may be quite different depending upon the setup of problems involved, especially upon the nature of the objectives.

We may be satisfied in setting up the control command in the so-called *open-loop* fashion (see Fig. 3.1a) without feedback, that is, in terms of an a priori given function $\bar{u}(t)$ of time only, thus ignoring a need to check the effects or whether the objective is actually being achieved the way we want it to be. In controlling a manipulator, this may, for instance, be the following case. We want the gripper to follow a *planned path* (model) in the base space. Using inverse kinematics, we calculate $\bar{q}_m(t)$ in Δ_q and its derivatives and substitute all this in a motion equation, say (1.3.15). Then, assuming for simplicity that Q_i^F is linear in u_i, that is, $Q_i^F(\bar{q}, \dot{\bar{q}}, u_i) = B_i(\bar{q}, \dot{\bar{q}})u_i$, $B_i(\bar{q}, \dot{\bar{q}}) \neq 0$, we have

$$u_{mi}(t) = \frac{1}{B_i}\left[\sum_j a_{ij}\ddot{\bar{q}}_{mj} + {}_iK_i(\bar{q}_m, \dot{\bar{q}}_m) - Q_i^D(\bar{q}_m, \dot{\bar{q}}_m) - Q_i^P(\bar{q}_m)\right], \qquad (3.1.1)$$

which is, through the known $\bar{q}_m(t)$, a given time function or an open-loop control force or torque. Applied to our manipulator, it should theoretically produce the desired path. In some cases it may indeed do so, for instance, when the constraints are such that the motion is totally predictable and the motion itself is so slow that there are no dynamic effects of structural forces, or in most of the robot "teaching" cases (see Markov, Zamanov, and Nenchev [1]).

Such an approach is central to the so-called *direct methods of manipulator control*; see Hanafi, Wright, and Hewit [1] for a very good classification. The technique is based upon complex computational algorithms which, as a rule, must involve derivatives of the Cartesian or generalized paths or both. So the

3.1 FROM CLASSICAL TO RECENT PROGRAMS AND OBJECTIVES

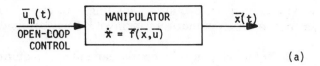

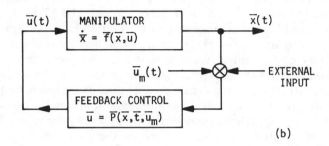

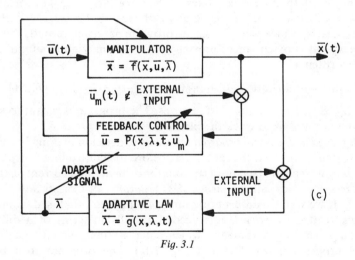

Fig. 3.1

main problems here are concerned with fast real-time calculation results. Hence the invention of auxiliary tabular look-up schemes for the computer; hence also the attempts to reformulate the models in order to avoid recursions and simplifications of the model by assuming low speeds, thus cutting off characteristics such as the Coriolis or centrifugal, etc. On the other hand, since this technique is open-loop, more than any other method it requires accurate models. In addition, the technique is intrinsically incapable of robustness against uncertainties, which are common in manipulator dynamics. We have already discussed the approach in Section 1.1, quoting some of the concerned literature. An interesting review of the problems involved may be found in Luh [1, 2].

At the present stage of development in robotic manipulators requiring both high accuracy of work and a high degree of robustness against uncertainty, we would want at least a current state related program for generating the current values of the control variable (see (1.5.2) and Fig. 3.1b). Then it becomes impossible to calculate the control force or torque in (1.3.17) *solely* on the basis of the inverse kinematics generated by the desired path. At each instant we have to get also *information* about the *actual state of the system*, which we can obtain only if the control variable based on the previous actual state is substituted into (1.3.15) and (possibly in the presence of all applied forces) the *actual* (real) trajectory of the system is either directly measured (output related control program) or obtained by analyzing solutions of the differential equations on-line, perhaps numerically or by any other method. Thus the overall control program (1.5.2) is based on the *feed-forward and feedback components*. The feed-forward is obtained as in (3.1.1), directly from a planned trajectory $\bar{q}_m(t)$, the feedback relates to the error $\bar{q}(t) - \bar{q}_m(t)$ in various ways that suit the case studies. Thus generally we have $\bar{u} = \bar{P}(\bar{q}, \dot{\bar{q}}, t, \bar{u}_m)$. Perhaps the most typical specification of such a program is given recently by Gusev, Timofeev, Yakubovich, and Yurevich [1], which in slightly simplified form may be written as

$$\bar{u}(t) = \bar{P}(\bar{q}(t), \dot{\bar{q}}(t), \ddot{\bar{q}}_m(t) + B_0(\dot{\bar{q}}(t) - \dot{\bar{q}}_m(t)) + B_1(\bar{q}(t) - \bar{q}_m(t))) \quad (3.1.2)$$

where B_0, B_1 are $m \times n$ matrix functions. The program is meant to stabilize the system as well as to track the planned $\bar{q}_m(t)$.

Such a *feedback control program* relies in its "decision making" on the state of the system, but has to be set *a priori* before the controlling process begins and thus must include fixed prescriptions for behavior at a current instant. It is therefore sometimes called a *memoryless program* (see Corless and Leitmann [1, 2, 3]). If we want to adjust our controller "on-line", that is, dynamically, a feedback related *adaptive law* is needed, usually in terms of a differential equation generating parameters which enter as time related signals to the control program (see (1.5.2) and Fig. 3.1c). The program then becomes feedback and *signal adaptive*. As the adaptive parameters may be a part of the manipulator structure, changing them "on-line," we not only adjust the program but may "reorganize" the structure. Unfortunately such reorganization could destabilize the system (see Barmish and Feuer [1]). Later in this text the role of adaptation will become evident and stabilization will be secured.

The *adaptive control of manipulators* represents an alternative class of methods to the mentioned direct methods. Adaptive methods are most likely to succeed in taking over since they are capable of the demanded robustness against uncertainty in the system parameter variations and thus improve the accuracy of work.

In adaptive control even more than in memoryless feedback control it is impossible to obtain the control force or torque by kinematic calculations from a given planned path, although such a path must quite obviously enter the program and the adaptive law as part of both. Those adaptive methods which deal with following a planned path by using the feedback program with signal synthesis adaptation only may be called *adaptive path tracking* methods. They already have a vast literature (see LeBorgue, Ibarra, and Espiau [1] for review and see also Abele and Sturz [1], Aksenov and Fomin [1], Brady, Hollerbach, and Johnson [1], Corless and Leitmann [1–5], Corless, Leitmann, and Ryan [1], Cvetkovic and Vukobratovic [1], Gusev and Yakubovich [1], Gusev, Timofeev, and Yakubovich [1], Hanafi, Wright, and Hewit [1], Horowitz and Tomizuka [1], Kulinich and Penev [1], Koivo and Paul [1], Luh and Lin [1], Ryan, Leitmann, and Corless [1], Slotine and Sastry [1], Stoten [1], Takegaki and Arimoto [1,2], Timofeev [2], Tkachenko, Brovinskaya, and Kondratenko [1], to name a few. The technique leaves a choice in the type of signal synthesis used, for example, local parametric optimization rules (see Asher and Matuszewski [1], Barnard [1], Kulinich and Penev [1]), Liapunov design (see Lindorf and Carrol [1]), or Popov hyperstability (see Jumarie [1]). This text deals mainly with the second type.

There is, however, a rapidly developing alternative to adaptive path tracking, still within the adaptive control of manipulators, namely utilization of the well-known technique of model reference adaptive control and identification (for a review see Landau [1, 2]). The path tracking is replaced by *dynamic model tracking*, which includes adaption of not only the control program (see Shaked [1]), but also directly the system parameters (see Fig. 3.1c), which must follow the parameters of the model. This is similar to the self-tuning control strategies which recently are evoking considerable interest in the process industries. The pioneering work in this method applied to manipulators belongs to Dubovsky and DesForges [1], with a sizeable number of publications following; see Balestrino, DeMaria and Sciavicco [1], Balestrino, DeMaria and Zinober [1], Corless, Goodal, Leitmann, and Ryan [1], Erzberger [1], Liegeois, Fournier, and Aldon [1], Skowronski and Pszczel [1], Skowronski [26–28].

In the model reference adaptive control, the Liapunov design technique dominates. In order to get the current state of the system and thus the actual deviation from the planned path or from a motion of the reference model, we need to either solve the motion equation or estimate the behavior of the solutions, possibly without solving the equation. This is where the Liapunov *direct method* is of such help because it tells us *wherefrom* in Δ *which task* can be achieved and thus saves a lot of money spent otherwise on numerical investigations. An answer to these questions is essentially what controllability is all about.

We turn now to the *objectives*. It is they which grow more and more complex, thus increasing the demand on control programs. Ever since its inception about 1960 when the first work appeared (see Kalman and Bertram [1], and Kalman [1]), the notion of controllability has been restricted to linear autonomous constant parameter systems, to the joy of people who like Laplace transformation, but not necessarily of those who would like the models to be justified physically. Moreover, the notion has been presented in a rather simplified and applicationally rigid form. Roughly speaking, it meant

the ability to transfer a system between two given states $\bar{x}^0$ and $\bar{x}^r = \bar{x}(t_r)$ in Δ, under specified control.

The associated *control problem* was to determine the set of such points $\bar{x}^0$ in Δ wherefrom the above transfer could be made under some control. Such a set was later called the region of controllability for the transfer concerned. Then, quite obviously, one was interested in specifying the control under which the transfer could be achieved from that region.

Even in this above-indicated simple transfer case, the control problem is not entirely trivial, particularly when the target $\bar{x}^r$ moves. We illustrate the point on the one-dimensional intuitively solvable situation in the following example.

Example 3.1.1. The gripper pursues an object on a conveyor. We ignore gravitation as well as any other applied forces except the control force, and let the masses of the gripper and the object plus conveyor both be equal to a unit. The conveyor moves along a straight line, say the ξ axis:

$$\xi = \xi^0 + 0.1t^3, \qquad \xi^0 > 0,$$

and the gripper follows $\xi(t)$, with the acceleration $\ddot{q}_G(t)$ equal to the controlling force: $\ddot{q}_G(t) = u(t), |u| \leq 1$. We choose states $q_G = x_1$ and $\dot{q}_G = x_2$; whence

$$\dot{x}_1 = x_2, \qquad \dot{x}_2 = u$$

become the state equations (1.5.1). Obviously the pursuit is made along the ξ axis and with the maximal force available $u = 1$, until the contact is made. The contact occurs when $x_1(t) = \xi(t)$ for some t, that is, assuming $x_1(0) = 0$, when $\frac{1}{2}t^2 = \xi^0 + 0.1t^3$. This, when solved for $t = t_r$, gives the time of contact. We can see that the contact is not possible if $\xi^0 > 1.85$, since there is no finite t_r; contact occurs at $t_r = 3.33$ if $\xi^0 = 1.85$ and at $t_r < 3.33$ for $\xi^0 < 1.85$. In the first of the three cases, the object cannot be reached. We also see that there is a set of initial positions ξ^0 wherefrom controllability is possible with the given constraint upon the control force: $|u| \leq 1$. Note also that in this case the trivial inverse kinematics $q_G(t) = \xi(t)$ has been used to calculate the open-loop control force. ∎

3.1 FROM CLASSICAL TO RECENT PROGRAMS AND OBJECTIVES

The linear controllability problem in the above or slightly modified sense was fully solved a long time ago, with simply expressed necessary and sufficient conditions given by Kalman himself at the first IFAC Congress in 1960 (see Barnett [1]). However, the manipulator is a nonlinear and often non-linearizable structure. Thus reviewing the further progress, we leave aside attempts to do the above controllability in terms of linearizable models (perturbed linear), mainly by Aronsson [1], Dauer [1–4], Lukas [1,2], Klamka [1,2], and others, and refer to the infrequent results without an attempt to linearize, like Markus [1], Gershwin and Jacobson [1], and later the Leitmann school, Sticht, Vincent, and Schultz [1], Grantham and Vincent [1], Vincent and Skowronski [1], Stonier [1], and Skowronski [19, 21].

Let us consider the system (1.5.3) and introduce the following:

Definition 3.1.1. The system is *controllable* at $\bar{x}^0$ *for reaching a given point* $\bar{x}^r$ in Δ, if there is a control program $P(\bar{x})$ and $t_r \geq 0$ such that $k(\bar{x}0, t_r) = \bar{x}^r$.

This $\bar{x}^0$ is called controllable for reaching $\bar{x}^r$, while the set of all such $\bar{x}^0$'s in Δ is termed the *region of controllability for reaching* $\bar{x}^r$, and designated Δ_r. The controllability is called *complete* if $\Delta_r = \Delta$. Any subset of Δ_r is called controllable for reaching $\bar{x}^r$. Given some set $\tilde{\Delta} \subset \Delta$, let $\tilde{\Delta} \cap \Delta_r = \tilde{\Delta}_r$ obviously controllable for reaching $\bar{x}^r$, then the system is said to be *controllable on* $\tilde{\Delta}_r$. The controllability of the above becomes $\bar{x}^0$-*uniform*, if t_r and $\bar{u}$ do not depend upon $\bar{x}^0$.

The concept of *reachability* is dual to controllability for reaching. The point $\bar{x}^r$ of Definition 3.1.1 is *reachable from* $\bar{x}^0$ and consequently from all subsets of Δ_r. In the case of the autonomous system, because of the group property (see Section 1.5) if $\bar{x}^r$ is reachable from $\bar{x}^0$ under some program P, it is also controllable for reaching $\bar{x}^0$ under the same P but in reverse time, that is, along trajectories with $\tau = -t$ substituted (retrograding). Also if all points of some subset $\Delta_0 \subset \Delta$ are both controllable for reaching and reachable with respect to any other point of that set, the set will be called *maneuverable* for our system (cf. Gayek and Vincent [1]). We shall return to these notions in more detail in Chapter 4.

Example 3.1.2. Consider the controllability for reaching the angle $x^r = 0$ by a gripper rotating with angle $x(t)$ and velocity

$$\dot{x} = \sin x - u \cos x,$$

with $\Delta : x(t) \in [-\pi, \pi]$, $U : u(t) \in [-1, 1]$. To achieve the transfer to $x^r = 0$ from $x^0 > 0$ we need $\dot{x} < 0$, and from $x^0 < 0$ we need $\dot{x} > 0$ (see Fig. 3.2).

Consequently, we use the control program defined by

$$u(t) = \begin{cases} 1, & x \in [0, \pi/4), \\ -1, & x \in (-\pi/4, 0], \end{cases}$$

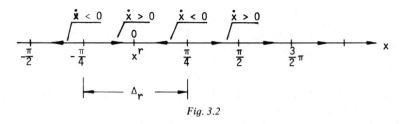

Fig. 3.2

to hold on $\Delta_r = (-\pi/4, 0] \cup [0, \pi/4)$. Observe that no $x^0 \notin \Delta_r$ is controllable for reaching x^r, no matter what u is applied, which defines Δ_r as the region of controllability for reaching $x^r = 0$. Hence there also is no need to define P outside Δ_r. ∎

The quoted Markus conditions for the controllability for reaching $\bar{x}^r$ are simply those for approaching $\bar{x}^r = 0$ as $t \to \infty$. We have

Theorem 3.1.1. *Let there be a function of class C^1 (continuously differentiable) $V(\cdot): R^N \to R$ and a C^1 program $\bar{P}(\cdot)$ defined on Δ and such that*

(i) $V(\bar{x}) > 0$ for $\bar{x} \neq 0$, $V(0) = 0$,
(ii) $V(\bar{x}) \to \infty$ as $|\bar{x}| \to \infty$,
(iii) $\nabla V(\bar{x}) \cdot f(\bar{x}, P(\bar{x})) < 0$ for $\bar{x} \neq 0$,

then the system (1.5.3) is completely controllable for reaching an arbitrarily small neighborhood of $\bar{x}^r = 0$.

Note that $\nabla V(\bar{x})$ means grad $V = ((\partial V/\partial x_1), \ldots, (\partial V/\partial x_N))$.

The proof is intuitively obvious; for details see Markus [1]. Heuristic justification follows from the fact that for the smooth function increasing from zero to infinity with $|\bar{x}|$, any stoppage in approaching $\bar{x} = 0$ along any trajectory would contradict (iii) which requires $\dot{V}$ to decay along all trajectories (see Fig. 3.3).

Let us now set up $V(\bar{x}) \equiv H(\bar{x})$ and let $\bar{x}^r = \bar{x}^e = 0$ be the basic energy level. Then recall our discussion in Section 2.3 on the energy flows within the local cups, in particular, the basic cup. The shape of $\mathscr{H}$ about $\bar{x}^e = 0$ guarantees that the energy levels Z_C nest about this equilibrium up to the threshold level, with overlapping interiors. By the condition (iii), given a suitable control program, the cup is covered with the strictly dissipative field $\mathscr{H}^-: f_0(\bar{x}, \bar{u}) = \dot{H}(\bar{x}(t)) < 0$ which makes all trajectories enter the interior of each level Z_C. Since the family of levels is continuous, all trajectories approach $\bar{x}^r = \bar{x}^e = 0$ asymptotically. Note that the converse is also true, whence in terms of energy the Markus conditions (i), (iii) are also necessary.

Observe that in Example 3.1.2 it suffices to choose $V = \frac{1}{2}x^2$ in order to satisfy Theorem 3.1.1.

3.1 FROM CLASSICAL TO RECENT PROGRAMS AND OBJECTIVES

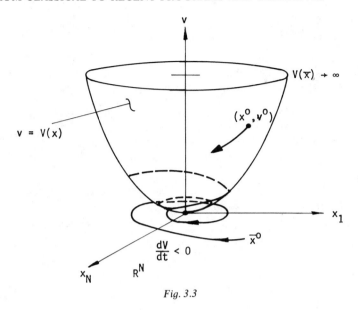

Fig. 3.3

Example 3.1.3. Consider the following simple system of the (1.5.16) type

$$\dot{x}_1 = -x_1 - x_2^2,$$
$$\dot{x}_2 = -x_2 + u,$$

and require complete controllability for reaching the small neighborhood of $(0, 0)$. Choosing $V = x_1^2 + x_2^2$ satisfies conditions (i) and (ii). To satisfy (iii) we select the program $P(x_1, x_2) = -x_1 x_2$ and calculate

$$\dot{V}(t) = 2x_1\dot{x}_1 + 2x_2\dot{x}_2 = 2x_1(-x_1 + x_2^2) + 2x_2(-x_2 + u),$$

which upon substituting $u = -x_1 x_2$ and rearranging gives $\dot{V}(t) = -2V(x_1(t), x_2(t))$, which obviously satisfies (iii) and secures the required controllability. ∎

The evident shortcoming of Theorem 3.1.1 is that $\bar{x}^e = 0$ is actually never reached, thus it is impossible to assess t_r, not to mention prespecify it, which usually is desired in manipulator work. Second (as the reader may have noticed), the energy surface $\mathscr{H}$ corresponding to the conditions concerned has only a single minimum (one equilibrium system), which does not mean that the system is linear or linearizable, but nevertheless considerably narrows the class of nonlinear systems available to study. With all the above, Markus was the first who brought in conditions for controllability that did not utilize linearization in any way, and his result is well suited to indicate many aspects of the theory.

The time interval for reaching $\bar{x}^r$ has to be stipulated when the gripper is allowed only so much time for a given operation.

Definition 3.1.2. *The system (1.5.3) is controllable at $\bar{x}^0$ for reaching $\bar{x}^r$ in the stipulated time $t_r < \infty$ if and only if there is $\bar{P}(\bar{x})$ such that $k(\bar{u}, \bar{x}^0, t_r) = \bar{x}^r$.*

We let again $\bar{x}^r = 0$. Gershwin and Jacobsen [1] were historically the first to introduce controllability in stipulated time; see also Chukwu [1]. The conditions have been proved for the nonautonomous system (1.5.3′). We simplify them to our autonomous case in the following theorem.

Theorem 3.1.2. *Suppose there is a C^1 function $V(\cdot): R^N \to R$ and a control program $\bar{P}(\bar{x})$ defined on R^N such that*

(i) *for all continuous N-vector functions $s(t)$,*

$$\lim_{t \to t_r} s(t) \neq 0 \Rightarrow \lim_{t \to t_r} V(s(t)) = \infty;$$

(ii) $\nabla V(\bar{x}) \cdot \bar{f}(\bar{x}, \bar{P}(\bar{x})) \leq c < \infty$

for suitable constant c. Then the system (1.5.3) is controllable at $\bar{x}^0$ for reaching $\bar{x}^r = 0$ in time $t_r < \infty$, and $\bar{P}(\bar{x})$ accomplishes this objective.

The proof is immediate: for $t < t_r$, condition (ii) yields

$$V(\bar{x}(t)) = V(\bar{x}^0) + \int_{t_0}^{t} \dot{V} dt \leq V(\bar{x}^0) + ct.$$

Thus

$$\lim_{t \to t_r} V(\bar{x}(t)) \leq V(\bar{x}^0) + ct_f \neq \infty,$$

and so by (i),

$$\bar{x}(t_r) = \lim_{t \to t_r} \bar{x}(t) = 0.$$

A closed and bounded (i.e., compact) target set $\mathfrak{T}$ (see Section 1.4), may be considered instead of the target point $\bar{x}^r$. The transfer to $\mathfrak{T}$ may then mean reaching either during arbitrary but finite time or in stipulated $t_r - t_0$, not necessarily terminal $t_r \leq t_f$.

Definition 3.1.3. *The system (1.5.3) is controllable at $\bar{x}^0$ for reaching the target $\mathfrak{T}$ in Δ, if and only if there is $\bar{P}(\bar{x})$ such that $k(\bar{u}, \bar{x}^0, R_0^+) \cap \mathfrak{T} \neq \emptyset$.*

The definition means that there is a $t_r > t_0 \geq 0$ such that $k(\bar{u}, \bar{x}^0, t_r) \in \mathfrak{T}$. We also have

3.1 FROM CLASSICAL TO RECENT PROGRAMS AND OBJECTIVES

Definition 3.1.3'. The system (1.5.3) is controllable at $\bar{x}^0$ for reaching $\mathfrak{T}$ during stipulated $t_r - t_0$ if there are $\bar{P}(\bar{x})$ and $t' \leq t_r$ such that $k(\bar{u}, \bar{x}^0, t') = \bar{x}^r \in \mathfrak{T}$.

The notions of the region of controllability, controllable sets, etc., remain unchanged in their definitions. However, denoting the region Δ_{rt} we observe that there is either $\Delta_{rt} \subset \Delta_r$ or $\Delta_r \subset \Delta_{rt}$ depending upon whether the stipulated t_r is larger or smaller than the undefined t_r following Definition 3.1.3 (see Section 4.4).

As mentioned previously, to solve the controllability problem is to determine the corresponding region of controllability in Δ. For this task one would at least like to have conditions that are necessary and sufficient for a particular controllability. As a rule, these conditions are very difficult to find and thus so is the region. But do we really need the region? If we have our manipulator arm at a specific point $\bar{x}^0$, or within a specific set $\Delta_0 \subset \Delta$, which is almost always the case, it is irrelevant to know *all* the initial states, that is, from where else the controllability is achievable, unless one needs this information for design purposes. Then, however, there will be state constraints to consider which will narrow the region to a specific subset Δ_0 of Δ anyway. Even if it is as large as Δ itself, it is still given. Thus, quite often it proves totally *satisfactory to have sufficient conditions for particular controllability satisfied on a specific set* $\Delta_0 \subset \Delta$, thus proving it to be a subset of the region of controllability. Obviously, in the case of our present controllability for reaching $\mathfrak{T}$ we must have $\Delta_0 \supset \mathfrak{T}$, since $\mathfrak{T}$ belongs trivially to Δ_r. Then also the intersection $\partial \mathfrak{T} \cap \Delta_0 \triangleq \partial' \mathfrak{T}$ is the *usable part* of the boundary $\partial \mathfrak{T}$.

Now, we specify Δ_0 as follows. Given a C^1 function $g(\cdot): \Delta \to R$, let $\mathring{\Delta}_0 \triangleq \{\bar{x} \in \Delta \mid g(\bar{x}) < 0\}$ and $\partial \Delta_0 \triangleq \{\bar{x} \in \Delta \mid g(\bar{x}) = 0\}$. We have the following theorem proved by Sticht, Vincent, and Schultz [1] in 1975.

Theorem 3.1.3. *Let there be two functions: a C^1 function $V(\cdot): \mathring{\Delta}_0 \to R$ with $\mathfrak{T} \triangleq \{\bar{x} \in \Delta \mid V(\bar{x}) \leq 0\} \neq \emptyset$ and $\bar{P}(\cdot): \overline{C\mathfrak{T}} \to U$, where $\overline{C\mathfrak{T}} = \Delta_0 - \mathfrak{T}$, such that*

(i) $V(\bar{x}) > 0, x \notin \mathfrak{T}$,
(ii) $V(\bar{x}) \to 0$ as $\bar{x} \to \partial \mathfrak{T}$,
(iii) $\nabla V(\bar{x}) \cdot \bar{f}(\bar{x}, P(\bar{x})) < 0$,
(iv) *for suitable $v_A \geq 0$ the closure $\bar{\Delta}_0$ encloses all levels $V(\bar{x}) = \text{const} \leq v_A$.*

Then the system (1.5.3) is $\bar{x}^0$ uniform controllable on Δ_0 for reaching $\mathfrak{T}$.

The proof may be found in the cited paper. Heuristically it is the same as for the Markus case. The obvious difference from Markus stems from the fact that $\mathfrak{T}$ is given about the origin and one must lower the surface $v = V(\bar{x})$ accordingly to get its minimum below zero (see Fig. 3.4). Here also the choice of $V(\bar{x})$ is restricted as $\partial \mathfrak{T}$ must be one of its levels. This is, however, not a

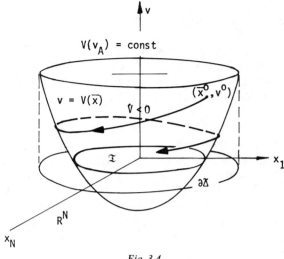

Fig. 3.4

serious restriction since, if the given $\partial \mathfrak{T}$ does not fit the desired V candidate, we may generate reaching something inside $\mathfrak{T}$ which does fit the shape of the V level, thus implementing reaching $\mathfrak{T}$ as well, see the target core in Fig. 1.2.

Example 3.1.4. For the state equations

$$\dot{x}_1 = -x_2 + ux_1^3, \qquad \dot{x}_2 = x_1 + ux_2^3,$$

we require controllability on $\Delta_0: x_1^2 + x_2^2 \leq 10$ for reaching $\mathfrak{T}: x_1^2 + x_2^2 \leq 1$. Setting up $V = x_1^2 + x_2^2 - 1$, $u = 1 - (x_1^2 + x_2^2)$ the objective is achieved. Indeed, conditions (i) and (ii) of Theorem 3.1.3 hold. To check (iii), we calculate

$$\dot{V}(t) = 2u(x_1^4 + x_2^4) = 2(x_1^4 + x_2^4)[1 - (x_1^2 + x_2^2)],$$

which is negative-definite for $x_1^2 + x_2^2 > 1$, that is, for $(x_1, x_2) \in C\mathfrak{T}$, and thus generates the desired controllability. ∎

The latter version of the set-to-set transfer provides the recent setup for the classical problem of controllability between two points. But, even in this relatively general version, it still does not cover the common manipulator problems, for instance, *path following* which requires not only reaching, but successive reachings in stipulated time instants or *staying within a moving target*. It also does not cover a requirement of a prolonged contact of gripper with a work piece. Such cases need the objective of capture introduced by Vincent and Skowronski [1].

Definition 3.1.4. The system (1.5.3) is *controllable at $\bar{x}^0$ for capture in $\mathfrak{T}$* if and

3.1 FROM CLASSICAL TO RECENT PROGRAMS AND OBJECTIVES

only if there are $\bar{P}(\bar{x})$ and a constant $t_c > 0$ such that $k(\bar{u}, \bar{x}^0, t) \in \mathfrak{T}$ for all $t \geq t_c$. When t_c is stipulated, we have *controllability for capture in $\mathfrak{T}$ during t_c*.

The state $\bar{x}^0$ is called controllable for capture and the set of all such $\bar{x}^0$ is the region of controllability for capture or, as we say it briefly, capturability, denoted Δ_c. Any subset of Δ_c is a set controllable for capture and if for some $\tilde{\Delta} \subset \Delta$, we have $\tilde{\Delta} \cap \Delta_c \triangleq \tilde{\Delta}_c \neq \emptyset$ then we say that the system is controllable for capture on $\tilde{\Delta}_c$. The controllability is $\bar{x}^0$ uniform if t_c and $\bar{P}$ of Definition 3.1.4 do not depend upon $\bar{x}^0$ but on $\tilde{\Delta}_c$.

Since capture implies reaching, we have $\Delta_c \subset \Delta_r$ and $\mathfrak{T} \subset \Delta_r$, but in general $\mathfrak{T} \not\subset \Delta_c$ (see Fig. 3.5). Thus, given $\bar{x}^0 \in \Delta_c$ there is a $\bar{P}$ such that for any $t \geq t_c$ the system is in $\mathfrak{T}$, but is not necessarily maintained in any subset of $\mathfrak{T}$. Furthermore, it does not follow that the system can even be maintained in a given subset of $\mathfrak{T} \cap \Delta_c$, since for certain points of $\mathfrak{T} \cap \Delta_c$ it may be possible that the system passes through the point, but under no $\bar{P}$ is it able to return to a neighborhood of that point.

Definition 3.1.5. A set $\tilde{M}$ is *positively invariant under the controlled* $k(\bar{u}, \bar{x}^0, \cdot)$ if and only if there is $\bar{P}$ such that $\bar{x}^0 \in \tilde{M}$ implies $k(\bar{u}, \bar{x}^0, R_0^+) \subset \tilde{M}$.

It follows from Definitions 3.1.4 and 3.1.5 that there is at least one nonvoid positively invariant-under-control subset of $\mathfrak{T}$. Any such open subset is called *capture subtarget* $\mathfrak{T}_c$. Let $\Delta_0 \subset \Delta$, $C\mathfrak{T}_c = \Delta_0 - \mathfrak{T}_c$ and let us agree upon the notation V (set) $= \{v = V(\bar{x}) | \bar{x} \in \text{set}\}$ for any set in R^N. Vincent and Skowronski [1] introduced the following sufficient conditions for the controllability in question.

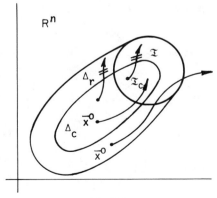

Fig. 3.5

Theorem 3.1.4. *Given* Δ_0, $\mathfrak{T}_c$ *suppose there is a* C^1 *function* $V(\cdot):D \to R$, *D (open)* $\supset C\mathfrak{T}_c$ *and a program* $\bar{P}$ *defined on* $C\mathfrak{T}_c$ *such that for all* $\bar{x} \in D$

(i) $V(\partial \mathfrak{T}_c) \leq V(\bar{x}) \leq V(\partial \Delta_0)$;
(ii) *there is a constant* $c > 0$ *such that* $\bar{u} = \bar{P}(\bar{x})$ *gives*

$$\nabla V(\bar{x}) \cdot \bar{f}(\bar{x}, \bar{u}) \leq -c.$$

Then the system is controllable on Δ_0 for capture in $\mathfrak{T} \supset \mathfrak{T}_c$.

Proof. Suppose that, given $\bar{x}^0 \in \Delta_0$, some trajectory from this point leaves Δ_0. Then there is $t_1 > 0$ such that $k(\bar{u}, \bar{x}^0, t_1) = \bar{x}^1 \in \partial \Delta_0$. By (i), $V(\bar{x}^1) \geq V(\bar{x}^0)$ which contradicts (ii). Hence

$$k(\bar{u}, \bar{x}^0, R_0^+) \subset \Delta_0, \quad \forall \bar{x}^0 \in \Delta_0. \tag{3.1.3}$$

Now consider $k(\bar{u}, \bar{x}^0, R_0^+)$, $\bar{x}^0 \in C\mathfrak{T}_c$. By (ii), we have $\dot{V}(\bar{x}(t)) \leq -c$, $\bar{x}(t) = k(\bar{u}, \bar{x}^0, t)$ for all $t > 0$. Integrating,

$$t \leq (1/c)[V(\bar{x}^0) - V(\bar{x}(t))]. \tag{3.1.4}$$

Denote $v_{\mathfrak{T}} = \inf V(\bar{x}) | \bar{x} \in \partial \mathfrak{T}_c$, $v_0 = \sup V(\bar{x}) | \bar{x} \in \partial \Delta_0$. By (i), $V(\bar{x}) - v_{\mathfrak{T}} \geq 0$ and $V(\bar{x}^0) - v_0 \leq 0$. Hence $V(\bar{x}) - v_{\mathfrak{T}} \geq V(\bar{x}^0) - v_0$ or $V(\bar{x}^0) - V(\bar{x}) \leq v_0 - v_{\mathfrak{T}}$ and (3.1.4) becomes

$$t \leq (1/c)(v_0 - v_{\mathfrak{T}}) = t_c \tag{3.1.5}$$

for an arbitrary trajectory in $C\mathfrak{T}_c$. Consequently, for $t > t_c$ the trajectory leaves $\overline{C\mathfrak{T}_c}$. By (3.1.3) it may not leave Δ_0; thus it enters $\mathfrak{T}_c$. A return to $C\mathfrak{T}_c$ is not possible, since then (i) and (ii) contradict by the same argument that led to (3.1.3), which completes the proof.

By specifying c from (3.1.5), we obtain

Corollary 3.1.4. *The system is controllable for capture in $\mathfrak{T}$ during (given) t_c if the conditions of Theorem 3.1.4 hold with (ii) replaced by*

$$\nabla V(\bar{x}) \cdot \bar{f}(\bar{x}, \bar{u}) \leq -(v_0 - v_{\mathfrak{T}})/t_c. \tag{3.1.6}$$

Note that capture follows also from Markus' theorem, although this was never postulated by Markus.

Example 3.1.5. Consider the system of Example 3.1.4, with the same target $\mathfrak{T}: x_1^2 + x_2^2 \leq 1$ and $\Delta_0: x_1^2 + x_2^2 \leq 10$. The controllability on Δ_0 for capture in $\mathfrak{T}$ is secured by the choice of $V = x_1^2 + x_2^2$ and $\bar{P}$, which follows from the calculation below. Indeed condition (i) of Theorem 3.1.4 holds in an obvious way. In order to check (ii), we let $c = (10 - 1)/t_c$, where t_c is a suitable interval and calculate $\dot{V}(t) = 2u(x_1^4 + x_2^4)$. Condition (ii) requires

$$\dot{V}(t) + 9/t_c = 2u(x_1^4 + x_2^4) + 9/t_c \leq 0;$$

3.1 FROM CLASSICAL TO RECENT PROGRAMS AND OBJECTIVES

whence $\bar{P}$ is given by a choice within the constraints

$$u(t) \leq -9/2t_c(x_1^4 + x_2^4), \qquad t_c > 0,$$

for $(x_1^2 + x_2^2) \in [1, 10]$. ∎

As mentioned, capture covers the case of prolonged contact with a workpiece. After state variables have been transformed into coordinates relative to the target, capture may as well cover the planned path tracking. But even the combination of successive reachings and captures (see Chapter 4) by no means exhausts our required objectives. Since the first developments in the 1960s, we have learned that for many applications, in particular, for the manipulator control, controllability is needed in a much wider sense than that described above. For instance, it is obvious that it is not only a planned path to be captured but sometimes a dynamic model, that our control needs not only reaching and capture but also avoidance, or a combination of both, and with respect to different objects at the same time or the same object at different times—what especially refers to objects on conveyors.

In general, the applications of control theory require much more flexible understanding of the idea of controllability. Again speaking roughly, it must be considered.

the ability to attain a prescribed objective regarding a behavior of the state trajectories under some control program.

The objective may be *qualitative* like stabilization, reaching, rendezvous, capture, or avoidance by the trajectories of certain sub-sets in Δ, rejection, periodic or quasiperiodic behavior of the trajectories, their convergence to a specified path or dynamical model, etc. The objective may also be *optimal* like extremizing a certain cost functional along the trajectories (see Section 3.5). The associate *controllability problem* is as before—to determine the region of controllability for the objective concerned—but in practical terms this set may be replaced by a candidate for a suitable controllable set. This candidate is then confirmed by checking against some sufficient conditions.

Exercises

3.1.1 Using the Markus theorem (Theorem 3.1.1), show the controllability on $\Delta = R^2$ for reaching a small neighborhood of $(0, 0) \in \Delta$ for the system

$$\dot{x}_1 = -x_1^3 + ux_2^2, \qquad \dot{x}_2 = -4x_1^2 x_2 - x_2^3.$$

The Liapunov function is $V(x_1, x_2) = 4x_1^2 + x_2^2$ and the control program is $P(x_1, x_2) = x_1$.

3.1.2 For which a, b is there a control u such that the system

$$\dot{x}_1 = ax_1 + bx_2 + u, \qquad \dot{x}_2 = -ax_2 - bx_1$$

is controllable on $\Delta_0: x_1^2 + x_2^2 < 10$ for reaching the target $\mathfrak{T}: x_1^2 + x_2^2 \leq 1$? Determine the feedback program.

3.1.3 For the system of Exercise 3.1.2 and a, b determined above, check conditions for controllability on the same Δ_0 for capture in the same $\mathfrak{T}$ and determine T_c with a stipulated rate of approach $c = 5$.

3.1.4 Consider $\ddot{q} + u + aq + bq^3 = F(q, \dot{q}, t)$, $a > b > 1$, $q, \dot{q} \in R$, $|F(q, \dot{q}, t)\dot{q}| < N$, with N a given constant. Determine conditions on $a, b, F(\cdot)$ for controllability on $\Delta_0: (aq - bq^3)q \geq 0$ for capture in $\mathfrak{T}: q^2 + \dot{q}^2 \leq 1$. [*Hint:* Use $u = l\dot{q}|\dot{q}|$, $l > 0$.]

3.1.5 Consider the system

$$\dot{x}_1 = x_1 - x_1^3 + u, \qquad \dot{x}_2 = -x_1^2 x_2,$$

and let $\mathfrak{T}: x_1^2 + x_2^2 \leq 1$, $\Delta_0: x_1^2 + x_2^2 \leq 10$. Using Theorem 3.1.4, find a control program securing capture in $\mathfrak{T}$ from Δ_0.

3.1.6 Given $\partial \mathfrak{T}_c, \partial \Delta_0$ as levels of a Liapunov function, show that condition (ii) of Theorem 3.1.4 is also necessary. Then given the rate c of change of $V(\bar{x})$, determine the region of controllability.

3.1.7 Consider the system of Exercise 3.1.1. Assuming that $\mathfrak{T}_c: 4x_1^2 + x_2^2 \leq 1$, find a suitable control program for controllability of capture on $\Delta_0: 4x_1^2 + x_2^2 \leq 20$, in stipulated time $T_c = 10$.

3.1.8 Basing on the proof of Theorem 3.1.2, how could one relax the conditions without jeopardizing the hypothesis?

3.1.9 For the system

$$\dot{x}_1 = -x_1^3 + u^2(x_1 x_2), \qquad \dot{x}_2 = -4x_2 - \tfrac{1}{2} x_1^4 x_2 u,$$

with $\Delta_0: x_1^2 + 2x_2^2 - 3 \leq 20$, $\mathfrak{T}: x_1^2 + 2x_2^2 - 3 \leq 0$, check the conditions of Theorem 3.1.3.

3.2 STABILITY, BOUNDEDNESS, AND STABILIZATION

The reader might already have observed that controllability problems may be cast one way or another into stability-type studies, in particular, via the Liapunov direct method. As we now pass over to broader objectives, it seems useful to begin with the objective that generated the method, i.e., stability.

Recall the state equation

$$\dot{\bar{x}} = \bar{f}(\bar{x}, \bar{u}, t) \qquad (1.5.3')$$

of the manipulator subject to external perturbation still without uncertainty, and with the control $\bar{u}$ as well as adaptation parameter $\bar{\lambda}$ again passive, (given functions of time or zero) while $k(\bar{u}, \bar{x}^0, t_0, \cdot): R_0^+ \to \Delta \times R$ is a motion through $(\bar{x}^0, t_0) \in \Delta \times R$ (see Section 1.5).

3.2 STABILITY, BOUNDEDNESS, AND STABILIZATION

Consider a set $M \subset \Delta \times R$, and let $M(t)$ denote the t section of M such that $M(t) \times R = M$. If there is a closed set $\Delta_M \subset \Delta \subset R^N$ such that $M(t) \in \Delta_M$ for all t, M is bounded. Then we let the distance from $\bar{x}(t) = k(\bar{u}, \bar{x}^0, t_0, t)$ to $M(t)$ in Δ be $\rho(\bar{x}(t), M(t)) = \inf \rho(\bar{x}, \bar{y}) \mid \bar{y} \in M(t)$, where $\rho(\bar{x}, \bar{y})$ denotes the distance of any two points $\bar{x}, \bar{y}$ in Δ. Given $\bar{u}(t)$, $\bar{\lambda}(t)$, the following definitions specify basic notions of Liapunov stability.

Definition 3.2.1. $\bar{M}$ is *stable* with respect to motions of (1.5.3′), if and only if for any $\varepsilon > 0$, $x^0 \in \Delta$, $t_0 \in R$, there is $\delta > 0$ such that $\rho(\bar{x}^0, \bar{M}(t_0)) < \delta$ implies $\rho(k(\bar{u}, \bar{x}^0, t_0, t), \bar{M}(t)) < \varepsilon$ for all $t \geq t_0$. When δ does not depend upon $\bar{x}^0$, $\bar{M}$ is *equistable*; when δ does not depend upon $\bar{x}^0$, t_0, $\bar{M}$ is *uniformly stable*. $\bar{M}$ is *unstable*, if and only if it is not stable.

Definition 3.2.2. $\bar{M}$ is an *attractor (repeller)* for motions of (1.5.3′) if and only if for each $\bar{x}^0$, $t_0 \in \Delta_0 \times R$ there is δ such that $\rho(\bar{x}^0, M(t_0)) < \delta$ implies $\rho(k(\bar{u}, \bar{x}^0, t_0, t), M(t)) \to 0$ as $t \to \infty$ ($t \to -\infty$). We have *equiattraction (repelling)* when δ does not depend on $\bar{x}^0$ and *uniform attraction (repelling)* if δ does not depend on $\bar{x}^0$, t_0.

Note that M being an attractor does not imply stability, because the motion may escape to infinity and return before approaching M. That is why attraction is also called *quasiasymptotic stability* (see Yoshizawa [1] and Antosiewicz [1]). The set of all $\bar{x}^0$ in Δ wherefrom motions are attracted to M is called the *region of attraction*, denoted by Δ_a.

Definition 3.2.3. $\bar{M}$ is *asymptotically stable* under the motions of (1.5.3′) if and only if it is a stable attractor. It is *equistable* or *uniformly asymptotically stable*, if and only if it is an equistable or uniformly stable attractor, respectively. Then also $\bar{M}$ is *completely unstable* if it is an unstable repeller.

Definition 3.2.4. M is *positively* (or *negatively*) *invariant* under $k(\bar{u}, \bar{x}^0, t_0, \cdot)$ if and only if there is $\bar{P}$ such that $(\bar{x}^0, t_0) \in M$ implies that $k(\bar{u}, \bar{x}^0, t_0, R_0^+) \subset M$ or $k(\bar{u}, \bar{x}^0, t_0, R_0^-)$, respectively (see Definition 3.1.5). If both, that is, if $k(\bar{u}, \bar{x}^0, t_0, R) \subset M$, then M is *invariant* under $k(\bar{u}, \bar{x}^0, t_0, \cdot)$. We call M *minimum invariant*, if it does not contain any other invariant set.

A nonvoid, minimum invariant M can be identified with a motion that starts at M:

$$\{k(\bar{u}, \bar{x}^0, t_0, R) \mid (\bar{x}^0, t_0) \in M\} = M, \tag{3.2.1}$$

(see Nemitzky and Stepanov [1]). An obvious subcase of (3.2.1) is an equilibrium [see (1.5.22), (1.5.27)]. The reference to M may also be used in a wider sense, namely, to investigate the stabilities with respect to a planned path by studying relative variables, in a manner *similar* to that in the case of controllability for capture. However note here that $\mathfrak{T}(R_0^+)$ does not satisfy (3.2.1) since generally it is not invariant; moreover the planned path in $\Delta \times R$ may not be a motion.

On the other hand, *given a motion* of (1.5.3') in the form of the curve $\bar{x}_p(t)$, $t \geq t_0$ in $\Delta \times R$, we may set up $\bar{z} = \bar{x} - \bar{x}_p(t)$, which transforms (1.5.3) into

$$\dot{\bar{z}} = \bar{f}(z + \bar{x}_p(t), \bar{u}, t) - \bar{f}(\bar{x}_p(t), \bar{u}, t).$$

If we now temporarily denote the right-hand side of the above equation by $\bar{f}^z(z, \bar{u}, t)$, clearly $\bar{f}^z(0, \bar{u}, t) \equiv 0$ and the zero solution or zero equilibrium $z(t) \equiv 0$ of the transformed equation corresponds to $\bar{x}_p(t)$. Therefore it is sufficient to discuss the stability of $z(t) \equiv 0$ instead of that of $M: \bar{x}_p(t), t \geq t_0$. Our system continues to be of the type (1.5.3') but with $f(0, 0, t) \equiv 0$. In particular, if $\bar{x}_p(t) = \text{const} = \bar{x}^e$ we may translate any local study to that of the basic energy cup. Thus we may investigate the stability of the zero equilibrium $\bar{x}(t) \equiv 0$ in lieu of any other $\bar{x}^e$, as long as the investigation is local.

Sufficient conditions for particular stabilities are given in Theorems 3.2.1–3.2.3, which were adapted from Yoshizawa [1].

Theorem 3.2.1. *Given $\bar{u}(\cdot)$, the equilibrium $\bar{x}(t) \equiv 0$ is equistable if there exists a C^1 function (continuously differentiable) $V(\cdot): \Delta \times [0, \infty) \to R$, $V(0, t) \equiv 0$ such that*

(i) $a(\|x\|) \leq V(\bar{x}, t)$, *where $\|\cdot\|$ is any norm in R^N, $a(r) > 0$ is a continuously increasing function;*

(ii) $(\partial V/\partial t) + \nabla V(x, t) \cdot \bar{f}(\bar{x}, \bar{u}, t) \leq 0.$ \hfill (3.2.2)

Proof. Indeed, since $V(\cdot)$ is continuous and $V(0, t_0) = 0$, we can find $\delta(t_0, \varepsilon) > 0$ such that $\|\bar{x}^0\| < \delta$ implies

$$V(\bar{x}^0, t_0) < a(\varepsilon) \tag{3.2.3}$$

(see Fig. 3.6). Suppose $k(\bar{u}, \bar{x}^0, t_0, R_0^+)$ crosses $\|\bar{x}\| = \varepsilon$. Then there is a $t_1 > t_0$ such that $k(\bar{u}, \bar{x}^0, t_0, t_1) = \bar{x}^1 = \varepsilon$. But then condition (ii) implies $V(x^1, t_1) \leq V(x^0, t_0)$, while by (3.2.3), $V(x^1, t_1) < a(\varepsilon)$, contradicting (i). We conclude that $k(\bar{u}, \bar{x}^0, t_0, R_0^+)$ may not cross $\|\bar{x}\| = \varepsilon$ which proves the hypothesis.

Theorem 3.2.2. *Given $\bar{u}(\cdot)$, the equilibrium $\bar{x}(t) \equiv 0$ is uniformly stable if the conditions of Theorem 3.2.1 hold, with condition (i) extended to*

(i') $a(\|\bar{x}\|) \leq V(\bar{x}, t) \leq b(\|\bar{x}\|),$
where $b(r) > 0$, is a continuously increasing function.

Choosing a $\delta(\varepsilon) > 0$ such that $b(\delta) < a(\varepsilon)$, by the same argument as for Theorem 3.2.1, we can show that if $\|\bar{x}^0\| < \delta(\varepsilon)$ and $t \in R$, then $\|k(\bar{u}, \bar{x}^0, t_0, t)\| < \varepsilon$ for all $t \geq t_0$, which is uniform stability.

3.2 STABILITY, BOUNDEDNESS, AND STABILIZATION

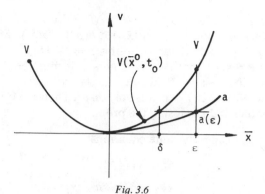

Fig. 3.6

Theorem 3.2.3. *Given $\bar{u}(\cdot)$, the equilibrium $\bar{x}(t) \equiv 0$ is uniformly asymptotically stable, if the conditions of Theorem 3.2.2 hold with condition (ii) extended to*

(ii') $\quad (\partial V/\partial t) + \nabla_x V(\bar{x}, t) \cdot \bar{f}(\bar{x}, \bar{u}, t) \leq -c(\|\bar{x}\|)$ \hfill (3.2.4)

for all $\|\bar{x}\| \neq 0$, where $c(r) > 0$ is a continuous function.

Proof. The proof is instructive for our further discussion. Under the conditions of Theorem 3.2.2, which are assumed, $\bar{x}(t) \equiv 0$ is uniformly stable, meaning that there is some subset $\Delta_0 \times R$ in $\Delta \times R$, possibly $\Delta \times R$ itself, of initial states $(\bar{x}^0, t_0)$ which is positively invariant (see Fig. 3.7). We want to show now that for each $\varepsilon > 0$ there is $\delta_1 > 0$ such that $(\bar{x}^0, t_0) \in \Delta_0 \times R$ implies $\|k(\bar{u}, \bar{x}^0, t_0, t)\| < \delta_1(\varepsilon)$ at some t.

Let us specify $\Delta_0 \triangleq \{\bar{x} \in \Delta \,|\, \|\bar{x}\| < \rho^0\}$ and start from above $\delta_1(\varepsilon)$, as shown in Fig. 3.7. Condition (ii') gives $V(\bar{x}, t) - V(\bar{x}^0, t_0) \leq -c(t - t_0)$, generating the estimate

$$t \leq t_0 + c^{-1}[V(\bar{x}^0, t_0) - V(\bar{x}, t)]. \tag{3.2.5}$$

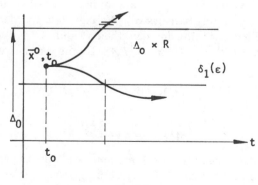

Fig. 3.7

By (i), $V(\bar{x}^0, t_0) \le b(\rho^0)$, $V(\bar{x}, t) \ge a(\delta_1)$ wherefrom $V(\bar{x}^0, t_0) - b(\rho^0) \le 0 \le V(\bar{x}, t) - a(\delta_1)$ or $V(\bar{x}^0, t_0) - V(\bar{x}, t) \le b(\rho^0) - a(\delta_1)$ and by (3.2.5), $t \le t_0 + c^{-1}[b(\rho^0) - a(\delta_1)]$. Thus, denoting $T = c^{-1}[b(\rho^0) - a(\delta_1)]$, we obtain that at some t, such that $t_0 \le t_1 \le t_0 + T$ it must be $\|k(\bar{u}, \bar{x}^0, t_0, t)\| < \delta_1(\varepsilon)$. Therefore, if $t > t_0 + T$, we have $\|k(\bar{u}, \bar{x}^0, t_0, t)\| < \varepsilon$, which was required to show. Note that T depends only on ρ^0. Then $\varepsilon \to 0$ as $t \to \infty$, which implies the hypothesis, thus closing the proof.

Example 3.2.1. Consider the state equations

$$\dot{x}_1 = -x_1^3 + x_1 x_2^3,$$
$$\dot{x}_2 = -x_2 - u^2, \qquad u = x_1 x_2,$$

and take $V(x_1, x_2) = x_1^2 + x_2^2$, which immediately satisfies condition $V(0) = 0$ and (i'). Differentiating,

$$\dot{V}(x_1(t), x_2(t)) = (\partial V/\partial x_1)\dot{x}_1 + (\partial V/\partial x_2)\dot{x}_2$$
$$= 2x_1(-x_1^3 + x_1 x_2^3) + 2x_2(-x_2 - x_1^2 x_2^2)$$
$$= -2(x_1^4 + x_2^2),$$

satisfying (ii') and thus implying uniform asymptotic stability of $x_1(t) \equiv 0$, $x_2(t) \equiv 0$. Observe that the region of attraction covers $\Delta \subset R^2$, as the conditions (i'), (ii') hold for all x_1, x_2. ∎

The following example will be useful later.

Example 3.2.2. Consider the linear N-dimensional system with constant coefficients

$$\dot{\bar{x}} = A\bar{x}$$

where A is $N \times N$ system matrix, and let the Liapunov function be the square form $V = \bar{x}^T P \bar{x}$, where $P = (p_{ij})$ is a symmetric positive definite $N \times N$ matrix. Such $V(\cdot)$ certainly satisfies (i'). We calculate the derivative $\dot{V}(t) = \dot{\bar{x}}^T P \bar{x} + \bar{x}^T P \dot{\bar{x}}$. When A is nonsingular, then P and APA^{-1} have the same eigenvalues. Hence $\dot{V}(t) = \bar{x}^T A^T P \bar{x} + \bar{x}^T P A \bar{x}$. By the symmetry of P, $\bar{x}^T P \dot{\bar{x}} = (P^T \bar{x})^T \dot{\bar{x}}$, whence

$$\dot{V}(t) = \bar{x}^T [A^T P + PA] \bar{x} = -\bar{x}^T Q \bar{x},$$

where

$$A^T P + PA = -Q,$$

which is called the *Liapunov Matrix Equation*, with Q also a symmetric matrix (quite often just a unit matrix). It follows that, given suitable A and positive-definite Q, we obtain $\dot{V}(t)$ negative-definite, thus securing the asymptotic

3.2 STABILITY, BOUNDEDNESS, AND STABILIZATION

stability required. On the other hand, again given suitable A, the matrix P could be obtained as a positive-definite solution to the Liapunov matrix equation, say with $Q = I$, securing the asymptotic stability through the corresponding Liapunov function. The suitable A is called *stable*, and it satisfies the above described role, if and only if its characteristic roots have negative real parts. ∎

It is to be expected that conditions sufficient for instability are somehow opposite to those for stability, though we may not get them by direct contradiction.

Theorem 3.2.4. *Given $\bar{u}(\cdot)$, the equilibrium $\bar{x}(t) \equiv 0$ is unstable if there is a C^1 function $V : \Delta \times R \to R$ such that*

(i) $V(\bar{x}, t) \to 0$, as $x(t) \to 0$ for all $t \in R$;
(ii) $V(\bar{x}, t) > 0$, for all $\bar{x} \neq 0$, $t \in R$;
(iii) $\partial V/\partial t + \nabla V(\bar{x}, t) \cdot \bar{f}(\bar{x}, \bar{u}, t) \geq 0$. (3.2.6)

The theorem is proved by Roxin [3].

Consider now (1.5.3′) covering both (1.5.4) and (1.5.16′), and take $V(\bar{x}, t) \equiv H(\bar{x})$. Then for $\bar{x}$ in any Δ_H^k, $k = 0, 1, \ldots, m$ we have the condition (i′) of Theorem 3.2.2 and the conditions (i), (ii) of Theorem 3.2.4 satisfied. Then, by (2.3.15′), we conclude that the power balance decides about the condition (ii′) of Theorem 3.2.3 or condition (iii) of Theorem 3.2.4. Thus with f_0 specified by (2.3.15′) in view of (2.3.19) we have

Proposition 3.2.1. (a) Dissipative field covering the energy cup $\Delta_H^k \times R$, generates uniform stability of the equilibrium in each $\Delta_H^k \times R$, $k = 0, 1, \ldots, m$.

(b) Strictly dissipative field covering $\Delta_H^k \times R$ generates uniform asymptotic stability of the equilibrium in each Δ_H^k, with the region of attraction enclosing at least Δ_H^k.

(c) Accumulative field generates instability on any subset of $\Delta_H^k \times R$ it covers, $k = 0, 1, \ldots, m$.

Ultimate boundedness of motions is a finite time and nonlocal alternative to stability, useful in stabilization.

Let ρ^0, β, B be positive numbers and $\Delta_0 \triangleq \{\bar{x} \in \Delta \,|\, \|\bar{x}\| < \rho^0\}$, $\Delta_B \triangleq \{\bar{x} \in \Delta \,|\, \|x\| < B\}$.

Definition 3.2.5. Given $\bar{u}(\cdot)$, a motion $k(\bar{u}, \bar{x}^0, t_0, R_0^+)$ of (1.5.3′) is *equibounded*, if and only if for any ρ^0 and $t_0 \in R$ there is $\beta > 0$ such that $\bar{x}^0 \in \Delta_0$ yields $\|k(\bar{u}, \bar{x}^0, t_0, t)\| < \beta$ for all $t \geq t_0$. The motion is *uniformly bounded*, if β does not depend upon t_0.

Definition 3.2.6. Given $\bar{u}(\cdot)$, motions of (1.5.3′) are *equi-ultimately bounded for bound B* if and only if there is $B > 0$ and if for any ρ_0 and $t_0 \in R$ there is

$T > 0$ such that $\bar{x}^0 \in \Delta_0$ implies that $k(\bar{u}, \bar{x}^0, t_0, t) \in \Delta_B$ for all $t \geq t_0 + T$. The motions are *uniformly ultimately bounded* if and only if T does not depend on t_0.

It is obvious that equi-ultimate boundedness implies equi-boundedness. Observe that in uniform ultimate boundedness T depends only on ρ^0, that is, on the size of the starting region for all motions involved; thus it does not depend upon individual motions.

Definition 3.2.7. If the bound $B < \rho^0$ in uniform ultimate boundedness of Definition 3.2.6 is given a priori, the set Δ_B is called a *finite time attractor* for the motions concerned, and the set Δ_0 is the *region of finite time attraction* (see Fig. 3.8). If T is also prespecified, we call Δ_B the *practical finite attractor* for the motions. It obviously means that Δ_0 is prescribed as well.

Observe that $\Delta_B \times R$ is positively invariant under the motions of (1.5.3′). The above concepts have been introduced by Yoshizawa [1], and adapted to mechanical systems by Skowronski [1, 4, 5, 12, 24], and Skowronski and Ziemba [2, 3].

Define $\Delta_R \triangleq \{\bar{x} \in \Delta \mid \|\bar{x}\| \geq R\}$, with $R > 0$ suitably small. Adapting the sufficient conditions proved by Yoshizawa [1], we obtain that the motions of (1.5.3′) are uniformly bounded if there is a C^1 function $V(\cdot): \bar{\Delta}_R \times R \to R$ such that

$$a(\|\bar{x}\|) \leq V(\bar{x}, t) \leq b(\|\bar{x}\|), \tag{3.2.7}$$

where $a(r)$ and $b(r)$ are continuously increasing and $a(r) \to \infty$ as $r \to \infty$;

$$\partial V/\partial t + \nabla V(\bar{x}, t) \cdot \bar{f}(\bar{x}, \bar{u}, t) \leq 0. \tag{3.2.8}$$

He also proves that the motions are uniformly ultimately bounded, if (3.2.7) holds with (3.2.8) replaced by

$$\partial V/\partial t + \nabla V(\bar{x}, t) \cdot \bar{f}(\bar{x}, \bar{u}, t) \leq -c(\|\bar{x}\|), \tag{3.2.9}$$

where $c(r) > 0$ is a continuous function.

Letting $V(\bar{x}, t) \equiv H(\bar{x})$, we see that in view of (2.3.23) and (2.3.15′), the dissipative field $\mathscr{H}^{0-}$ with $f_0(\bar{x}, \bar{u}, t) \leq 0$ outside $\Delta_L \times R$ secures (3.2.7) and (3.2.8) in $C_\Delta \Delta_L \times R$, that is, outside the cup in-the-large, thus yielding uniform boundedness of motions of (1.5.3′).

Now observe that there is no motion possible in the hyperplane $\dot{\bar{q}} = 0$ and that at all regular points of Δ (that is, except the equilibria) all the motions of (1.5.3′) cross this hyperplane instantaneously. Moreover, by the continuity of the function $f_0(\bar{x}, \bar{u}, t)$ in $\bar{x}$, upon crossing the hyperplane, the motions maintain the sign of the power balance. Hence $f_0(\bar{x}, \bar{u}, t) < 0$, $\dot{\bar{q}} \neq 0$ suffices to satisfy (3.2.9) everywhere in $C_L \Delta_L \times R$, thus producing uniform ultimate boundedness of the motions with a bound B.

3.2 STABILITY, BOUNDEDNESS, AND STABILIZATION

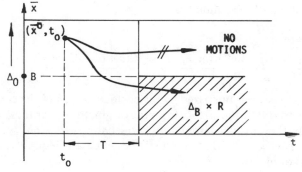

Fig. 3.8

Alternatively, by the same argument, for a given B such that $\Delta_B \triangleq \Delta_L$, the strictly dissipative field $\mathscr{H}^-: f_0(\bar{x}, \bar{u}, t) < 0$, $\dot{\bar{q}} \neq 0$ on $C_\Delta \Delta_L \times R$ produces the finite time attraction to $\Delta_L \times R$.

By a similar argument, we may have uniform ultimate boundedness or finite time attraction to some subsets of the local cups $\Delta_H^k \times R$, $k = 0, 1, \ldots, m$, provided there is a strictly dissipative field cover in neighborhoods of such subsets (see Skowronski [1, 24] and Skowronski and Ziemba [2, 3]). In conclusion, we have

Proposition 3.2.2. Dissipative field covering $C_\Delta \Delta_L \times R$ secures uniform boundedness of motions of (1.5.3'). When it becomes strictly dissipative everywhere for $\dot{\bar{q}} \neq 0$, we have uniform ultimate boundedness, and if the bound B is given by $\partial \Delta_L$, the cup in-the-large is a finite time attractor and positively invariant. The above holds if $C_\Delta \Delta_L$ is replaced by the complement to Δ_H^k of any subset of such a local cup, $k = 0, 1, \ldots, m$.

Example 3.2.3. Let us return to the RP manipulator of Section 1.1. The kinetic energy was

$$T(\bar{q}, \dot{\bar{q}}) = \tfrac{1}{2}(m_1 r_1^2 + m_2 q_2^2)\dot{q}_1^2 + \tfrac{1}{2}m_2 \dot{q}_1^2. \tag{1.1.18}$$

Letting $\sin q_1 = q_1 - \tfrac{1}{6}q_1^3$, the potential energy (1.1.17) becomes

$$\mathscr{V}(\bar{q}) = \frac{1}{2}aq_1^2 - \frac{9.81}{6}m_1 r_1 q_1^3 + \frac{1}{4}bq_1^4 + 9.81 m_2 q_1 q_2 - \frac{9.81}{6}m_2 q_2 q_1^3. \tag{3.2.10}$$

To obtain the free autonomous system equilibria, we need

$$\begin{aligned} \frac{\partial \mathscr{V}}{\partial q_1} &= aq_1 - \frac{9.81}{2}m_1 r_1 q_1^2 + bq_1^3 + 9.81 m_2 q_2 - \frac{9.81}{2}m_2 q_2 q_1^2 = 0, \\ \frac{\partial \mathscr{V}}{\partial q_2} &= 9.81 m_2 q_1 - \frac{9.81}{6}m_2 q_1^3 = 0. \end{aligned} \tag{3.2.11}$$

Solving, we obtain three equilibria:

$$q_1^e = 0, \quad \sqrt{6}, \quad -\sqrt{6},$$

$$q_2^e = 0, \quad \frac{\sqrt{6}(a+6b) - 3 \cdot 9.81 m_1 r_1}{2 \cdot 9.81}, \quad \frac{-\sqrt{6}(a+6b) - 3 \cdot 9.81 m_1 r_1}{2 \cdot 9.81} \quad (3.2.12)$$

which are obviously parametrized by the coefficients a, b of the compensating spring force and will be the farther from the origin, the larger these parameters are. Naturally, they must be large enough to secure $\sqrt{6}(a + 6b) \geq 3 \cdot 9.81 m_1 r_1$ in order to have the second q_2^e nonnegative. In case of equality, all three q_2^e reduce to the origin.

In view of (1.1.18) and (3.2.10) the total energy is

$$H(\bar{x}) = \frac{1}{2}(m_1 r_1^2 + m_2 q_2^2)\dot{q}_1^2 + \frac{1}{2} m_2 \dot{q}_2^2 + \frac{1}{2} a q_1^2 - \frac{9.81}{6} m_1 r_1 q_1^3 + \frac{1}{4} b q_1^4$$

$$+ 9.81 m_2 q_1 q_2 - \frac{9.81}{6} m_2 q_2 q_1^3. \quad (3.2.13)$$

We have a single Dirichlet stable equilibrium $(0,0,0,0)$ and a threshold specified by

$$q_1^e = \sqrt{6}, \quad q_2^e(a,b,m_1,m_2,r_1) = \frac{\sqrt{6}(a + 6b) - 3 \cdot 9.81 m_1 r_1}{2 \cdot 9.81 m_2}$$

with the Z_H: $H(\sqrt{6}, q^e(a,b,m_1,m_2,r_1),0,0) = \text{const} = h_H$, which defines $C_\Delta \Delta_L$ as much as $\Delta_L = \Delta_H$. Simple calculation yields

$$h_H = 3a - 9.81\sqrt{6} m_1 r_1 + 9b, \quad (3.2.14)$$

which is also parametrized by a, b, m_1 but is independent of the load m_2. To satisfy Proposition 3.2.2 we need the dissipative field for $h \geq h_H$, which for (1.1.19) means

$$f_0(\bar{x}, \bar{u}) = [Q_1^F(\bar{q}, \dot{\bar{q}}, u_1)\dot{q}_1 - \lambda_1 |\dot{q}_1|\dot{q}_1] + [Q_2^F(\bar{q}, \dot{\bar{q}}, u_2)\dot{q}_2 - \lambda_2 \dot{q}_2] \leq -c(\|\bar{x}\|), \quad (3.2.15)$$

which is obviously satisfied when for $\dot{q}_i \neq 0$, $i = 1,2$, the power balances $Q_1^F(u_1) < \lambda_1 |\dot{q}_1|\dot{q}_1$, $Q_2^F(u_2) < \lambda_2 \dot{q}_2$ hold, so that in our simple case, the controller has to offset the damping only. Note that the same conditions render the dissipative field within the energy cup below h_H given by (3.2.14), thus producing asymptotic stabilization with regard to the zero equilibrium of our manipulator. ∎

Manipulators often work as actuator controlled inverted pendulums (see Section 1.1), which is the typical example for Dirichlet instability (see Section 2.2). Thus often we are interested in the uniform ultimate boundedness of motions in a neighborhood of an energy threshold equilibrium rather than

3.2 STABILITY, BOUNDEDNESS, AND STABILIZATION

within an energy cup. On the other hand, Definitions 3.2.5–3.2.7, together with the sets Δ_0, Δ_B and sufficient conditions (3.2.7)–(3.2.9) are referred to $\bar{x}^e = 0$. Fortunately, as mentioned before when dealing with stability, it is a simple matter of replacing $\|\bar{x}\|$ by $\|\bar{x} - \bar{x}^e\|$ if we want to refer the mentioned notions to any other equilibrium $\bar{x}^e \neq 0$, including our present threshold equilibrium. Assume this to be done, and let the Liapunov function in the so modified conditions (3.2.7)–(3.2.9) be $V(\bar{x}, t) = |H(\bar{x}^e) - H(\bar{x})|$.

Then, with $H(\bar{x}^e) = \text{const}$, (2.3.23) still yields (3.2.7) since $|H(\bar{x}^e) - H(\bar{x})|$ decreases without any extrema as $\bar{x} \to \bar{x}^e = \text{const}$. For (3.2.9) we have $\dot{V}(\bar{x}(t), t) = -\dot{H}(\bar{x}(t)) = -f_0(\bar{x}, \bar{u}, t)$ for all points $(\bar{x}, t)$ below the threshold specified by $h = H(\bar{x}^e)$ and $\dot{V}(\bar{x}(t), t) = \dot{H}(\bar{x}(t)) = f_0(\bar{x}, \bar{u}, t)$ for all points $(\bar{x}, t)$ above this threshold, while the $\bar{x}$ are the regular points with $\dot{\bar{q}} \neq 0$ in some semineighborhood $\mathcal{N}(\partial \Delta_0) \cap \Delta_0$, where $\mathcal{N}(\partial \Delta_0)$ is a neighborhood of the boundary of a set $\Delta_0 = \{\bar{x} \in \Delta \,|\, \|\bar{x} - \bar{x}^e\| < \rho^0\}$ that surrounds our threshold equilibrium. Hence to produce the modified (3.2.9) on this semineighborhood, we need $f_0(\bar{x}, \bar{u}, t) \geq c(\|\bar{x} - \bar{x}^e\|)$ below the threshold energy level $Z_H \times R$ in order to raise the representing point $(\bar{x}, t)$, $\bar{x}(t) = k(\bar{u}(\cdot), \bar{x}^0, t_0, t)$ to this level, and we need $f_0(\bar{x}, \bar{u}, t) \leq -c(\|\bar{x} - \bar{x}^e\|)$ above $Z_H \times R$ in order to lower the point $(\bar{x}, t)$ to this level. The reader is asked now to recall our discussion in Section 2.2 on the role of energy flow and the power characteristics f_0 (see Fig. 2.5).

According to our modification, if we now let

$$\Delta_B \triangleq \{\bar{x} \in \Delta \,|\, \|\bar{x} - \bar{x}^e\| < B\} \subset \Delta_0,$$

in view of the above, we may state the following:

Proposition 3.2.3. If there is Δ_0 enclosing an energy threshold equilibrium $\bar{x}^e$ such that regular ($\dot{\bar{q}} \neq 0$) points $\bar{x} \in \mathcal{N}(\partial \Delta_0) \cap \Delta_0$ above the threshold energy level $Z_H^k \times R$ are covered by a strictly dissipative field $\mathcal{H}^-$ and points below this level are covered by an accumulative field $\mathcal{H}^+$, then the motions of (1.5.3') from Δ_0 are uniformly ultimately bounded for a bound $B < \rho^0$ with the time interval $T(\rho^0) > 0$ specified by Δ_0. Given B, the set $\Delta_B \times R$ becomes a finite attractor if the above assumptions hold on $\mathcal{N}(\partial \Delta_0) \cap \Delta_0 \triangleq \Delta_0 - \Delta_B$. The set $\Delta_B \times R$ becomes a practical finite attractor if either Δ_0 or $T(\rho^0)$ is given *a priori*.

Now we make the control program (1.5.2) active again, in particular memoryless but nonstationary: $\bar{P}(\bar{x}, t)$, in order to produce controllability for the discussed stability objectives which we may call briefly *stabilization*. We shall discuss one type of stabilization only, just to show the technique. For the others, the reader is referred to the literature, with an up-to-date review by Leitmann [2, 3] and Corless and Leitmann [2, 4, 5].

Definition 3.2.8. The system (1.5.3') is *stabilizable*, if and only if there is a program $\bar{P}(\bar{x},t)$ such that the motions $k(\bar{u}(\cdot),\bar{x}^0,t_0,R_0^+)$ from some $\Delta_0 \times R \subset \Delta \times R$ are uniformly ultimately bounded for bound B, referring to either $\bar{x}^e = 0$ or to any other equilibrium. When a desired $B > 0$ is given a priori, we have stabilization in the given $\Delta_B \times R$. The set $\Delta_s \subset \Delta$ of all $\bar{x}^0$ from which the stabilization or the stabilization in given $\Delta_B \times R$ may be achieved is called the *region of stabilization* or of stabilization in given $\Delta_B \times R$, respectively.

When Δ_B is simply connected and bounded, it may be feasible to specify the norm $\|\cdot\|$ in terms of energy, as, for instance, within the cups

$$\Delta_B = \{\bar{x} \in \Delta_H^k \,|\, H(\bar{x}) \leq h_B = B\} \qquad (3.2.16)$$

or, similarly, in $C\Delta_L$, that is, about the cup in-the-large

$$\Delta_B = \{\bar{x} \in \Delta \,|\, H(\bar{x}) \leq h_B = B, h_B \geq h_L\}. \qquad (3.2.17)$$

Then the stabilization in given $\Delta_B \times R$ may be called the *stabilization under given level* $Z_B \times R$.

Clearly $\Delta_0 \subset \Delta_s$, as the region of stabilization is by definition the maximal Δ_0 available. Note also that the time interval in the ultimate boundedness concerned (see Definition 3.2.6) becomes the time interval for stabilization and stabilization in given $\Delta_B \times R$. It will be denoted T_s. Observe that it depends only on ρ^0 in Δ_0 and not on particular motions, and may be calculated similarly to T in the proof of Theorem 3.1.3. The method for doing it will be introduced later in terms of general controllability. Similarly to Δ_B, the set Δ_0 may also be given in terms of energy. For instance, for the cups:

$$\Delta_0 = \{\bar{x} \in \Delta_H^k \,|\, H(\bar{x}) \leq h_\rho = \rho^0\}. \qquad (3.2.18)$$

We may attempt to estimate the region Δ_s a posteriori, which often proves a difficult task, or we may give a priori a desired Δ_0, in which case $T_s(\rho^0)$ will also be given (see Skowronski [24]).

Definition 3.2.9. The stabilization, or stabilization in Δ_B is called *practical* if a desired Δ_0 is given a priori. When the norm $\|\cdot\|$ is specified by the energy, we have practical stabilization from the corresponding level $Z_0 \times R$, with Z_0 specified by h_ρ.

Obviously T_s will be thus given either in terms of ρ^0 or h_ρ. Note also that then the region of stabilization Δ_s becomes irrelevant.

Definition 3.2.10. The system (1.5.3) is *asymptotically stabilizable about an equilibrium* $\bar{x}^e$, if and only if it is stabilizable and the corresponding program $\bar{P}$ is such as to generate uniform asymptotic stability of $\bar{x}^e$ under the motions of (1.5.3'). For $\bar{x}^e = 0$ we can have *asymptotic stabilization about the basic energy level*. The set Δ_{as} of all $\bar{x}^0$ where from the asymptotic stabilization is achievable under some $\bar{P}$ is called the *region of asymptotic stabilization*. If a specific

3.2 STABILITY, BOUNDEDNESS, AND STABILIZATION

$\Delta_0 \subset \Delta_{as}$ is given a priori, we have the *practical asymptotic stabilization* on Δ_0, and if Δ_0 is expressed in terms of energy, *practical asymptotic stabilization from the level* $Z_0 \times R$.

Here again for practical asymptotic stabilization, the region Δ_{as} is irrelevant.

The above stabilization concepts represent slight modification of those used by the Leitmann school (see Skowronski [1, 3, 4, 5, 7, 8, 24], Barmish, Corless and Leitmann [1], Barmish and Leitmann [1], Barmish, Petersen and Feuer [1], Corless and Leitmann [1–5]).

To focus attention on some cases of the above stabilization properties, let us consider the stabilization about the *cup in-the-large*, thus referring to the basic equilibrium and sets $\Delta_0 = \{\bar{x} \in \Delta | H(\bar{x}) \le h_\rho\}$ with $h_\rho = \rho^0 \ge h_L$, and (3.2.18) (see Fig. 3.9).

Recall now that the power characteristics $f_0(\bar{x}, \bar{u}, t)$ are defined by (2.3.15') and note that for $\dot{\bar{q}} \ne 0$, the damping power $\bar{Q}^{DD}(\bar{q}, \dot{\bar{q}})\dot{\bar{q}}$ is negative by (2.3.2), and decreasing in $|\bar{x}|$ by (2.3.5) and (2.3.6), while the powers $\bar{Q}^{DA}(\bar{q}, \dot{\bar{q}})\dot{\bar{q}}$, $\bar{Q}^F(\bar{q}, \dot{\bar{q}}, \bar{u})\dot{\bar{q}}$, and $\bar{Q}^R(\bar{q}, \dot{\bar{q}}, t)\dot{\bar{q}}$ are bounded independently of their sign with the bound $N < \infty$. Hence for some $\bar{P}(\bar{x}, t)$ there are values $B_1, \ldots, B_N$ of $q_1, \ldots, q_n$, $\dot{q}_1, \ldots, \dot{q}_n$, depending on N, such that for $\|\bar{x}\| > B$, $B \triangleq |\bar{B}|$, and $\bar{B} \triangleq (B_1, \ldots, B_{2n})$, the power characteristic f_0 becomes negative for $\dot{\bar{q}} \ne 0$, so that Proposition 3.1.2 for uniform ultimate boundedness is satisfied, generating stabilization. Hence we have

Proposition 3.2.4. Given $N < \infty$, there is a control program $\bar{P}(\cdot)$ such that for some $B(N)$ the system (1.5.3') is stabilized for the level $Z_B \times R$ in $C\Delta_L \times R$.

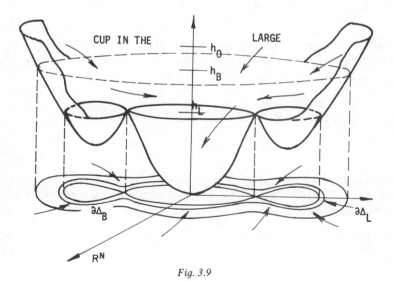

Fig. 3.9

Obviously, if we want stabilization in a given Δ_B or under a given $Z_B \times R$, then either the controller is selected accordingly to kill the excess in accumulation power or the bound N may perhaps be adjusted (if it is adjustable at all). Indeed, in $(\Delta - \Delta_B) \times R$ we want

$$f_0(\bar{x}, \bar{u}, t) \leq -c(\|\bar{x}\|), \quad \dot{\bar{q}} \neq 0. \tag{3.2.19}$$

It is economic to have $c(\|\bar{x}\|)$ as small as possible. By (2.3.2), (2.3.5), and (2.3.6), the damping power $\bar{Q}^{DD}(\bar{q}, \dot{\bar{q}})\dot{\bar{q}}$ is negative and continuously decreasing in $|\bar{x}|$. We may thus choose $c(\|\bar{x}\|) = -\varepsilon \bar{Q}^{DD}(\bar{q}, \dot{\bar{q}})\dot{\bar{q}}$, where $\varepsilon > 0$ but suitably small. Then, by (2.3.15'), condition (3.2.19) holds if

$$\bar{Q}^F \dot{\bar{q}} + \bar{Q}^R \dot{\bar{q}} + \bar{Q}^{DA} \dot{\bar{q}} \leq (\varepsilon - 1)\bar{Q}^{DD}\dot{\bar{q}}, \quad \dot{\bar{q}} \neq 0. \tag{3.2.20}$$

Denote

$$N_B \triangleq \inf_{\Delta - \Delta_B} [(\varepsilon - 1)\bar{Q}^{DD}(\bar{q}, \dot{\bar{q}})\dot{\bar{q}}],$$

and observe from the above and (2.3.14') that for $N < N_B$ there is no need to control at all since the system is self-stabilizing with $\bar{Q}^F(\bar{q}, \dot{\bar{q}}, \bar{u}) \equiv 0$. It is only for $N > N_B$ that the controller must cancel the excess in accumulation power, that is, (3.2.20) must be made to hold. Observe that the larger ε is, the less positive is the right-hand side of (3.2.20); thus the stronger controller is needed to reduce the positive accumulative powers of $\bar{Q}^R$ and $\bar{Q}^{DA}$. To produce a practical algorithm for the control program, given ε, we set up the control condition as

$$\bar{Q}^F(\bar{q}, \dot{\bar{q}}, \bar{u})\dot{\bar{q}} + \bar{Q}^R(\bar{q}, \dot{\bar{q}}, t)\dot{\bar{q}} + \bar{Q}^{DA}(\bar{q}, \dot{\bar{q}})\dot{\bar{q}} = (\varepsilon - 1)\bar{Q}^{DD}(\bar{q}, \dot{\bar{q}})\dot{\bar{q}}, \quad \dot{\bar{q}} \neq 0. \tag{3.2.21}$$

Unfortunately this is as close as we can get to $\bar{u}$ without specifying the transmission force $\bar{Q}^F(\cdot)$. If, as it is often the case, we may have $Q_i^F = B_i(\bar{q}, \dot{\bar{q}})\bar{u}_i$, $B_i(\bar{q}, \dot{\bar{q}}) \neq 0$, $i = 1, \ldots, n$, then (3.2.21) is satisfied by

$$u_i(t) = \begin{cases} \dfrac{-1}{B_i(\bar{q}, \dot{\bar{q}})}[Q_i^R(\bar{q}, \dot{\bar{q}}, t) + Q_i^{DA}(\bar{q}, \dot{\bar{q}}) + (1 - \varepsilon)Q_i^{DD}(\bar{q}, \dot{\bar{q}})], & \dot{q}_i \neq 0, \\ C, & \dot{q}_i = 0, \end{cases} \tag{3.2.22}$$

which determines the program $\bar{P}$. Here C is a suitable constant.

For our RP manipulator of Section 1.1, continued in Example 3.2.2, the control condition (3.2.21) reduces to [see (3.2.15)]

$$Q_1^F(\bar{q}, \dot{\bar{q}}, u_1)\dot{q}_1 \leq (1 - \varepsilon)\lambda_1 |\dot{q}_1|\dot{q}_1^2, \quad Q_2^F(\bar{q}, \dot{\bar{q}}, u_2)\dot{q}_2 \leq (1 - \varepsilon)\lambda_2 \dot{q}_2^2, \quad \dot{q}_i \neq 0, \tag{3.2.23}$$

3.2 STABILITY, BOUNDEDNESS, AND STABILIZATION

which, with the above specification of $\bar{Q}^F$, gives

$$u_1(t) = \frac{(1-\varepsilon)\lambda_1|\dot{q}_1|\dot{q}_1}{B_1(\bar{q},\dot{\bar{q}})}, \quad u_2(t) = \frac{(1-\varepsilon)\lambda_2\dot{q}_2}{B_2(\bar{q},\dot{\bar{q}})}, \quad \dot{q}_i \neq 0, \quad (3.2.24)$$

and

$$u_1(t) = u_2(t) = 0 \quad \text{for} \quad \dot{q}_i = 0, \quad i = 1, 2.$$

The latter is obviously a simplified version of (3.2.22) with $Q_i^R \equiv 0$, $Q_i^{DA} \equiv 0$. The actuators work in this case for compensation of the damping in joints and other energy losses in transmission.

When practical stabilization is attempted, the range of (3.2.20) holding is narrowed from $\Delta - \Delta_B$ to $\Delta_0 - \Delta_B$, but the same controller works. Similar investigations may be made for the other types of stabilization utilizing conditions on $f_0(\bar{x}, \bar{u}, t)$ suitable for Propositions 3.2.1–3.2.3. Asymptotic stabilization is obviously tailored to stabilize motions in cups, that is, about the Dirichlet stable equilibria.

Exercises

3.2.1 Determine the equilibrium point other than the origin of the system

$$\dot{x}_1 = x_4 - 2x_1x_2, \quad \dot{x}_2 = -2x_2 + x_1x_2.$$

Apply a transformation of coordinates which moves this point to the origin and find the new system.

3.2.2 Show by solving the equations explicitly and applying Definition 3.2.1 that the origin is a stable equilibrium point for the system

$$\dot{x}_1 = ax_1 - x_2, \quad \dot{x}_2 = x_1 + ax_2,$$

with $a \leq 0$, $x_1(t_0) = \alpha$, $x_2(t_0) = \beta$.

3.2.3 Convert the system $\ddot{q} + a\dot{q} + bq = 0$ to the state space form. Using $V = \bar{x}^T P \bar{x}$ obtain the necessary and sufficient conditions $a > 0$, $b > 0$ for asymptotic stability.

3.2.4 Find the region of asymptotic stability of the origin for the following systems:

(a) $\dot{x} = \frac{1}{2}x(x-2)$;
(b) $\dot{x}_1 = x_2 - x_1^3, \dot{x}_2 = -x_1 - x_2$.

3.2.5 Consider the system $\dot{x}_1 = x_1(x_2^2 - 1)$, $\dot{x}_2 = x_2(x_1^2 - 1)$. Using $V = \frac{1}{2}(x_1^2 + x_2^2)$ estimate the behaviour of trajectories with respect to the Liapunov level curves. Show that the circle $V(x_1, x_2) = 1$ is positively invariant. Sketch the trajectories.

3.2.6 Consider the system $\ddot{q} + l\dot{q} + \sin q = 0$, $l > 0$, $q(t) \in R$. Using the function $V = \frac{1}{2}(\sqrt{l}q + (1/\sqrt{l})\dot{q})^2 + (1/l)(q^2 + \dot{q}^2)$, show asymptotic stability of $(0, 0)$ and determine its region.

3.2.7 Using the function $V = 5x_1^2 + 2x_1x_2 + 2x_2^2$ show that the origin of $\dot{x}_1 = x_2$, $\dot{x}_2 = -x_1 - x_2 - (x_1 + 2x_2)(x_2^2 - 1)$ is asymptotically stable. Find the region of asymptotic stability.

3.2.8 Write the system $\ddot{q} + \dot{q} + q + q^3 = u$ in the state space form, then show that the destabilizing program $u = 2\dot{q}$ prevents stability of the origin $(q, \dot{q}) = (0, 0) \in R^2$.

3.2.9 Write the system $\ddot{q} - q^2\dot{q} + u + q + q^3 = 0$ in the state form, then find a stabilizing program $u = P(q, \dot{q})$ which asymptotically stabilizes the system about $(q, \dot{q}) = (0, 0) \in R^2$ from $\Delta = R^2$.

3.3 CONTROLLABILITY UNDER UNCERTAINTY

So far in our model, both the inertial and the applied forces have been assumed known and well defined. As already mentioned, the manipulator may work in difficult, unpredictable, or only partially predictable conditions and the forces may be at least uncertain. The mass, inertia, shape or orientation of the handled objects, friction coefficients and elasticity coefficients of transferred movement mechanisms, parameters of drives and sensor transducers, damping in joints, and wear in every element of the system: all of these factors contribute to the fact that for large displacements the system parameters that enter the forces mentioned change considerably causing or contributing to the uncertainty as much as does the environment where the robot works. The total effect is to render useless any model sensitive to the above-mentioned uncertainties.

Moreover, wherever else the uncertainty appears, the inertia matrix, being uncertain, pollutes all the characteristics. We have to make our control study robust against all that noise, and we can do it provided the noiseband is bounded. We do it via the so-called *worst case design* (also called game against nature), killing the noiseband with a controller (see Sections 3.3–3.5) or more economically via *adaptive control* (see Section 3.6). Later in the text, we shall attempt to *identify the noise adaptively* (see Chapter 6).

Let $\bar{r}(t) = (r_1(t), \ldots, r_s(t))^T$ be a *vector of uncertain parameters* bounded for all $t \geq t_0$ in some known, bounded band $\Re_1 \subset R^s$ and defined by an unknown function $\bar{r} = \bar{r}(\bar{q}, \dot{\bar{q}}, t)$ or $\bar{r}(\bar{q}, \bar{p}, t)$ depending upon the model. Moreover, let $\bar{Q}^r = \bar{Q}^r(\bar{q}, \dot{\bar{q}}, t)$ or $\bar{Q}^r = \bar{Q}^r(\bar{q}, \bar{p}, t)$, $\bar{Q}^r = (Q_1^r, \ldots, Q_n^r)^T$, be *unknown environmental perturbation forces* bounded by a known norm:

$$\bar{Q}^r \in \Re_2 = \{\bar{Q}^r \mid \|\bar{Q}^r(\bar{x}, t)\| \leq \rho(\bar{x}, t), \forall (\bar{x}, t) \in \Delta \times R\}.$$

3.3 CONTROLLABILITY UNDER UNCERTAINTY

Then according to what we said before, (1.3.15′) becomes

$$\sum_i a_{ij}(\bar{q},\bar{r})\ddot{q}_j + K_i(\bar{q},\dot{\bar{q}},\bar{r}) - Q_i^D(\bar{q},\dot{\bar{q}}) - Q_i^P(\bar{q},\bar{r}) = Q_i^F(\bar{q},\dot{\bar{q}},u_i) + Q_i^r(\bar{q},\dot{\bar{q}},t),$$

$$i = 1,\ldots,n, \quad (1.3.15'')$$

or in the inertially decoupled form

$$\ddot{q}_i + \Gamma_i(\bar{q},\dot{\bar{q}},\bar{r}) + D_i(\bar{q},\dot{\bar{q}},\bar{r}) + G_i(\bar{q},\bar{r}) + \Psi_i(\bar{q},\bar{r}) = F_i(\bar{q},\dot{\bar{q}},u_i,\bar{r}) + R_i^r(\bar{q},\dot{\bar{q}},\bar{r},t),$$

$$i = 1,\ldots,n, \quad (1.3.24'')$$

with $R_i^r(\cdot)$ being the perturbation characteristics from $Q_i^r(\bar{q},\dot{\bar{q}},t)$.

Since the inertia matrix is noise polluted we write $A(\bar{q},\bar{r})$, similarly for the kinetic energy $T(\bar{q},\bar{p},\bar{r})$, which together with $\bar{Q}^P(\bar{q},\bar{r})$ makes the total energy $H(\bar{x},\bar{r})$ uncertain. Hence the Hamiltonian form of state equations (1.5.16″) becomes

$$\dot{q}_i = \frac{\partial H(\bar{q},\bar{p},\bar{r})}{\partial p_i},$$

$$\dot{p}_i = -\frac{\partial H(\bar{q},\bar{p},\bar{r})}{\partial q_i} + Q_i^D(\bar{q},\bar{p}) + \bar{Q}_i^F(\bar{q},\bar{p},u_i) + Q_i^r(\bar{q},\bar{p},t).$$

$$(1.5.16'')$$

The advantage with respect to the state equations generated by (1.3.24″) is that even in this inertia-decoupled form neither the damping (positive or negative) nor the control transmitting forces are polluted by the uncertainty.

Forming now the $s + n$ vector

$$\bar{w}(t) = (\bar{r}(\bar{q}(t),\dot{\bar{q}}(t),t), \bar{Q}^r(\bar{q}(t),\dot{\bar{q}}(t),t))^T \in \Re_1 \times \Re_2,$$

with $W \triangleq \Re_1 \times \Re_2 \subset R^{s+n}$, we may now write the general form corresponding to (1.5.1′) as

$$\dot{x}_i = f_i(\bar{x},\bar{u},t,\bar{w}), \quad i = 1,\ldots,N. \quad (1.5.1'')$$

and the vector form as

$$\dot{\bar{x}} = \bar{f}(\bar{x},\bar{u},t,\bar{w}). \quad (1.5.3'')$$

Obviously solutions to (1.5.3″) will now depend on an unpredictable function $\bar{w}(\cdot)$ and in general may not exist at all, let alone be unique. If that was not enough, we have another reason for their unpredictability and that is the control program (1.5.2) which was assumed to be substituted, may not be suitable to produce $f_i(\bar{x},\bar{u},t,\bar{w})$ locally Lipschitz continuous even without $\bar{w}$ inserted. Indeed, in cases of optimal control, such a program itself may be discontinuous (bang-bang) and possibly set-valued. Then there are no unique solutions to (1.5.1′). For instance, the simple system $\dot{x} = u\sqrt{x}$, $x(t) \in R$, with the program defined by $u = \pm\sqrt{x}$, produces solutions $x(t) = x^0 e^{\pm t}$ obviously with two branches. The motion discontinuities have a real, not only academic meaning: They may produce dangerous vibrations (see Rees and Jones [1]).

Let us introduce the *set-valued control program* which accomodates these cases in terms of the function $\mathscr{P}: \Delta \times \Lambda \times R \to$ all subsets of U, defined by (see Fig. 3.1c),

$$\bar{u}(t) \in \mathscr{P}(\bar{x}(t), \bar{\lambda}(t), t). \tag{1.5.2'}$$

Now we can write the general form of our system as the *contingent equation* (see Roxin [1, 2])

$$\dot{\bar{x}} \in \{\bar{f}(\bar{x}, \bar{u}, t, \bar{w}) \,|\, \bar{u} \in \mathscr{P}(\bar{x}, \bar{\lambda}, t), \bar{w} \in W\}. \tag{3.3.1}$$

The vector function $\bar{f}$ of above is called the *selector* from within the *orientor field* described by the right-hand side of (3.3.1) for each $t \in R_0^+$. In the case for which $\mathscr{P}(\cdot)$ reduces to the single valued $\bar{P}(\bar{x}, \bar{\lambda})$ and $w(\cdot)$ is given, the equation (3.3.1) reduces to (1.5.3''), which is thus called a *selecting equation*. Until we consider the adaptive control in Section 3.4, the argument $\bar{\lambda}(t)$ in (1.5.2) is passive and may be dropped out.

Now, if we are interested in solutions to (3.3.1), we have the following notions to investigate: Given any event $(\bar{x}^0, t_0) \in \Delta \times R$, a solution to (3.3.1) is an absolutely continuous function $k(\bar{x}^0, t_0, \cdot): R_0^+ \to \Delta$, identified at $k(\bar{x}^0, t_0, t_0) = \bar{x}^0$, that when substituted satisfies (3.3.1) almost everywhere on R_0^+. We designate the family of such solutions by $\mathscr{K}(\bar{x}^0, t_0)$. The set of events $(\bar{x}, t) \in \Delta \times R$ such that $\bar{x} = k(\bar{x}^0, t_0, t)$, $t \in R_0^+$, represents a unique curve in $\Delta \times R$. We denote it by $k(\bar{x}^0, t_0, R_0^+)$ and call it the motion from $(\bar{x}^0, t_0)$ (see Fig. 3.10).

Given t, the set $K(\bar{x}^0, t_0, (\cdot), t) \triangleq \{k(\bar{x}^0, t_0, t) \,|\, k(\cdot) \in \mathscr{K}(\bar{x}^0, t_0)\}$ in Δ (t-section of $\Delta \times R$) is called the *attainable set* at t from $(\bar{x}^0, t_0)$. Then the set of events

$$K(\bar{x}^0, t_0, \mathscr{P}(\cdot), [t_0, t]) \triangleq \bigcup_\tau K(\bar{x}^0, t_0, \mathscr{P}(\cdot), \tau) \,|\, \tau \in [t_0, t]$$

in Ω is called the *reachable set at* t from $(\bar{x}^0, t_0)$, while the expression $K(\bar{x}^0, t_0, \mathscr{P}(\cdot), R_0^+)$ designates the *reachable cone* from $(\bar{x}^0, t_0)$ in $\Delta \times R$.

If we choose to leave our model autonomous, that is, let $Q_i^r(\bar{q}, \dot{\bar{q}}, t) \equiv 0$, the explicit t vanishes from under the function $\bar{f}(\bar{x}, \bar{u}, \bar{r})$ in (1.5.3'') and we will have the same notions of motion, attainable set, and reachable set, but in Δ instead of $\Delta \times R$.

Returning to (3.3.1), there are at least several choices of conditions sufficient to the existence of solutions in the above sense, that is, to the fact that $\mathscr{K}(x^0, t_0)$ is nonvoid on $\Delta \times R$ (see Davy [1]). Filippov [1] requires that the set on the right-hand side of (3.3.1) be obtained by a compact set-valued function which is continuous and bounded on $\Delta \times R$. Obviously, suitable conditions must be imposed on $\mathscr{P}(\cdot)$ in order to imply the above. We shall call such a program *admissible*. Then the controlling agent has the convenience of picking up any

3.3 CONTROLLABILITY UNDER UNCERTAINTY

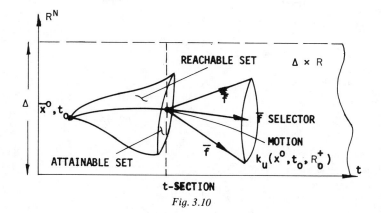

Fig. 3.10

value of $\bar{u}(t)$ from the corresponding set $\mathscr{P}(\bar{x}, t)$ and may use any motion from $\mathscr{K}(\bar{x}^0, t_0)$ to do the job for him. On the other hand, Filippov also showed that, if $\bar{f}(\cdot)$ is continuous and $\mathscr{P}(\cdot)$ is upper-semicontinuous, there are measurable functions $\bar{u}(\cdot)$, $\bar{w}(\cdot)$ such that $\dot{k}(t) = f(k(t), t, \bar{u}(t), \bar{w}(t))$ for almost all $t \in R_0^+$. Hence the agent can always find control functions that allow one to represent $k(\bar{x}^0, t_0, \cdot)$ as unique solutions to the selecting equations (1.5.3″), and thus (3.3.1) as a family of (1.5.3″). Thus, we have $\mathscr{K}(\bar{x}^0, t_0)$ nonvoid and equipped with a definite differential structure. Let us illustrate this with two simple examples.

Example 3.3.1. (Stonier [1]). Consider (3.3.1) specified as

$$\dot{x} \in \{z = -wu \,|\, w \in [1, 2]\}$$

with $x(t) \in R$, $\mathscr{P}(x, t) \equiv x$. The attainability set is

$$K(x^0, t_0, P(\cdot), t) = \begin{cases} (x^0 e^{-2(t-t_0)}, & x^0 e^{-(t-t_0)}), & x \geq 0, \\ (x^0 e^{-(t-t_0)}, & x^0 e^{-2(t-t_0)}), & x < 0. \end{cases}$$

Considered for all $t \geq 0$, it gives the reachability cone. ∎

Example 3.3.2. (Hajek [1]). More generally, we may consider the N-dimensional system

$$\dot{\bar{x}} \in \{\bar{z} = A\bar{u} - \bar{w} \,|\, \bar{w} \in W\}$$

with $\bar{x}(t) \in R^N$, A given $N \times N$ matrix, and $\mathscr{P}(\bar{x}, t) \equiv \bar{x}$. Then the reachability set at t is

$$K(\bar{u}, \bar{x}^0, t_0, \mathscr{P}(\cdot), [t_0, t]) = \int_{t_0}^{t} e^{-As} W \, ds,$$

where

$$\int_{t_0}^{t} e^{-As} W\, ds = \left\{ \int_{t_0}^{t} e^{-As} w(s)\, ds \,\middle|\, w(t) \in W \right\}.$$

For any $w(\cdot)$, the motion is defined by

$$k(\bar{u}, \bar{x}^0, t_0, t) = e^{At} \left[\bar{x}^0 - \int_{t_0}^{t} e^{-As} w(s)\, ds \right],$$

which is obtainable from the selector equation $\dot{\bar{x}} = A\bar{u} - \bar{w}$ by the classical formula of variation of parameters. ∎

The contingent form (3.3.1), particularly with all the characteristics substitued, may not seem too simple at first glance. To reconcile this with the reader, let us say that we will deal de facto with the selecting equations considered for all $\bar{u} \in \mathcal{P}(\bar{x}, \bar{\lambda}, t)$ and all $\bar{w}(t) \in W$. Thus, the meaning of (3.3.1) for all practical purposes is symbolic, but it should help us to understand the problem.

The objectives listed so far must be adjusted to cover the uncertainty problem; the new objectives coming must include the uncertainty directly. It is convenient and instructive now to set up a generally applicable *technical pattern* for the controllabilities concerned, if only to avoid repetitions later. Although some conclusions follow immediately from such a pattern, they will be discussed only if needed for our case study. Generalization for its own sake is not our purpose.

As mentioned briefly at the end of Section 3.1, the objectives may be *qualitative* or *optimal* (quantitative) or both. We discuss the first type now, leaving the second to Section 3.4.

A qualitative objective is of the YES or NO type with respect to obtaining a certain *objective property* Q of the motions of (3.3.1) on some subset of $\Delta \times R$ called the Q zone. This property may in fact be anything specifying the pattern of motions, as long as it is well defined. We attempt to achieve Q in spite of all possible "effort" of the uncertainty, which plays the role of an *opponent* with the objective $\varnothing$ in our game against uncertainty, or as it is more often called, game against nature. Such an opponent plays blindly—it has no control program of its own to select $\bar{w}(t)$ but is *assumed to choose the worst* for us on the principle that if something wrong can happen, it will. Such an assumption is obviously penalizing economically (even if the cost of insurance is substracted), and we shall try to overcome it later by adaptive control, or still better by identification of $w(t)$. The Q zone may cover the entire Δ, be empty, or be anything between these two limits.

Example 3.3.3. Consider the system similar to that in Example 3.1.2, but

3.3 CONTROLLABILITY UNDER UNCERTAINTY

with the noise

$$\dot{x} = u\cos x + w\sin x$$

on $\Delta = \{x \in R \,|\, 0 \leq x \leq \tfrac{1}{2}\pi\}$, with $U: |u| \leq 1$, $W: |w| \leq 1$. Our objective property Q is to reach the target $\mathfrak{T} = \{0\}$, while our opponent, the noise, wants to prevent it. Hence for any $x^0 \in \Delta$ we want $x(t)$ to decrease ($\dot{x}(t) < 0$) against all the noiseband $-1 \leq w \leq 1$, see Fig. 3.11. We thus use $P: u = -1$, $\forall x$. Our opponent does his worst, attempting to increase $x(t)$ ($\dot{x}(t) > 0$) against all $-1 \leq u \leq 1$. Thus he uses $w = 1$ for all x. There is a rest point $\dot{x} = 0$ at $\cos x = \sin x$ wherefrom nobody can move the x, but at any point below it, that is, over Δ_c, we win against the noise, while anywhere above it, the noise wins. The rest point $\dot{x} = 0$ represents the boundary $\partial \Delta_c$ of the region of controllability for capture, separating it from the rest of Δ (see Fig. 3.11). Isaacs [1] called it a *barrier*, with obvious reason. Our conclusions follow immediately from Theorem 3.1.4 if we let $V = x$, implying $\dot{V} = \dot{x}(t)$, and choose the control program as indicated. In terms of energy, Δ_c corresponds to the basic cup Z_H, with $f_0 = \dot{x}$ representing the power balance between actuators and the noise in almost every force concerned. ∎

Considering the controllability, we must first make sure that the objective Q is achievable at all without the opposition of the noise. Such a problem is called *weak* and in fact does not differ from the controllabilities defined in Section 3.1.

Definitions 3.3.1. The system (3.3.1) is *controllable* at $(\bar{x}^0, t_0) \in R$ *for* Q if and only if there is a control program $\mathscr{P}(\bar{x}, t)$ and an uncertainty function $w(\cdot)$ such that at least one motion of the corresponding $\mathscr{K}(\bar{x}^0, t_0)$ has the property Q. The state $\bar{x}^0$ is then called controllable for Q and the set of all such $\bar{x}^0$ is the *region of controllability for* Q, denoted by Δ_q. The latter may also be briefly called the *region for* Q. The controllability is *complete* if $\Delta_q = \Delta$. Any subset of Δ_q is controllable for Q.

Our standard procedure will remain the same as before. We will choose a candidate $\tilde{\Delta}$ for a subset of Δ_q and then check it against sufficient conditions for the controllability concerned. Making $\tilde{\Delta} \cap \Delta_q \triangleq \tilde{\Delta}_q$ controllable, we will

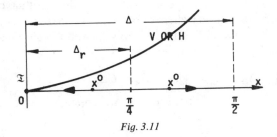

Fig. 3.11

have the controllability $\bar{x}^0$ uniform. In turn, if $\tilde{\Delta}_q \times R$ is controllable, the controllability is $\bar{x}^0$, t_0 uniform, or simply *uniform on* $\tilde{\Delta}_q \times R$.

Note that there is *nothing exclusive about the controllability* for Q in that at the same $\bar{x}^0$, t_0 there may be another trajectory which is controllable for $\mathcal{Q}$ by our opponent noise. Granted that Δ_q is not void, the next question is who is the winner (see Example 3.3.3) and where in Δ the winning may be implemented. This problem is called *strong*. We have

Definitions 3.3.2. The system (3.3.1) is *strongly controllable* at $(\bar{x}^0, t_0) \in \Delta \times R$ *for* Q if and only if there is $\mathcal{P}(\bar{x}, t)$ such that all motions of the corresponding $\mathcal{K}(\bar{x}^0, t_0)$ have Q. The state $\bar{x}^0$ is then called strongly controllable for Q, and the set of all such states is the *region of strong controllability for* Q, denoted by Δ_Q. It may be also called the *winning region* for Q or the *strong region* for Q. The diameter of Δ_Q gives the *degree of strong controllability*. The strong controllability is *complete* if $\Delta_Q = \Delta$, and any subset $\tilde{\Delta} \subset \Delta_Q$ is strongly controllable for Q.

Again, in a standard manner, we choose a candidate $\tilde{\Delta}$ for a subset in Δ_Q suitable for our case assignment and check it against some sufficient conditions yielding the strong controllability for Q. The uniformity in $\bar{x}^0$, t_0 is defined in the same way as for the weak problem, but now the implementation of strong controllability depends also on $k(\cdot)$ in $\mathcal{K}(\bar{x}^0, t_0)$. Thus the strong controllability becomes k *uniform* if the latter does not occur. Moreover, $\mathcal{P}$ is called the *winning program*, and we assume that it is applied whenever available.

In Example 3.3.3, the region Δ_c of controllability for capture is also the region Δ_C of strong controllability for capture with $\partial \Delta_c = \partial \Delta_C : \dot{x}(t) \equiv 0$.

In its attempt to do the worst for us, the uncertainty opposition will aim at the following:

(a) at least to prevent the strong controllability for Q,
(b) if still possible, to prevent the controllability for Q.

Securing (a), it assures the controllability for $\mathcal{Q}$. Indeed, contradicting the strong controllability for Q means that, given $\bar{x}^0$ for each $\mathcal{P}$, there is $w(\cdot)$ such that some motion of the corresponding $\mathcal{K}(\bar{x}^0, t_0)$ does not have Q, thus must have $\mathcal{Q}$. As this happens everywhere outside Δ_Q, we have $\Delta - \Delta_Q \supset \Delta_{\mathcal{Q}}$. On the other hand, Definitions 3.3.1 and 3.3.2 imply $\Delta_Q \cap \Delta_{\mathcal{Q}} = \varnothing$. Then, defining the semineutral set $C\Delta_Q \triangleq \Delta - \Delta_Q$, we have

Property 3.3.1. $\Delta_Q \cap \Delta_{\mathcal{Q}} = \varnothing, \quad \Delta_{\mathcal{Q}} \subset C\Delta_Q.$

In Example 3.3.3, Δ_Q, $\Delta_{\mathcal{Q}}$ partition Δ. If Δ_Q is open, we conclude that $C\Delta_Q$ is closed whence

Property 3.3.2. Open Δ_Q implies $\partial \Delta_Q \subset C\Delta_Q$.

3.3 CONTROLLABILITY UNDER UNCERTAINTY

Also since strong controllability for an objective implies the corresponding controllability, we have (see Fig. 3.12)

Property 3.3.3. $\Delta_Q \subset \Delta_q$.

By the same argument, we have

Property 3.3.4. If Q' implies Q, then $\Delta_{q'} \subset \Delta_q$, $\Delta_{Q'} \subset \Delta_Q$.

Let us now introduce the following:

Definition 3.3.3. Suppose there is a hypersurface in Δ separating its neighborhood into two disjoint open sets. It is a *semibarrier for* Q, denoted B_Q, if and only if, given $(\bar{x}^0, t_0) \in B_Q \times R$, there is $\tilde{\bar{w}}(\cdot)$ such that for all $\mathcal{P}$ no motion of (3.3.1) may enter $\Delta_Q \times R$.

Clearly the property defining the semibarrier B_Q implies the conditions defining $C\Delta_Q$, whence $B_Q \subset C\Delta_Q$. Thus one of the sets separated by B_Q includes Δ_Q. We call it an *interior set* (INT) while calling the other *exterior* (EXT) (see Fig. 3.12). Let B_Q be a smooth surface with a nonzero normal $\bar{n} = (n_1, \ldots, n_N) \in \Delta$ directed positively toward EXT. By Definition 3.3.2, each of the motion selectors $\bar{f}(\bar{x}, \bar{u}, \tilde{\bar{w}}, t)$ within the orientor field (3.3.1) makes the sharp angle $\gamma \leq 90°$ with $\bar{n}$ (see Fig. 3.12). This is equivalent to saying that there is $\tilde{\bar{w}}(\cdot)$ such that

$$\bar{n} \cdot \bar{f}(\bar{x}, \bar{u}, \tilde{\bar{w}}, t) \geq 0 \tag{3.3.2}$$

for all $\bar{u}(t) \in U$.

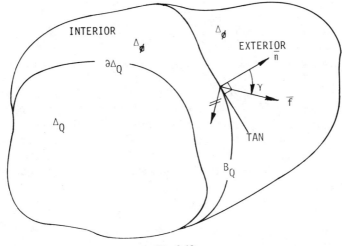

Fig. 3.12

The reader may have guessed already that B_Q estimates the size of Δ_Q and the estimate becomes perfect when it becomes the boundary $\partial\Delta_Q$. This may happen if $C\Delta_Q$ could be covered by condition (3.3.2). Observe, however, that (3.3.2) is only a necessary condition for B_Q, and we may need sufficient conditions to confirm the candidate $\partial\Delta_Q = B_Q$. For instance, in Example 3.3.3, $\partial\Delta_Q : \dot{x}(t) \equiv 0$ is such a candidate, but so is any other point to the right of it. Skowronski and Stonier [2] introduced sufficient conditions for the semi-barrier which may be adapted to our case as follows.

Theorem 3.3.1. *A surface $\mathscr{S}$ that separates its neighborhood into two disjoint open sets, with the normal vector directed to the exterior, becomes a semibarrier B_Q if there is a C^1 function $V(\cdot)$: $\overline{\text{INT}} \to R$ such that*

(i) $V(\text{INT}) < V(\mathscr{S})$,
(ii) *there is $\bar{w}(\cdot)$ generating*

$$\nabla V(\bar{x}) \cdot \bar{f}(\bar{x}, \bar{u}, \bar{w}, t) \geq 0 \qquad (3.3.3)$$

for all $\bar{u}(t) \in U$, $\bar{x} \in \mathscr{S}$, $t \in R_0^+$.

Proof. Proof is immediate. Given $\bar{x}^0, t_0 \in \mathscr{S} \times R \subset \Delta_q \times R$, suppose some motion crosses into $\text{INT} \supset \Delta_Q$. Then there is $t_1 > t_0$ such that $k(\bar{u}, \bar{x}^0, t_0, t_1) = x^1 \in \text{INT}$; whence by (i), $V(\bar{x}^0) > V(x^1)$, contradicting (ii), which operates there by continuity of $\dot{V}(t)$.

Corollary 3.3.1. *When $\mathscr{S}$ is smooth and specified as a $V(\bar{x})$ level, $\nabla V(\bar{x}) = \bar{n}$, condition (3.3.3) also becomes necessary.*

Proof follows immediately from (3.3.2).

The above gives a method for defining Δ_Q if the case investigation requires it instead of assuming a prespecified $\tilde{\Delta}$, as mentioned before. Observe, however, that (3.3.2) being an inequality is not very convenient for use as an algorithm. If the uncertainty acts in a more definite or stronger way, we may obtain a sharper condition. To this end, and for the sake of completing the study, let us now consider task (b) of the two aims of uncertainty posed preceding Property 3.3.1. The aim has been to prevent controllability for Q. If the aim succeeds, then there is $\bar{w}(\cdot)$ such that for all available $\mathscr{P}$ we lose, that is, all motions became $\emptyset$. Holding the pessimistic view that if the worst can happen, it will, the uncertainty will choose such a $\bar{w}(\cdot)$ and we have strong controllability for $\emptyset$ with a nonempty $\Delta_\emptyset$ (see Fig. 3.13).

Now, let us introduce the neutral set

$$\Delta_N \triangleq \Delta - (\Delta_Q \cup \Delta_\emptyset),$$

which by Property 3.3.4 yields

3.3 CONTROLLABILITY UNDER UNCERTAINTY

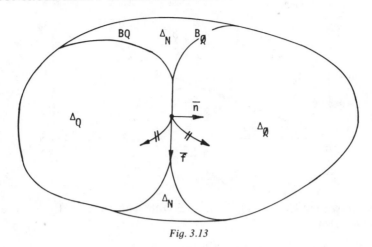

Fig. 3.13

Property 3.3.5. $\Delta_N = C\Delta_Q \cap C\Delta_{\varnothing}$.

Then we define the surface $\mathfrak{B} \triangleq B_Q \cap B_{\varnothing}$ and call it a *barrier* (see Fig. 3.13). Obviously $\mathfrak{B} \subset \Delta_N$, but also, obviously, it may be empty. Maintaining the direction of $\bar{n}$, the inequality in (3.3.2) for $B_{\varnothing}$ must be inverted, and we obtain the necessary condition for $\mathfrak{B}$ specified by the requirement that for some $\tilde{\tilde{u}}, \tilde{\tilde{w}}$,

$$\bar{n} \cdot \bar{f}(\bar{x}, \tilde{\tilde{u}}, \tilde{\tilde{w}}, t) = 0 \tag{3.3.4}$$

at all $(\bar{x}, t) \in \mathfrak{B} \times R$. Following Isaacs [1], a hypersurface in Δ defined by (3.3.4) is called *semipermeable* and is denoted by S. Consequently, B is semipermeable, but obviously there may be many other S surfaces in Δ. Again, all S's may represent candidates for $\mathfrak{B}$, but it is the S which satisfies sufficient conditions for the barrier that becomes $\mathfrak{B}$.

Now consider the choice of $\mathfrak{B}$ candidates. It is obvious that all the S surfaces must be located in Δ_N. Thus if there is a single S in Δ_N, it is $\mathfrak{B}$ and there is no need to check it against any sufficient conditions. This is for instance the case when we know or can guess $\partial \Delta_Q$, $\partial \Delta_{\varnothing}$ and if $\partial \Delta_Q \cap \partial \Delta_{\varnothing} \neq \varnothing$ is S. Otherwise the problem is open and we must use sufficient conditions. The obvious way is to apply Theorem 3.3.1 twice, simultaneously for B_Q and $B_{\varnothing}$, taking two S surfaces as candidates. For more details, see Skowronski and Stonier [1, 2].

In Example 3.3.3 the obvious candidate for $\mathfrak{B}$ is $S = \{\frac{1}{4}\pi\}$. It is the closest point to $\partial \Delta_C$ in Δ_N with $\mathfrak{B} = \partial \Delta_C$, $\partial \Delta_C$ being a semipermeable set and boundary of the region of strong controllability for capture Δ_C. Owing to the specific objective there, target capture, we may also point out another method of selecting a suitable candidate for both B_C and $\mathfrak{B}$, namely, that which is closest to the target. Note that this selection can be made without knowing $\partial \Delta_C$ a priori, or even estimating it.

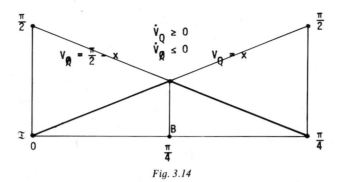

Fig. 3.14

Now, if we let $V_Q = x$, $V_{\cancel{Q}} = \frac{1}{2}\pi - x$, with S defined as $V_Q = x = \text{const}$, whence $n = 1$, we have

$$\dot{V}_Q = \dot{x} = u \cos x + w \sin x,$$
$$\dot{V}_{\cancel{Q}} = -\dot{x} = -u \cos x - w \sin x.$$

The state equation in Example 3.3.3 gives S defined by $\dot{x} = 0$, that is, $x = \frac{1}{4}\pi$. Condition (i) of Theorem 3.3.1 holds for each $V_{Q,\cancel{Q}}$, and condition (ii) follows from our discussion in Example 3.3.3 regarding $\dot{x}(t)$ (see Fig. 3.14). Hence $\mathfrak{B} = \partial \Delta_Q \cap \partial \Delta_{\cancel{Q}} = \{\frac{1}{4}\pi\}$ is confirmed.

Exercises and Comments

3.3.1 Consider the selecting equation $\dot{x} = u\sqrt{x} + w$, $x(t) \in R$ and let $\mathscr{P}(x)$ be defined by $u = \pm\sqrt{x}$, and $W = [-1, 1]$. Write the contingent form (3.3.1) and determine the reachability sets.

3.3.2 Verify that $x(t) = x^0 \pm t$ are two motions of the contingent equation $\dot{x} \in [-1, 1]$. Are there any other motions?

3.3.3 Consider the system

$$\dot{x}_1 = x_2, \qquad \dot{x}_2 = -x_1 - ux_2 + wx_1^3,$$

with $u \in [0, 1]$, $w \in [1, 2]$. Find equilibria and design the feedback control program $\mathscr{P}(x_1 x_2)$ for strong controllability for capture in $x_1^2 + x_2^2 \leq r$, where r is a suitable small constant. Specify the region.

3.3.4 Consider the RP manipulator of Section 1.1 with the equilibria and energy thresholds calculated in Example 3.2.3. Suppose the right-hand sides of (1.1.27) are perturbed and rearranged to become $B_i(\bar{q}, \dot{\bar{q}})u_i + w_i$, $i = 1, 2$, $w(t) \in [0, 1]$. Find a control program $\mathscr{P}(\bar{q}, \dot{\bar{q}})$ such that the threshold (3.2.14) becomes the barrier for asymptotic stabilization in Δ_H.

3.3.5 Give an example of (3.3.1) in which the attainability sets at least on some time interval are simply connected. Would the latter be possible if W

3.4 STRONG AND ADAPTIVE STABILIZATION

were disjoint? For a general answer to this problem and on geometry of reachable sets as a whole, the reader is advised to consult Roxin [2].

3.3.6 Do you know sufficient conditions for the existence of motions of (3.3.1) other than those of Philippov?

3.4 STRONG AND ADAPTIVE STABILIZATION

We shall now illustrate our discussion in Section 3.3 on an immediate application to the objective property Q specified in terms of stabilization, introduced without uncertainty in Section 3.2. Considering now the uncertain system (3.3.1), we must produce a corresponding *strong stabilization* below some energy level.

Definition 3.4.1. The system (3.3.1) is strongly controllable on a set $\Delta_0 \times R$ in $\Delta \times R$ for stabilization, briefly *strongly stabilizable* on $\Delta_0 \times R$, if and only if there is a program $\mathscr{P}$ and two constants: B and (for each ρ_0) some $T_S(\rho_0)$ such that $(\bar{x}^0, t_0) \in \Delta_0 \times R$ implies $K(\bar{x}^0, t_0, \mathscr{P}(\cdot), t) \in \Delta_B$, $t \geq t_0 + T_S$, with $\Delta_B \triangleq \{\bar{x} \in \Delta \,|\, \|\bar{x}\| < B \leq \rho^0\}$ (see Definition 3.2.6). The union of all such Δ_0 is the *region* of strong stabilization Δ_S in Δ.

Recall $\Delta_R \triangleq \{\bar{x} \in \Delta \,|\, \|\bar{x}\| > R\}$, $\rho^0 > R > 0$ suitably small, from Section 3.2. We have

Theorem 3.4.1. *The system (3.3.1) is strongly stabilizable on $\Delta_0 \times R$ if there are C^1 functions $V(\cdot): \bar{\Delta}_R \times R \to R$ and $\mathscr{P}$ defined on $\Delta_R \times R$ such that*

(i) $a(\|\bar{x}\|) \leq V(\bar{x}, t) \leq b(\|\bar{x}\|)$, (3.2.7')

where $a(r), b(r)$ are continuously increasing, $a(r) \to \infty, r \to \infty$;

(ii) *for each $\bar{u} \in \mathscr{P}(\bar{x}, t)$,*

$$\partial V/\partial t + \nabla V(\bar{x}, t) \cdot \bar{f}(\bar{x}, \bar{u}, \bar{w}, t) \leq -c(\|\bar{x}\|), \quad (3.2.9')$$

for all $\bar{w}(t) \in W$, with $c(r) > 0$ a continuous function.

Proof. By the same argument as that in the proof of Theorem 3.2.1, condition (i) implies that no motion of $\mathscr{K}(\bar{x}^0, t_0)$ with $(\bar{x}^0, t_0) \in \Delta_0 \times R$ leaves $\Delta_0 \times R$ (see Fig. 3.7). Given $\bar{x}^0 \in \Delta_0 \cap \Delta_{\mathscr{R}}$, $t \geq t_0$ and $\mathscr{K}(\bar{x}^0, t_0)$, consider an arbitrary motion of this class. By (ii), $V(\bar{x}, t) - V(\bar{x}^0, t_0) \leq -c(t - t_0)$, yielding

$$t \leq t_0 + c^{-1}[V(\bar{x}^0, t_0) - V(\bar{x}, t)].$$

By (i), $V(\bar{x}^0, t_0) \leq b(\rho^0)$, $B(\bar{x}, t) \geq a(B)$, where from $t \leq t_0 + c^{-1}[b(\rho^0) - a(B)] = t_0 + T_S$. Thus for $t > t_0 + T_S$ we have $\|k(\bar{x}^0, t_0, t)\| < B$. The motion may not leave $\Delta_B \times R$ by the same argument which secured invariance of $\Delta_0 \times R$ above, which concludes the proof.

We designate the *Liapunov derivative* by

$$\mathscr{L}'(\bar{x}, t, \bar{u}, \bar{w}) = \partial V/\partial t + \nabla V(\bar{x}, t) \cdot \bar{f}(\bar{x}, \bar{u}, \bar{w}, t). \tag{3.4.1}$$

Theorem 3.3.2 has the following practical deduction.

Corollary 3.4.1. *Given* $(\bar{x}, t) \in \Delta_R \times R$, *if there is a pair* $\tilde{\bar{u}}(\cdot), \tilde{\bar{w}}(\cdot)$ *such that*

$$\mathscr{L}'(\bar{x}, t, \tilde{\bar{u}}, \tilde{\bar{w}}) = \min_{\bar{u}} \max_{\bar{w}} \mathscr{L}'(\bar{x}, t, \bar{u}, \bar{w}) \leq -c(\|\bar{x}\|), \tag{3.4.2}$$

then the condition (3.2.9′) is met with $\tilde{\bar{u}} \in \mathscr{P}(\bar{x}, t)$.

Proof. The proof follows immediately from the fact that

$$\min_{\bar{u}} \max_{\bar{w}} \mathscr{L}'(\bar{x}, t, \bar{u}, \bar{w}) \geq \min_{\bar{u}} \mathscr{L}'(\bar{x}, t, \bar{u}, \bar{w})$$

for all $\bar{w}(t) \in W$.

This means that it may be possible to deduce the winning program from (3.4.2). In particular, let the selecting equation (1.5.3″) have a special shape

$$\dot{x} = \bar{f}^0(\bar{x}, t) + B(\bar{x}, t)(\bar{u} + \bar{w}), \tag{3.4.3}$$

with the uncontrolled and well-defined $\bar{f}^0(\bar{x}, t)$ generating uniform asymptotic stability of the zero motion, $B(\cdot)$ continuous $N \times n$ matrix function, and $W = \{\bar{w} \mid \|\bar{w}\| \leq \rho(x, t)\}$. Moreover, let $V_0(\bar{x}, t)$ be the Liapunov function generating the said asymptotic stability, with the negative-definite Liapunov derivative $\mathscr{L}'_0(\bar{x}(t), t)$. Then

$$\mathscr{L}'(\bar{x}, t, \bar{u}, \bar{w}) = \mathscr{L}'_0 + \nabla V(\bar{x}, t) \cdot B(\bar{x}, t)(\bar{u} + \bar{w}), \tag{3.4.4}$$

and from (3.4.2) we require

$$\min_{\bar{u}} \max_{\bar{w}} \nabla V(\bar{x}, t) \cdot B(\bar{x}, t)(\bar{u} + \bar{w}) \leq 0. \tag{3.4.5}$$

Then we can use the Leitman–Gutman control program, called briefly the *L–G controller* $\mathscr{P}$ defined by

$$u(t) = \begin{cases} -\rho(\bar{x}, t) \dfrac{\alpha(\bar{x}, t)}{\|\alpha(\bar{x}, t)\|}, & \alpha(\bar{x}, t) \neq 0, \\ \{\bar{u} \in R^n \mid \|\bar{u}\| \leq \rho(\bar{x}, t)\}, & \alpha(\bar{x}, t) = 0, \end{cases} \tag{3.4.6}$$

where $\alpha(\bar{x}, t) \triangleq B^{\mathrm{T}}(\bar{x}, t) \cdot \nabla_x^{\mathrm{T}} V(\bar{x}, t)$. Simple calculation shows (see Gutman [2] and Gutman and Palmor [1]) that (3.4.6) substituted into (3.4.4) makes $\mathscr{L}'(\bar{x}, t, \bar{u}, \bar{w}) \leq \mathscr{L}'_0(\bar{x}, t)$, thus satisfying (3.4.5).

Note that $\mathscr{P}$ is unique and smooth for $\alpha \neq 0$, but it is a set for $\alpha = 0$. It means that for $\alpha = 0$, $\bar{u}$ chooses its values over the said set, but the choice depends upon the uncertainty $\bar{w}(\cdot)$ which is not known. However, since $\mathscr{L}'(t) < 0, \alpha = 0$ for any $\bar{u}$ in the set, say $\bar{u} = 0$, for a short time interval $V(\cdot)$ decreases. Then, we

3.4 STRONG AND ADAPTIVE STABILIZATION

may specify (3.4.6) as

$$u(t) = \begin{cases} -\rho(\bar{x},t)\dfrac{\alpha(\bar{x},t)}{\|\alpha(\bar{x},t)\|}, & \alpha(\bar{x},t) \neq 0, \\ 0, & \text{or any other admissible value for a short time as } \alpha(\bar{x},t) = 0. \end{cases} \quad (3.4.7)$$

A modified L–G controller was introduced in Corless and Leitmann [5], making $\mathscr{P}(\cdot)$ consist of branches $P(\bar{x},t)$ which are continuous in $\bar{x}$ and measurable in t, that is, are admissible in our sense. The branches are unique and are applicable to the system (3.4.3) relaxed to the case of several bounded uncertainties represented by state dependent functions, with constant and variable bounds.

Let us apply Theorem 3.4.1 to strongly stabilizing the system under the *same* level $Z_B \times R$ in the complement to the cup-in-the-large. Here, however, we have an immediate problem to solve: Since the total energy $H(\bar{x},\bar{r})$ is noise polluted, the reference frame $\mathscr{H}$ is not well defined. Fortunately, there is a relatively simple solution to the problem. Since $\bar{r}$ is bounded in the closed $\mathfrak{R}_1$, we choose the Liapunov function $V(\bar{x},t) \equiv \tilde{H}(\bar{x}) \triangleq \min_{\bar{r}} H(\bar{x},\bar{r})$, and adjust our investigation to the corresponding surface $\mathscr{H}: \tilde{h} = \tilde{H}(\bar{x})$.

The levels $\tilde{Z}_C: \tilde{H}(\bar{x}) = \tilde{h}_C$ may be then adjusted within the band $\delta\tilde{h} = \max_{\bar{r}} H(\bar{x},\bar{r}) - \tilde{h}_C$ covering the uncertainty in energy. All the notions of Chapter 2 may be similarly rearranged. In particular, the equilibria are defined by [see (1.5.27)]

$$\min_{\bar{w}} \bar{f}(\bar{x}^e, 0, \bar{w}, t) = 0, \qquad \forall t \geq t_0, \quad (3.4.8)$$

with an uncertainty band $\delta\bar{x}^e$ depending on W. As we may assume

$$T(\bar{q}, 0, \bar{r}) \equiv 0, \quad (3.4.9)$$

and

$$Q^D(\bar{q}^e, 0, \bar{r}) \equiv 0, \quad (3.4.10)$$

the equilibria will be defined by [see (2.2.8)]

$$\dot{\bar{q}}_i^e = 0, \qquad \min_{\bar{r}} Q_i^P(\bar{q}^e, \bar{r}) = 0, \qquad i = 1, \ldots, n, \quad (3.4.11)$$

with accuracy to the corresponding band $\delta\bar{q}^e$.

Now, for the sake of the strong stabilization in-the-large, it suffices to use $\tilde{Z}_B$, as *it makes it hold for any level above*, thus also covering the band $\delta\tilde{h}_B$ (see Fig. 3.15). We have to use the system with uncertainty (1.3.15″), see the opening of this section, and consequently we have the balance of power (2.3.15) becoming

$$f_0(\bar{x}, \bar{u}, \bar{Q}^r, t) = \bar{Q}^{DA}(\bar{q}, \dot{\bar{q}})\dot{\bar{q}} + \bar{Q}^F(\bar{q}, \dot{\bar{q}}, \bar{u})\dot{\bar{q}} + \bar{Q}^r(\bar{q}, \dot{\bar{q}}, t)\dot{\bar{q}} + \bar{Q}^{DD}(\bar{q}, \dot{\bar{q}})\dot{\bar{q}}, \quad (3.4.12)$$

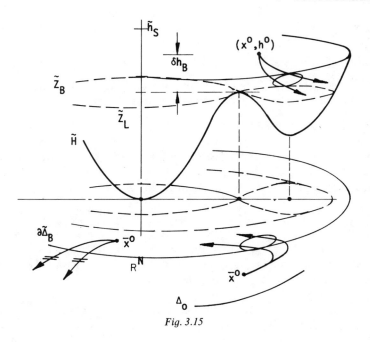

Fig. 3.15

with the axiom of bounded accumulation (2.3.14′) adjusted to

$$\max_{\bar{Q}^r} |\bar{Q}^{DA}(\bar{q},\dot{\bar{q}})\dot{\bar{q}} + \bar{Q}^F(\bar{q},\dot{\bar{q}},\bar{u})\dot{\bar{q}} + \bar{Q}^r(\bar{q},\dot{\bar{q}},t)\dot{\bar{q}}| \leq N < \infty.$$

Then Theorem 3.4.1 is satisfied by the same argument as for Proposition 3.2.4, except that (3.4.2) replaces (3.2.9′). This gives

Proposition 3.4.1. Given $N < \infty$, there is a control program $\mathcal{P}(\bar{x},t)$ such that for some $B(N)$ the system (3.3.1) is strongly stabilized for the level $\tilde{Z}_B \times R$ in $C_\Delta \Delta_L \times R$.

To discuss the control program for Proposition 3.4.1, we use Corollary 3.4.1. Let $V(\bar{x},t) \equiv \tilde{H}(\bar{x})$, whence $\mathcal{L}'(\bar{x},t,\bar{u},\bar{w}) = f_0(\bar{x},t,\bar{u},\bar{Q}^r)$ of (3.4.12). The condition (3.4.2) requires

$$\min_{\bar{u}} \max_{\bar{Q}^r} f_0(\bar{x},t,\bar{u},\bar{Q}^r) \leq \varepsilon \bar{Q}^{DD}(\bar{q},\dot{\bar{q}})\dot{\bar{q}}, \quad \dot{\bar{q}} \neq 0, \qquad (3.4.13)$$

where $\bar{Q}^{DD}(\bar{q},\dot{\bar{q}})\dot{\bar{q}}$ is negative and continuously decreasing with $|\bar{x}|$, and $\varepsilon > 0$, a suitably small constant. Observe that f_0 represents the time rate of the energy flow. Thus since we are in $C_\Delta \Delta_L \times R$, (3.4.13) means that strict dissipation of energy, as long as we are above $\tilde{Z}_B \times R$, produces the strong stabilization. It also means that at each level $h = H(\bar{x}) = $ const with negative f_0, the corresponding set of states lies in the region of strong stabilization Δ_S. On the

3.4 STRONG AND ADAPTIVE STABILIZATION

other hand, by (3.4.12), f_0 represents the power balance and it is easy to see that the control power $\bar{Q}^F(\bar{q}, \dot{\bar{q}}, \bar{u})\dot{\bar{q}}$ must drive f_0 to the negative values against the opposition of the positive powers $\bar{Q}^{DA} \cdot \dot{\bar{q}}$ and $\max \bar{Q}^r \cdot \dot{\bar{q}}$. Thus the game against uncertainty is played on the $\tilde{h}$ axis for $\tilde{h} \geq \tilde{h}_B$, and one may expect the semibarrier B_S determined by such a semipermeable point $\tilde{h}_S$ defined by

$$\min_{\bar{u}} \max_{\bar{Q}^r} f_0(\bar{x}, t, \bar{u}, \bar{Q}^r) = 0, \qquad (3.4.14)$$

which is closest to the level $\tilde{h}_B$. The set of values $\bar{x}$ corresponding to the level $\tilde{h}_S$ in Δ is the boundary $\partial \Delta_S$, provided that the interval $[\tilde{h}_B, \tilde{h}_S]$ is covered by (3.4.13) (see Fig. 3.15). To find the barrier, we would have to secure strong destabilization by the opposition, which in view of the difference between $\tilde{h}_S$ and the noisy h_S, $\tilde{h}_S < h_S$, would require introducing $\hat{H}(\bar{x}) = \max_r H(\bar{x}, \bar{r})$ with corresponding $\hat{h}_S \geq h_S$ defined by

$$\max_{\bar{Q}^r} \min_{\bar{u}} f_0(\bar{x}, t, \bar{u}, \bar{Q}^r) = 0, \qquad (3.4.15)$$

and checking whether $\tilde{h}_S = \hat{h}_S$.

In Example 3.3.3, the energy levels may simply be measured by the height of the rotating gripper (see also Example 3.1.2) or, equivalently, by the angle of rotation, which is $x(t)$ and which has been taken as the Liapunov function. Then (3.4.13) corresponds to the requirement that $\dot{x} < 0$, and semibarrier (3.4.14) is defined by $\dot{x} = 0$, yielding $B_Q = \{\frac{1}{4}\pi\}$. This point also corresponds to both the first and highest energy threshold, and the rotating gripper is asymptotically stabilizable at any x^0 below this threshold. However, in $C_\Delta \Delta_L : \frac{1}{2}\pi < x < \infty$, no admissible control can stabilize the gripper unless we allow higher control constraints, which is the same as raising the upper bound on the control power, so that the whole of $C_\Delta \Delta_L$ is covered by the region of strong controllability for destabilization.

For practical purposes of determining Δ_S and the semibarrier both (3.4.13) and (3.4.14) must be specified by (3.4.12).

Now, in order to actually find the winning program $\mathscr{P}$, we could use the method discussed in Section 3.2, but instead we follow the technique applied for the L–G controller. From (3.4.12),

$$\min_{\bar{u}} \max_{\bar{Q}^r} f_0(\bar{x}, t_0, \bar{u}, \bar{Q}^r) - \left[\min_{\bar{u}} \bar{Q}^F \dot{\bar{q}} + \max_{\bar{Q}^r} \bar{Q}^r \dot{\bar{q}} + \bar{Q}^{DA} \dot{\bar{q}} \right] = \bar{Q}^{DD} \dot{\bar{q}}. \qquad (3.4.16)$$

Since $\bar{Q}^{DD} \dot{\bar{q}}, \dot{\bar{q}} \neq 0$, is negative and decreasing with $|\bar{x}|$, (3.4.2) is satisfied if the following *control condition* holds:

$$\min_{\bar{u}} \bar{Q}^F(\bar{q}, \dot{\bar{q}}, \bar{u})\dot{\bar{q}} + \max_{\bar{Q}^r} \bar{Q}^r(\bar{q}, \dot{\bar{q}}, t)\dot{\bar{q}} \leq -\bar{Q}^{DA}(\bar{q}, \dot{\bar{q}})\dot{\bar{q}}, \qquad \dot{\bar{q}} \neq 0. \qquad (3.4.17)$$

Here, again, this is as close to $\bar{u}$ as we can get without having $\bar{Q}^F(\cdot)$ and $\bar{Q}^{DA}(\cdot)$ specified, which will be possible in our case studies. Note that this technique is obtained from that in Section 3.2 if we allow there the less economic choice of $\varepsilon = 1$.

Example 3.4.1. We return to the RP manipulator of Section 1.1 and Example 3.2.3. Let us rewrite Eqs. (1.1.19), inserting the uncertainties

$$(m_1 r_1^2 + m_2 q_2^2)\ddot{q}_1 + 2m_2 q_2 \dot{q}_1 \dot{q}_2 + \lambda_1 |\dot{q}_1|\dot{q}_2 + 9.81(m_1 r_1 + m_2 q_2)(1 - \tfrac{1}{2}q_1^2)$$
$$- 9.81 m_1 r_1 + aq_1 + bq_1^3 = B_1(\bar{q}, \dot{\bar{q}})u_1 + Q_1^{\mathrm{r}}(\bar{q}, \dot{\bar{q}}, t), \tag{3.4.18}$$
$$m_2 \ddot{q}_2 - m_2 q_2 \dot{q}_1^2 + \lambda_2 \dot{q}_2 + 9.81 m_2(q_1 - \tfrac{1}{6}q_1^3) = B_2(\bar{q}_1, \dot{\bar{q}})u_2 + Q_2^{\mathrm{r}}(\bar{q}, \dot{\bar{q}}, t),$$

where the payload is uncertain, making $m_2 \in \Re_1$ unknown, and where $Q_i^R \in \Re_2$, $i = 1, 2$, are uncertain perturbations. On the other hand, the total energy (3.2.13) as well as the equilibria (3.2.12) become uncertain due to m_2. Since m_2 is positive and bounded with a known bound, we may as well let $\Re_1 = \{m_2 | \underline{m} \leq m_2 \leq \tilde{m}\}$, and define

$$\tilde{H}(\bar{x}) \triangleq \frac{1}{2}(m_1 r_1^2 + \underline{m}q_2^2)\dot{q}_1^2 + \frac{1}{2}\underline{m}\dot{q}_2^2 + \frac{1}{2}aq_1^2 - \frac{9.81}{6}m_1 r_1 q_1^3 + \frac{1}{4}bq_1^4$$
$$+ 9.81 \min_{m_2}(m_2 q_1 q_2) - \frac{9.81}{6}\min_{m_2}(m_2 q_2 q_1^3).$$

Note that the threshold h_H [see (3.2.14)] is not affected, and neither is the basic equilibrium $(0, 0, 0, 0)$. So it is only the levels defined by $\tilde{h}^0$ and $\tilde{h}_B$ which have to be adjusted from h^0, h_B. We have already specified $\Delta_L = \Delta_H$, whence strong stabilization in-the-large is expected above the level defined by h_H [see (3.2.14)]. From (3.4.18) we have the power characteristic as

$$f_0(\bar{x}, t, \bar{u}, \bar{w}) = (B_1 u_1 + Q_1^{\mathrm{r}} - \lambda_1 |\dot{q}_1|\dot{q}_1)\dot{q}_1 + (B_2 u_2 + Q_2^{\mathrm{r}} - \lambda_2 \dot{q}_2)\dot{q}_2, \tag{3.4.19}$$

which should satisfy (3.4.13). It holds if [see (3.4.17)]

$$B_1 u_1 \dot{q}_1 + \max Q_1^{\mathrm{r}} \dot{q}_1 \leq (1 - \varepsilon)\lambda_1 |\dot{q}_1|\dot{q}_1^2,$$
$$B_2 u_2 \dot{q}_2 + \max Q_2^{\mathrm{r}} \dot{q}_1 \leq (1 - \varepsilon)\lambda_2 \dot{q}_1^2. \tag{3.4.20}$$

which represent the control conditions specifying the controller. Applying (3.4.19) on $\Delta - \Delta_H$, we aim at $h_B = h_H$—prespecified. Determination of the barrier $\tilde{h}_S$ from (3.4.19) is left to the reader. ∎

In considering (1.5.3″) so far, we have been assuming that the noise band W for $w(t)$ is known. The philosophy of the worst-case design or game against uncertainty, although penalizing and noneconomic, still works since it is based on the above assumption. Using an *adaptive signal* in the feedback controller would have been the alternative. This alternative, however, becomes a must if we *do not know the size* of W; in other words, if $w(t)$ is still bounded, but we do not know the bound. The model is more convenient for the case studies if we

3.4 STRONG AND ADAPTIVE STABILIZATION

do not alter $w(t)$ in (1.5.3″), but simply add another perturbation whose bound is not known. Consider

$$\dot{\bar{x}} = \bar{f}(\bar{x}, t, \bar{u}, \bar{w}) + B^A(\bar{x}, t)[\bar{u}^A - \bar{w}^A(\bar{x}, t)], \qquad (3.4.21)$$

where $\bar{f}(\cdot)$ is the right-hand side of (1.5.3″), $B^A(\bar{x}, t)$ is a known Lipschitz continuous $n \times n$ matrix function, and $\bar{w}^A(\bar{x}, t)$ is an unknown function bounded by an unknown bound $\beta = \text{const}$:

$$\|\bar{w}^A(\bar{x}, t)\| < \beta. \qquad (3.4.22)$$

We have added an adaptive control vector $\bar{u}^A(t) = (u_1^A(t), \ldots, u_n^A(t))^T$ to offset the unknown-bound uncertainty $\bar{w}^A(\cdot)$. The vector is to be selected by a program $\mathscr{P}^A(\bar{x}, \bar{\lambda}, t)$ defined by

$$\bar{u}^A(t) \in \mathscr{P}^A(\bar{x}(t), \bar{\lambda}(t), t), \qquad (3.4.23)$$

where $\bar{\lambda}(t) = (\lambda_1(t), \ldots, \lambda_l(t))^T \in \Lambda \subset R^l$ is a vector of adaptive parameters bounded in a known band $\Lambda \subset R^l$ and satisfying the *adaptive law*

$$\dot{\bar{\lambda}} = \bar{f}^A(\bar{x}, \bar{\lambda}, t), \qquad (3.4.24)$$

with initial conditions such that $\lambda(t_0) = \lambda^0 > 0$. 0.

The discussion to follow is a simplified version of what has been done by Corless and Leitmann [2, 3, 4]. Let us call the system (1.5.3″) represented in (3.4.21) by the vector $\bar{f}$ *nominal*, as opposed to the *perturbed* (3.4.21). Suppose that the stabilization problem specified by Definition 3.4.1 has been solved for the nominal system; that is, we have some nominal Liapunov pair $V_N(\bar{x}, t)$, $\mathscr{L}_N(\bar{x}, \bar{w}, t)$ and a nominal control program $\mathscr{P}^N(\bar{x}, t)$ such that Theorem 3.4.1 holds. Hence, the general control program $\mathscr{P}(\bar{x}, \bar{\lambda}, t)$ of (1.5.2′) is to be specified by the pair $(\mathscr{P}^N(\bar{x}, t), \mathscr{P}^A(\bar{x}, \bar{\lambda}, t))$ with the first element given, the second to be found.

Denote as before $\alpha^A(\bar{x}, t) \triangleq [B^A(\bar{x}, t)]^T \cdot \nabla^T V_N(\bar{x}, t)$, and specify the adaptive law (3.4.24) by

$$\dot{\bar{\lambda}} = \|\alpha^A(\bar{x}, t)\|. \qquad (3.4.24')$$

Introducing the constant vector $\bar{\beta} = (\beta_1, \ldots, \beta_{s+n}, 0, \ldots, 0)$ in R^l, $l \geq s + n$, such that $|\bar{\beta}| \triangleq \beta$, we denote $\bar{\mu}(t) = \bar{\lambda}(t) - \bar{\beta}$, wherefrom $\dot{\bar{\lambda}} = \dot{\bar{u}}, t \geq t_0$. Then we set up the Liapunov function for the perturbed system $V(\bar{x}, t) = V_N(\bar{x}, t) + \frac{1}{2}\bar{\mu}^2$ and attempt to find $\mathscr{P}^A(\bar{x}, \bar{\lambda}, t)$ such that the conditions of Theorem 3.4.1 hold for (3.4.21) with such $V(\bar{x}, t)$.

First, we observe that condition (i) of Theorem 3.4.1 holds if we add to each side of the inequalities involved the positive term $\frac{1}{2}\bar{u}^2$; thus it holds for $V(\bar{x}, t)$ as much as it did for $V_N(\bar{x}, t)$. Then let us take $\mathscr{P}^A$ defined by

$$\bar{u}^A(t) = -(\alpha^A/\|\alpha^A\|)\bar{\lambda}(t), \qquad (3.4.23')$$

where $\bar{\lambda}(t)$ is a solution of (3.4.24′), then substitute it into (3.4.21) and calculate the derivative of $V_N(\bar{x}, t)$ along the motions of such (3.4.21):

$$\dot{V}_N(\bar{x}(t), t) = \mathscr{L}'_N(\bar{x}, t, \bar{u}, \bar{w}) + [\alpha^A]^T[-(\alpha^A/\|\alpha^A\|)\bar{\lambda} - w^A(\bar{x}, t)]$$
$$\leq \mathscr{L}'_N(\bar{x}, t, \bar{u}, \bar{w}) - \|\alpha^A\|\bar{\lambda} + \|\alpha^A\|\beta$$
$$= \mathscr{L}'_N(\bar{x}, t, \bar{u}, \bar{w}) - \|\alpha^A\|\bar{\mu}.$$

Substituting the adaptive law $\dot{\bar{\mu}} = \dot{\bar{\lambda}} = \|\alpha^A\|$,

$$\dot{V}_N(\bar{x}, (t), t) \leq \mathscr{L}'_N(\bar{x}, t, \bar{u}, \bar{w}) - \bar{\mu}\dot{\bar{\mu}} = \mathscr{L}'_N(\bar{x}, t, \bar{u}, \bar{w}) - \frac{d}{dt}\left(\frac{1}{2}\bar{\mu}^2\right),$$

whence

$$\dot{V}_N(\bar{x}(t), t) + \frac{d}{dt}\left(\frac{1}{2}\bar{\mu}^2\right) \leq \mathscr{L}'_N(\bar{x}, t, \bar{u}, \bar{w}),$$

or

$$\dot{V}(\bar{x}(t), t) \leq \mathscr{L}'_N(\bar{x}, t, \bar{u}, \bar{w}).$$

Since by assumption $\mathscr{L}'_N(\bar{x}, t, \bar{u}, \bar{w}) \leq -c(\|\bar{x}\|)$, then $\dot{V}(x(t), t)$ satisfies condition (ii) of Theorem 3.4.1, and we have

Proposition 3.4.2. Given the function $V_N(\bar{x}, t)$ and the control program $\mathscr{P}^N(\bar{x}, t)$ satisfying Theorem 3.4.1 for the nominal system (3.3.1) on $\Delta^0 \times R$, the control program $\mathscr{P}^A$ defined by (3.4.23') and the function $V(\bar{x}, t) = V_N(\bar{x}, t) + \frac{1}{2}\bar{\mu}^2$ stabilize the perturbed system (3.4.21) on the same $\Delta_0 \times R$ and with the same bound B and time T_S.

Obviously, if the known bound uncertainty $\bar{w}(t)$ in the nominal system is known, the situation simplifies, and we do not need the contingent form (3.3.1) at all. Note that such a system (3.4.21) actually covers both cases as far as our manipulator is concerned, provided that $B^A(\bar{x}, t)$ is suitably chosen in terms of the applied forces.

Translating the above into our manipulator terms, we may investigate (1.3.15″) and (1.3.24″) or (1.5.16‴), both augmented by terms corresponding to $B^A(\bar{x}, t)(\bar{u}^A - \bar{w}^A)$. Let us take (1.3.15″) and assume that the uncertainty $\bar{r}$ is bounded with a known bound, that is, $\mathfrak{R}_1$ is given, while the environmental forces $\bar{Q}^r(\bar{q}, \dot{\bar{q}}, t)$ have an unknown bound and must be adaptively controlled. Thus, given rows B_i^A of B^A reduced to $n \times n$, we have

$$\sum_j a_{ij}(\bar{q}, \bar{r})\ddot{q}_j + K_i(\bar{q}, \dot{\bar{q}}, \bar{r}) - Q_i^{DD}(\bar{q}, \dot{\bar{q}}) - Q_i^{DA}(\bar{q}, \dot{\bar{q}}) - \bar{Q}_i^P(\bar{q}, \bar{r})$$
$$= B_i(\bar{q}, \dot{\bar{q}})\bar{u}_i + B_i^A(\bar{q}, \dot{\bar{q}})[\bar{u}_i^A - Q_i^r(\bar{q}, \dot{\bar{q}}, t)], \qquad (3.4.25)$$

assuming the actuator transmission functions to be linear with respect to the control force from the motor. The above is also under the assumption that

3.5 OPTIMAL CONTROLLABILITY

$\bar{w}(t) = \bar{r}(t) \in \mathfrak{R}_1 \subset R^S$ with the known bound and that $\|Q^r(\bar{q}, \dot{\bar{q}}, t)\| < \beta$, β unknown. Then the nominal control program $\mathscr{P}^N$ is defined from (3.4.17) by

$$u_i(t) = -\left(\sum_i Q_i^{DA}(\bar{q}, \dot{\bar{q}})\dot{q}_i\right) \Big/ B_i(\bar{q}, \dot{\bar{q}}), \qquad \dot{q}_i \neq 0, \qquad (3.4.26)$$

and $u_i(t) = 0$ for $\dot{q}_i = 0$, in order to offset the negative damping of high-speed manufacturing only. Hence, with the adaptive law

$$\dot{\bar{\lambda}} = \|[B^A(\bar{q}, \dot{\bar{q}}, t)]^T \nabla^T \tilde{H}(\bar{q}, \dot{\bar{q}})\|, \qquad \bar{\lambda}(t_0) > 0, \qquad (3.4.27)$$

the almost total (environmental) control is made by the adaptive controller $\bar{u}^A$ with the program $\mathscr{P}^A$ specified by

$$\bar{u}^A(t) = -\frac{[B^A(\bar{q}, \dot{\bar{q}}, t)]^T \nabla^T \tilde{H}(\bar{q}, \dot{\bar{q}})}{\|[B^A(\bar{q}, \dot{\bar{q}}, t)]^T \nabla^T \tilde{H}(\bar{q}, \dot{\bar{q}})\|} \bar{\lambda}(t). \qquad (3.4.28)$$

Hence, Propositions 3.4.1 and 3.4.2 imply

Proposition 3.4.3. The control programs (3.4.26) and (3.4.28), together with the adaptive law (3.4.27), make the uncertain system (3.4.25) strongly stabilized for the level $\tilde{Z}_B \times R \subset C\Delta_L \times R$.

Extension of system (3.4.18) to case (3.4.25) and adaptive laws (3.4.27), as well as the adaptive program (3.4.28), is almost immediate and is left to the reader.

Exercises and Comments

3.4.1 Consider the RP manipulator (3.4.18) and sketch the energy surfaces $\tilde{\mathscr{H}}, \mathscr{H}$ and point out the difference. Then find the min–max controller for stabilization in-the-large from (3.4.20), assuming $\underline{m} = 1$, $\tilde{m} = 10$, $\varepsilon = 0.9$, $\lambda_1 = \lambda_2 = 0.1$. Derive the barrier from (3.4.19).

3.4.2 Consider the system (3.4.25) and specify the controllers (3.4.26), (3.4.28) and adaptive laws (3.4.27) for the case of Exercise 3.4.1.

3.4.3 Write down several shapes of $\bar{Q}^F(\cdot)$ that are other than linear in $\bar{u}$ and that allow explicit calculation of $\bar{u}$ from (3.4.17).

3.5 OPTIMAL CONTROLLABILITY

Suppose the system (3.3.1) is strongly controllable at $(x^0, t_0) \in \Delta \times R$ for Q and moves along some of the motions of $\mathscr{K}(\bar{x}^0, t_0)$. Let us introduce a functional $\mathscr{V}(\cdot)$ which assigns a unique real number to such a motion. The number will be called the *cost* of the transfer from $(\bar{x}^0, t_0)$ to $K(\bar{x}^0, t_0, \mathscr{P}(\cdot), t_f)$ while achieving Q, and it must be consistent with Q during the time concerned for the functional to be called a (quantitative) *performance index*. The cost will be assumed to be dependent upon the corresponding $\mathscr{P}(\cdot)$ and possibly t_f. Although t_f must exist for each $k(\cdot)$ concerned, it may not be specified. Then we delete it from the arguments of the cost function, writing $\mathscr{V}(\bar{x}^0, t_0, k(\cdot), \mathscr{P})$. The cost is identified at some event

$$(\bar{x}^\sigma, t_\sigma): \mathscr{V}(\bar{x}^\sigma, t_\sigma, k(\cdot), \mathscr{P}(\cdot), t_\sigma) = 0 \qquad (3.5.1)$$

for all admissible $\mathscr{P}(\cdot)$ and all $k(\cdot) \in \mathscr{K}(\bar{x}^\sigma, t_\sigma)$, $x^\sigma = k(\bar{u}, \bar{x}^0, t_0, t_\sigma)$, the t_σ being either t_0 or t_f depending upon Q. For example, the cost of reaching is the smaller, the closer we are to a target and is zero at the target. The cost of avoidance behaves in the opposite manner. The cost is additive along the motion. For a more detailed and very illustrative description of the above notions, we recommend Leitmann [7].

The function $\mathscr{V}(\bar{x}^0, t_0, k(\cdot), \mathscr{P}, \cdot): t \to \mathscr{V}(t) \in R$ is called the *cost flow*, and it plays the same role in optimization as the energy flow in qualitative behavior. It is assumed to be C^1 and is written $\mathscr{V}(\bar{u}, \bar{x}^0, t_0, \cdot)$ or even $\mathscr{V}(\cdot)$ when the initial state and time are obvious or irrelevant.

A qualitative strong controllability with some adjoint quantitative objective will consist of two problems.

Problem 1. Given $(\bar{x}^0, t_0) \in \Delta \times R$, secure strong controllability for Q and then among winning programs find $\mathscr{P}_*$ such that

$$\mathscr{V}(\bar{x}^0, t_0, k(\cdot), \mathscr{P}_*(\cdot)) \leq \min_{\mathscr{P}} \max_{w(\cdot)} \inf_{k(\cdot)} \mathscr{V}(\bar{x}^0, t_0, k(\cdot), \mathscr{P}(\cdot)) \qquad (3.5.2)$$

for all admissible $\mathscr{P}(\cdot)$, $w(\cdot)$ and all $k(\cdot) \in \mathscr{K}(\bar{x}^0, t_0)$, with the right-hand side of (3.5.2) called the *upper value* of the game against uncertainty and denoted $\mathscr{V}^+ = \text{const}$. Here again we adopt the philosophy of Section 3.3 that w aims to make it the worst for us, that is, attempts to maximize our cost while we try to minimize it. The minimizing program $\mathscr{P}_*$ is called *optimal*.

The second problem reflects the situation from the viewpoint of the uncertainty, which aims at preventing our goal.

Problem 2. Given $(\bar{x}^0, t_0) \in \Delta \times R$, contradict strong controllability for Q (secure controllability for $\bar{Q}$), and if that is not possible, at least use $w^*(\cdot)$ such that

$$\mathscr{V}(\bar{x}^0, t_0, k(\cdot), \mathscr{P}(\cdot)) \geq \max_{\bar{w}(\cdot)} \min_{\mathscr{P}} \sup_{k(\cdot)} \mathscr{V}(\bar{x}^0, t_0, k(\cdot), \mathscr{P}(\cdot), \bar{w}(\cdot)) \qquad (3.5.3)$$

for all admissible $\mathscr{P}$, $\bar{w}(\cdot)$ and all $k(\cdot) \in \mathscr{K}(\bar{x}^0, t_0)$, with the right-hand side of (3.5.3) being a constant $\mathscr{V}^-$ and called the *lower value* of the game against uncertainty. Here again $\bar{w}^*(\cdot)$ is called *optimal* for the uncertainty.

3.5 OPTIMAL CONTROLLABILITY

In the case $\mathcal{V}^- = \mathcal{V}^+$ we say that there exists a *game value* $\mathcal{V}^*(\bar{x}^0, t_0) = \mathcal{V}^- = \mathcal{V}^+$. Let us write $\mathcal{V}(\bar{x}^0, t_0, \mathcal{P}(\cdot), \bar{w}(\cdot))$ for the cost to acknowledge the fact that for each pair $\mathcal{P}(\cdot)$, $\bar{w}(\cdot)$ we admit many motions $k(\bar{u}, \bar{x}^0, t_0, \cdot) \in \mathcal{K}(\bar{x}^0, t_0)$, each generating a (possibly different) cost. The pair $P_*(\cdot), \bar{w}^*(\cdot)$ such that the cost $\mathcal{V}(\bar{x}^0, t_0, \mathcal{P}_*(\cdot), \bar{w}^*(\cdot)) = \mathcal{V}^*(\bar{x}^0, t_0)$ is called *optimal* and so are the corresponding motions from some $\mathcal{K}^*(\bar{x}^0, t_0)$. Note that all the values $\mathcal{V}(\bar{x}^0, t_0, \mathcal{P}_*(\cdot), \bar{w}^*(\cdot))$ must be equal for given (x^0, t_0).

Theorem 3.5.1. *A game against uncertainty has the value if and only there are $\mathcal{P}_*(\cdot)$ and $\bar{w}^*(\cdot)$ such that, given $(\bar{x}^0, t_0)$,*

$$\mathcal{V}(\bar{x}^0, t_0, \mathcal{P}(\cdot), \bar{w}(t)) \leq \mathcal{V}(\bar{x}^0, t_0, \mathcal{P}(\cdot), \bar{w}^*(t)) \\ \leq \mathcal{V}(\bar{x}^0, t_0, \mathcal{P}(\cdot), \bar{w}^*(\cdot)) \tag{3.5.4}$$

for all $\mathcal{P}(\cdot)$ and $\bar{w}(\cdot)$.

The proof follows immediately from the fact that in view of (3.5.2) and (3.5.3), $\mathcal{V}^- = \mathcal{V}^+$ implies (3.5.4). The converse holds since always min max $\mathcal{V}(\mathcal{P}(\cdot), \bar{w}(\cdot)) \geq$ max min $\mathcal{V}(\mathcal{P}(\cdot), \bar{w}(\cdot))$. Condition (3.5.4) is called the *saddle condition* and is frequently used to determine the optimal program, in a way similar to that for (3.4.2) of Corollary 3.4.1. The game value $\mathcal{V}^*(\bar{x}^0, t_0) = \mathcal{V}(\bar{x}^0, t_0, \mathcal{P}_*(\cdot), \bar{w}_*(\cdot))$ is unique, that is, the $\mathcal{V}^*$'s along all $k(\cdot)$'s are equal.

Definition 3.5.1. A system (3.3.1) is *optimally controllable* at $(\bar{x}^0, t_0)$ *for* Q if and only if it is controllable for Q there and the corresponding $\mathcal{P}(\cdot)$, $\bar{w}(\cdot)$ are optimal. Then $\bar{x}^0$ is optimally controllable for Q, and Δ_q^* denotes the *region* of such controllability.

Definition 3.5.2. A system (3.3.1) is *strongly optimally controllable* at (x^0, t_0) *for* Q if and only if it is strongly controllable at this point for Q, and among the winning $\mathcal{P}$ there is $\mathcal{P}_*$ such that combined with some $\bar{w}^*(\cdot)$ they form an optimal pair. Then $\bar{x}^0$ is the region Δ_Q^* of strong optimal controllability for Q.

The remaining notions follow the pattern in Section 3.3. Note that strong optimal controllability solves Problem 1. On the other hand, optimal controllability for $\mathcal{Q}$ makes strong controllability for Q impossible, thus completely solving Problem 2 rather than solving only its optimizing part.

Definitions 3.5.1 and 3.5.2 yield

Property 3.5.1. $\Delta_q^* \subset \Delta_q, \Delta_Q^* \subset \Delta_Q$.

It follows that $\Delta_N^* \triangleq \Delta - (\Delta_Q^* \cup \Delta_{\mathcal{Q}}^*) \supset \Delta_N$. Moreover, if there is a nonempty barrier $\mathfrak{B}^* = B_Q^* \cap B_{\mathcal{Q}}^*$, then $\mathfrak{B}^* \subset \mathfrak{B} = B_Q \cap B_{\mathcal{Q}} \subset \Delta_N \subset \Delta_N^*$, whence

Property 3.5.2. $\mathfrak{B}^* \neq \varnothing \Rightarrow \mathfrak{B} \neq \varnothing$, which may help in determining $\mathfrak{B}$. On

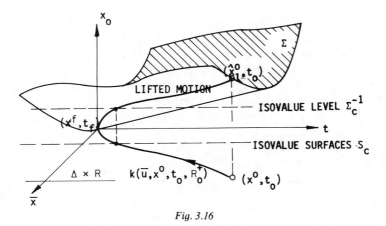

Fig. 3.16

the other hand, note that the $\mathcal{B}^*$ candidates coming possibly from necessary conditions for (3.5.4) may be confirmed to be $\mathcal{B}^*$ by twice using Theorem 3.3.1, which in this way produces sufficient conditions for optimization. Indeed, it is easy to check that Theorem 3.3.1 used twice as prescribed for determining $\mathcal{B}$ generates the Leitmann sufficient conditions for optimization (see Leitmann [1]).

Let us briefly investigate the necessary conditions. We said that the cost flow plays a role corresponding to that of the energy flow. To show it, we need a *cost surface* $\Sigma: x_0 = \mathcal{V}^*(\bar{x}, t)$ in the space R^{N+2} of vectors $(\hat{x}, t)$, where $\hat{x} = (x_0, x_1, \ldots, x_N)$ (see Fig. 3.16).

We let $\mathcal{V}^*(\bar{x}, t)$ be C^1 function, single sheeted due to the property that the value of the game is unique. It is often called the *game potential* (see Krassovski and Subbotin [1]), forming a reference frame for a conflict much as the energy surface does for qualitative control. In a way similar to that for the energy, we form a family of levels: $\Sigma(c): \mathcal{V}^*(\bar{x}, t) = \text{const} = c$, which, when projected isometrically onto $\Delta \times R$, produce a continuous family of surfaces S_c that fill up $\Delta \times R$ and are ordered by the parameter c. The motions cross the isocost levels in exactly the same way that they crossed the constant-energy levels H_C in Section 2.3 (see Fig. 2.6 and the accompanying discussion).

Let the investigated slice of Σ be regular (no extremal points) and such that the given S_c separates its neighborhood into two disjoint sets

$$\text{EXT}: \mathcal{V}^*(\bar{x}, t) > c, \tag{3.5.5}$$

$$\text{INT}: \mathcal{V}^*(\bar{x}, t) < c. \tag{3.5.6}$$

Since Σ is smooth, so is S_c, and at any $(x, t) \in S_c$ there is a gradient

$$\nabla_{x,t} \mathcal{V}^*(\bar{x}, t) \triangleq (\partial \mathcal{V}^*/\partial t, \partial \mathcal{V}^*/\partial x_1, \ldots, \partial \mathcal{V}^*/\partial x_N) \neq 0$$

3.5 OPTIMAL CONTROLLABILITY

directed toward EXT. Consider now the point $(\bar{x}, t)$ on a motion with the selector $\bar{f}(\bar{x}, \bar{u}, \bar{w}, t) \neq 0$, and let this point be regular on S_c (see Fig. 3.17 and compare Section 2.6). There obviously will be $\mathcal{K}(\bar{x}, t)$ of such motions through the same $(\bar{x}, t)$. From (3.5.4) we have at this point

$$\mathcal{V}(\bar{x}, t, \mathcal{P}_*(\cdot), w(\cdot)) \leq \mathcal{V}^*(\bar{x}, t) \quad (3.5.7)$$

for all $w(\cdot)$ and

$$\mathcal{V}(\bar{x}, t, \mathcal{P}(\cdot), w^*(\cdot)) \geq \mathcal{V}^*(\bar{x}, t) \quad (3.5.8)$$

for all $\mathcal{P}(\cdot)$. From (3.5.7) we conclude that no motion with the selector $\bar{f}(\bar{x}, \bar{u}_*, \bar{w}, t)$ may penetrate EXT no matter what $\bar{w}(\cdot)$ is used. This contradicts $\mathcal{V}^*(t) > 0$ along such motions, thus yielding $\mathcal{V}^*(t) \leq 0$, that is,

$$\partial \mathcal{V}^*/\partial t + \nabla_x \mathcal{V}^*(\bar{x}, t) \cdot \bar{f}(\bar{x}, \bar{u}_*, \bar{w}, t) \leq 0 \quad (3.5.9)$$

for all $\bar{u}_* \in \mathcal{P}(\bar{x}, \bar{\lambda}, t)$ and all $\bar{w}(\cdot)$. In turn, from (3.5.8) we conclude that no motion with $\bar{f}(\bar{x}, \bar{u}, \bar{w}^*, t)$ may penetrate INT no matter what $\mathcal{P}(\cdot)$ is used, yielding

$$\partial \mathcal{V}^*/\partial t + \nabla_x \mathcal{V}^*(\bar{x}, t) \cdot \bar{f}(\bar{x}, \bar{u}, \bar{w}^*, t) \geq 0 \quad (3.5.10)$$

for all $\bar{u} \in \mathcal{P}(\bar{x}, \bar{\lambda}, t)$. It follows that conditions (3.5.9) and (3.5.10) are necessary for the *saddle* (3.5.4).

Further implications lead to the following three cases.

(A) Motions with selectors $f(\bar{x}, \bar{u}_*, \bar{w}, t)$ enter INT possibly with sliding upon S_c (cf. $\gamma \geq 90°$ in Fig. 3.17). We call $(\bar{x}, t)$ an *entry point* with contact, or if $\gamma > 90°$, a *strict entry point*. The motions then are *cost dissipative* or strictly dissipative, respectively, which is the objective of our control.

(B) Motions with selectors $f(\bar{x}, \bar{u}, \bar{w}^*, t)$ leave INT possibly with sliding upon S_c (cf. $\gamma \leq 90°$ in Fig. 3.17). Then we call our point $(\bar{x}, t)$ an *exit point* with

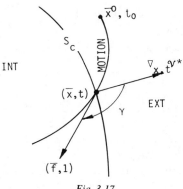

Fig. 3.17

contact, or if $\gamma < 90°$, a *strict exit point*. The latter means that the motions are *cost accumulative* as desired by the uncertainty opposition.

(C) The motions with selectors $f(\bar{x}, \bar{u}_*, \bar{w}^*, t)$ stay on S_c, and with (3.5.9), (3.5.10) applied together we have

$$\partial \mathcal{V}^*/\partial t + \min_{\bar{u}} \max_{\bar{w}} [\nabla_{\bar{x}} \mathcal{V}^*(\bar{x}, t) \cdot \bar{f}(\bar{x}, \bar{u}, \bar{w}, t)] = 0 \quad (3.5.11)$$

for all $\bar{u} \in \mathcal{P}(\bar{x}, \bar{\lambda}, t)$, and all $\bar{w} \in W$, which defines S_c. The surface is thus *cost conservative* or cost positively invariant, and the forces forming $\bar{f}(\bar{x}, \bar{u}_*, \bar{w}^*, t)$ may be called cost potential forces. Thus this immediately suggests the fact that Eq. (1.5.3″) with right-hand side $\bar{\psi}(\bar{x}, t) \triangleq \bar{f}(\bar{x}, t, \bar{u}_*, \bar{w}^*)$ should be integrable and that $\mathcal{V}^*(\bar{x}, t) = $ const is a first integral (see Krassovski and Subbotin [1]), so that the analogy to energy is complete.

Recall now that (3.5.9) follows from preventing penetration into EXT and (3.5.10) follows from preventing penetration into INT [cf. (3.3.2)], which results in (3.5.11) being an analog to (3.3.4). This makes S_c a cost semipermeable surface and an obvious candidate for both semibarriers as well as a barrier, after confirmation by Theorem 3.3.1.

Now let $\bar{x}^e$ be an extremal point for the surface $\Sigma(t)$ such that $\Sigma \triangleq \Sigma(t) \times R$, and let us take the cost level $S_c = \{\bar{x}^e\} \times R$ with the gradient vanishing. This extremal point is either a local minimum or maximum (threshold). Consider the minimum first. The neighborhood is a well EXT with INT collapsed to an empty set. As the gradient vanishes, we work directly from (3.5.7) and (3.5.8), which obviously still hold. As before, motions with $\bar{f}(\bar{x}^e, \bar{u}_*, \bar{w}, t)$ cannot penetrate EXT, but since INT $= \emptyset$, the best the uncertainty can do is to force the motions to rest at $\{\bar{x}^e\} \times R$; that is,

$$\bar{f}(\bar{x}^e, \bar{u}_*, \bar{w}, t) \equiv_t 0 \quad (3.5.12)$$

for all $\bar{u}_* \in \mathcal{P}(\bar{x}, \bar{\lambda}, t)$, $\bar{w} \in W$, $t \in R_0^+$, which thus replaces (3.5.9) as a necessary condition for the minimum. By symmetric argument with EXT $= \emptyset$, we obtain

$$\bar{f}(\bar{x}^e, \bar{u}, \bar{w}^*, t) \equiv_t 0$$

for all $\bar{u} \in \mathcal{P}(\bar{x}, \bar{\lambda}, t)$, all $\mathcal{P}$, and $\bar{w} \in W$, as a necessary condition for $\bar{x}^e$ being a local maximum. As a result, the motions with $\bar{f}(\bar{x}, \bar{u}_*, \bar{w}^*, t)$ cannot leave $\{\bar{x}^e\} \times R$ at all, whence

$$\bar{f}(\bar{x}^e, \bar{u}_*, \bar{w}^*, t) \equiv_t 0 \quad (3.5.13)$$

is necessary for $\bar{x}^e$ to be an equilibrium [see (3.5.8)].

Apart from the extrema $\bar{x}^e$, the surface Σ may have discontinuities in differentiability—sets of points where the gradient is not defined. The corresponding projection into R^N forms what Isaacs calls a *singular surface*, an $(N-1)$-dimensional manifold on which the regular behavior of motions fails. These surfaces may attract, repel, or be neutral, much as the extrema might.

3.5 OPTIMAL CONTROLLABILITY

As a practical conclusion of our analogy between the energy and cost surfaces, we can have that if the cost is considered as energy, the stabilization of Section 3.4 applies directly. When the cost is something else, Definition 3.5.1, Theorem 3.5.1, and Proposition 3.5.1 apply as well but must be translated into the cost language. Then we talk about *strong cost stabilization* rather than strong stabilization. On the other hand, when the objectives are mixed, qualitative and quantitative, e.g., we have both energy and other cost stabilization, the name is *strong optimal stabilization*. The translations are immediate and are left to the reader.

The necessary conditions (3.5.11) are useful in many ways, but to us mainly to produce semipermeable candidates for strong controllability, not necessarily optimal, for various objectives—in particular capture (see Section 4.4). The practical means of applying (3.5.11) is through the so-called Pontriagin principle, which we outline here very briefly for further use.

Let us specify the cost as the integral

$$\mathcal{V} = \int_{t_0}^{t_f} f_0^V(\bar{x}, t, \bar{u}, \bar{w}) \, dt \qquad (3.5.14)$$

and write the cost flow as

$$x_0(t) = x_0(t_0) + \int_{t_0}^{t} f_0^V(\bar{x}, \tau, \bar{u}, \bar{w}) \, d\tau, \qquad x_0(t_0) = 0,$$

which in terms of the vector $\hat{\bar{x}} = (x_0, x_1, \ldots, x_N)^T$ (see Fig. 3.16) produces the *cost-state equations*

$$\dot{\hat{\bar{x}}} = \hat{f}(\hat{\bar{x}}, t, \bar{u}, \bar{w}), \qquad (3.5.15)$$

where $\hat{\bar{f}} = (f_0^V, \bar{f})^T$, with cost-state motions $\hat{k}(\cdot)$ in $\Delta \times \{x_0\} \times R$ governed by the augmented selectors $\hat{\bar{f}}$. Then (3.5.11) becomes

$$\partial \mathcal{V}^*/\partial t + \min_{\bar{u}} \max_{\bar{w}} [\nabla_{\hat{\bar{x}}} \mathcal{V}^*(\bar{x}, t) \cdot \hat{\bar{f}}(\hat{\bar{x}}, t, \bar{u}, \bar{w})] = 0, \qquad (3.5.11')$$

where

$$\nabla_{\hat{\bar{x}}} \mathcal{V}^*(\bar{x}, t) = \left(\frac{\partial \mathcal{V}^*}{\partial x_0}, \ldots, \frac{\partial \mathcal{V}^*}{\partial x_N} \right) \qquad \text{with} \qquad \frac{\partial \mathcal{V}^*}{\partial x_0} = 1.$$

Following Isaacs, we call (3.5.11') the *main equation*. Recall now the vector $\bar{n}$ of (3.3.4) and let $\hat{\bar{n}} = (n_0, n_1, \ldots, n_{N+1})^T$ be defined by $n_0 = \partial \mathcal{V}^*/\partial x_0 = 1$, $n_i = \partial \mathcal{V}^*/\partial x_i$, $i = 1, \ldots, N$, $n_{N+1} = \partial \mathcal{V}^*/\partial t$. We shall refer to $\hat{\bar{n}}$ as *adjoint* or *costate vector*. Then (3.5.11') becomes

$$\min_{\bar{u}} \max_{\bar{w}} \hat{\bar{n}} \cdot \hat{\bar{f}}(\hat{\bar{x}}, t, \bar{u}, \bar{w}) = 0, \qquad (3.5.16)$$

with the geometric interpretation as in Fig. 3.17 and the corresponding discussion (see also Section 3.3). Let us introduce now Pontriagin's

Hamiltonian $\mathfrak{H}(\hat{\bar{x}}, t, \bar{u}, \bar{w}, \hat{\bar{n}}) \triangleq \hat{\bar{n}}(t) \cdot \hat{\bar{f}}(\bar{x}, t, \bar{u}, \bar{w})$ and rewrite (3.5.16) as

$$\min_{\bar{u}} \max_{\bar{w}} \mathfrak{H}(\hat{\bar{x}}, t, \bar{u}, \bar{w}) = 0. \tag{3.5.17}$$

The selecting costate equations (3.5.15); and the so called *adjoint equations* producing $\hat{\bar{n}}$ are

$$\dot{x}_i = \frac{\partial \mathfrak{H}}{\partial n_i}, \qquad i = 0, 1, \ldots, N, \tag{3.5.18}$$

$$\dot{n}_i = -\frac{\partial \mathfrak{H}}{\partial x_i} = \sum_{j=0}^{N} n_j \frac{\partial f_i}{\partial x_i}, \qquad i = 0, 1, \ldots, N. \tag{3.5.19}$$

Conditions (3.5.17)–(3.5.19) form what is known as the Pontriagin *min–max principle*. When the cost is time: $f_0^V(\bar{x}, t, \bar{u}, \bar{w}) \equiv 1$, all the above simplifies. Indeed, $x_0(t) = x_0(t_0) + (t_f - t_0), \hat{\bar{f}} = (1, f)$ with

$$\mathfrak{H} = n_0 + \sum_{i=1}^{N} n_i f_i(\bar{x}, t, \bar{u}, \bar{w}) + \frac{\partial \mathcal{V}^*}{\partial t}. \tag{3.5.20}$$

3.6 STABILIZATION SYNTHESIS

In the adaptive stabilization of Section 3.4, we already had the case in which parameter adjustment helped the controller to secure stabilization (see also Carrol [1] and Nguen [1]). Such help in general terms of controllability for an objective may be expected, perhaps even more effectively, from adjustable parameters outside the controller, that is, embedded directly in the manipulator structure (see Section 4.7). Still more generally, the *structure could be made adjustable* (see Furasov [1], Zinober, El-Gesawi, and Billings [1]). To be more specific, we may have all manipulator characteristics adjustable. The general theory of such synthesis is far beyond the scope of this book. Interested readers are referred to a wide variety of literature (Grayson [1,2], Porter [1]) for review. In particular, the Liapunov formalism based *synthesizable stability* was introduced and developed in general terms in Skowronski [5, 7–11, 24], Bogusz and Skowronski [1, 2], Olas, Ryan, and Skowronski [1].

We refer to case studies dealing with characteristics whose synthesis is suitable for manipulation. Mainly, it means adding spring forces or reducing elasticity and thus also influencing equilibria. However, even for such a reduced scope, the variety of used approaches is so large that no attempt can be made here to give their description or classification. We narrow our study to a few specific cases—to show how the synthesis procedure can be used. In particular, we shall discuss the interface between the gravity–spring equilibria design and the influence of active actuator controllers.

3.6 STABILIZATION SYNTHESIS

By now it is obvious to the reader that a suitably designed controller, given enough brute force, could push the manipulator everywhere on $\mathscr{H}$. This is, however, for a given manipulator structure, that is, for an existing real manipulator. It is perhaps equally obvious that when designing a manipulator, we may apply a more intelligent solution to the problem of achieving a class of specified objectives—namely, to adjust the structure, that is, adjust $\mathscr{H}$ to a given task, thus saving perhaps a lot in actuators power. Since the kinetic energy is a well-established square form that does not influence the shape of $\mathscr{H}$ except by narrowing it to a canyonlike type, we have at our disposal the potential energy and potential forces and thus either *gravity* or *spring compensation* for the controller.

In order to relate directly to energy, let us consider the Hamiltonian equations of motion, first in the form (1.5.16'), that is, without uncertainties. We rewrite these equations here as

$$\dot{q}_i = \frac{\partial T}{\partial p_i} = \sum_j a_{ij}^{-1}(\bar{q}) p_j,$$

$$\dot{p}_i = -\frac{\partial T}{\partial q_i} + Q_i^P(\bar{q}) + Q_i^D(\bar{q},\bar{p}) + Q_i^F(\bar{q},\bar{p},u_i), \quad i = 1,\ldots,n. \quad (3.6.1)$$

Assuming the kinetic forces and damping fixed, we may want the sum $Q_i^P(\bar{q}) + Q_i^F(\bar{q},\bar{p},u_i) = -\partial\mathscr{V}/\partial q_i + Q_i^F(\bar{q},\bar{p},u_i)$ to be synthesized. Let $\mathscr{V}^*(\bar{q})$ be the desired or objective potential energy function and let the *potential compensation force* $Q_i^C = -\partial\mathscr{V}^C/\partial q_i, i = 1,\ldots,n$, be defined from

$$\frac{\partial\mathscr{V}^*}{\partial q_i} \triangleq \frac{\partial\mathscr{V}}{\partial q_i} + \frac{\partial\mathscr{V}^C}{\partial q_i}, \quad (3.6.2)$$

so that the sum $Q_i^P + Q_i^F$ in (3.6.1) becomes

$$-\frac{\partial\mathscr{V}}{\partial q_i} - \frac{\partial\mathscr{V}^C}{\partial q_i} + Q_i^F = -\frac{\partial\mathscr{V}^*}{\partial q_i} + Q_i^F \quad (3.6.3)$$

as desired. For instance, in Example 3.2.3, the potential energy (1.1.17) of our RP manipulator, after substituting $\sin q_1 = q_1 - \frac{1}{6}q_1^3$, produces the energy (3.2.10), which may generate too high an energy threshold for getting the manipulator stabilized in Δ_H (see the calculations of h_H in Chapter 4, Example 4.3.1). If this is the case (and in Example 4.3.1 we will see that indeed it perhaps is), then instead of producing an enormously powerful accumulative controller to get over this threshold, we may lower the threshold. It is obviously done by adjusting the $\mathscr{V}(\bar{q})$ to the desired $\mathscr{V}^*(\bar{q})$ with a lower threshold, that is, according to (3.6.2), by adding $-\bar{Q}_i^C, i = 1,\ldots,n$, to $\partial\mathscr{V}/\partial q_i$, which makes $q_1 - \frac{1}{6}q_1^3$ into a function with a lower maximum at the equilibrium $q_1^e = \sqrt{6}$, $q_2^e = (\sqrt{6}(a + 6b) - 3 \cdot 9.81 m_1 r_1)/2 \cdot 9.81 m_2$. Obviously, such compensation changes q_1^e, q_2^e as well. With such an arrangement, it is hoped that we may attain the desired value of the sum (3.6.3) without excessive power adjustment of $Q_i^F(\bar{q}, \bar{p}, u_i)$.

We may now rewrite (3.6.1) into the system

$$\dot{q}_i = \sum_j a_{ij}^{-1}(\bar{q}) p_j,$$
$$\dot{p}_i = -\frac{\partial H}{\partial q_i} + Q_i^D(\bar{q}, \bar{p}) + Q_i^F(\bar{q}, \bar{p}, u_i) + Q_i^C(\bar{q}), \tag{3.6.4}$$

where Q_i^C plays the role of a potential compensation control force. There is an obvious way in which Q_i^C can cooperate with any of the controllers we introduced in this chapter, including the min–max, the L-G, and the adaptive controllers. Takegaki and Arimoto [2], for a slightly simplified case of (3.6.1), discuss at length such cooperation for a controller representing positive damping, linear in velocity $\bar{u} = Q\dot{\bar{q}}$, where Q is the positive-definite matrix of the Liapunov matrix equation (see Example 3.2.2).

Example 3.6.1. Consider the single DOF manipulator (one link arm) described by the motion equation

$$\ddot{q} + u\dot{q} + 2q + 3q^2 + cq^2 = 0, \tag{3.6.5}$$

where $Q^P = 2q + 3q^2$ and $Q^C = cq^2$, c being the compensation coefficient. In the state form, we have

$$\dot{x}_1 = x_2, \qquad \dot{x}_2 = -ux_2 - 2x_1 - 3x_1^2 - cx_1^2 \tag{3.6.6}$$

with the total energy

$$H(x_1, x_2) = \tfrac{1}{2} x_2^2 + x_1^2 + x_1^3 + \tfrac{1}{3} c x_1^3, \tag{3.6.7}$$

and equilibria $x_1^e = 0$ or $-2/(3 + c)$, $x_2^e = 0$, the negative equilibrium being a saddle (inflection with respect to x_1). The energy level through the latter equilibrium gives the so-called conservative separatrix which specifies the boundary $\partial \Delta_H$. Indeed, if we let $x_2^2/2 + x_1^2 + x_1^3 + \tfrac{1}{3} c x_1^3 = $ const through $x_1^e = -2/(3 + c)$, $x_2^e = 0$, we obtain

$$\text{const} = h_H = \frac{4(3 + c) - 8(1 + \tfrac{1}{3} c)}{(3 + c)^3},$$

which yields the separatrix defined by (see Fig. 3.18)

$$\partial \Delta_H : \frac{1}{2} x_2^2 + x_1^2 + x_1^3 + \frac{1}{3} c x_1^3 = \frac{4(3 + c) - 8(1 + \tfrac{1}{3} c)}{(3 + c)^3}. \tag{3.6.8}$$

Observe that without potential compensation, $c = 0$, we have an established position of the equilibria and thus a specified size of Δ_H. With a suitable controller $u = l(r_1) r_2$, $l(r_1) > 0$, implying that $\dot{H}(x_1(t), x_2(t)) = f_0(x_1, x_2) =$

3.6 STABILIZATION SYNTHESIS

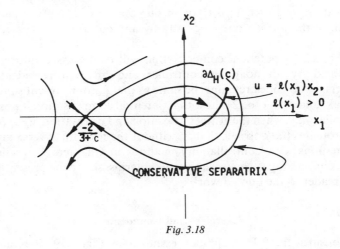

Fig. 3.18

$-l(x_1)x_2^2 < 0$, we may asymptotically stabilize the arm to $(0,0)$ at least from everywhere in Δ_H. On the other hand, to do so from behind the threshold h_H calculated in this example, we may need an accumulative controller for part of the trajectory or compensation $c > 0$ selected such that it enables us to include the (x_1^0, x_2^0) concerned into $\Delta_H(c)$, now depending upon c as specified by $\partial \Delta_H$ of (3.6.8). Sufficiently large c may push the second equilibrium beyond the boundary of the working region generating $\Delta_H = \Delta$.

Naturally, the compensation cq^2 may be obtained either by adding a spring $Q^s = cq^2$ (which is obviously our present case) or by using gravity balancing (*counterweight*) about the axis of rotation. ∎

Turning now to the perturbed case with uncertainty, that is, to the form (1.5.16″) in Section 3.3, we may observe that a suitably selected compensation force $\bar{Q}^c(\bar{q})$ may balance the uncertainty in $\bar{Q}^P(\bar{q}, \bar{r})$, $\bar{r} \in \mathfrak{R}_1$, thus freeing us from the need of adjusting the energy surface $\mathcal{H}$ either to $\mathcal{H}$ or $\mathcal{H}$. It is clearly visible from Example 3.6.1. Consider (3.6.5) augmented as follows:

$$\ddot{q} + u\dot{q} + 2rq + 3rq^2 + cq^2 = w \tag{3.6.9}$$

with $r \in [\underline{r}, \tilde{r}]$, $w \in [0,1]$, which within Δ_H would require $\tilde{\mathcal{H}} : \tilde{h} = \min_r H(q, \dot{q}, r)$ for the asymptotic stabilization aimed at in Example 3.6.1. Choosing c as desired for the initial conditions (x_1^0, x_2^0) but not smaller than required by

$$H(\bar{x}, r, \bar{Q}^c) \le \tilde{H}(\bar{x}), \tag{3.6.10}$$

we cancel the effects of r. Condition (3.6.10) holds generally. In our case we want

$$rx_1^2(1 + x_1) + \tfrac{1}{3}cx_1^3 \le \underline{r}x_1^2(1 + x_1). \tag{3.6.10'}$$

Compensation against $Q'(\bar{q}, \dot{\bar{q}}, t)$, or in our case w in (3.6.9), is usually not needed since the job is done efficiently by any controller discussed in this chapter.

A wider case of potential compensation will be discussed in Section 6.4, accompanied by an adaptive controller and an adaptive identifier of parameters, but the idea remains the same as that for our present simple case.

Takegaki and Arimoto [2] show that potential compensation is possible, in a very similar way, when the gripper's position and orientation are specified in base coordinates (task space), that is, without the need for inverse kinematic transformations. Our controllability methods in Chapters 4–6 will give a means of considering such a case in general terms, so for the present time we refer the reader to the quoted work.

Exercises and Comments

3.6.1 Consider the RP manipulator shown in Fig. 3.19. It is a slightly modified version of the case in Section 1.1. The compensating springs are as marked in the figure with c_i, $i = 1, 2$, measuring the distance from the coordinate origin, $k_i > 0$, $i = 1, 2$, constant coefficients. We also have damping in the joints: linear viscous $Q_1^D = l_1 \dot{q}_1$, $l_1 = \text{const} > 0$, and Coulomb friction $Q_2^D = l_2 \mu(\dot{q}_2) \operatorname{sign} \dot{q}_2$, $l_2 = \text{const} > 0$, μ-unit step function.

Derive the equations of motion in both Newtonian and Hamiltonian forms, rewrite them as state equations, and form the equations for equilibria in terms of c_1, c_2, k_1, k_2. Discuss the possible shapes of $\mathscr{V}$ and choose c_1, c_2 to provide for the shape which gives the fastest descent to the origin under the asymptotic stabilization condition given by the damping and the two actuator controls u_1, $u_2 \in [-1, 1]$.

3.6.2 Make m_2 uncertain in Exercise 3.6.1 and such that $m_2 \in [\underline{m}_2, \tilde{m}_2]$ ($\underline{m}_2, \tilde{m}_2$ given) is positive. Using (3.6.10), compensate for the uncertainty by producing suitable c_1, c_2.

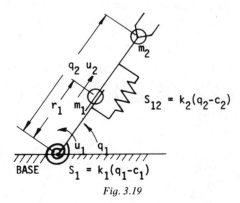

Fig. 3.19

Chapter 4

Reaching and Capture

4.1 REACHING AND RELATED OBJECTIVES

As the reader might have deduced from Section 3.1, reaching and capture represent quite different objective properties for the manipulator in that they have different applications even if attained together. Reaching is satisfactory for point-to-point or set-to-set control and when we attempt a very short contact. Punching holes for riveting may require reaching only, but inserting the rivets or welding along a planned path requires more. When a longer operation is required, we need at least a temporary rendezvous, possibly in stipulated time, or capture (see Section 4.4). When tracking is required, we may get away with successive reachings at stipulated instants (see Section 4.3), but a better job is done by the arm staying continuously with a moving target, which means either successive rendezvous in stipulated time intervals (see Section 4.5) or capture in relative coordinates referenced to the target, which may then be stationary (see Section 4.4). There are obviously also some *direct methods* designed for tracking (see Section 4.5), such as stabilization about a moving target as introduced recently by Corless, Leitmann, and Ryan [1] or convergence to the diagonal in the product of state spaces of the manipulator and the moving target.

Before discussing any such combined objectives, we must analyze the basic reaching and basic capture for our nonautonomous uncertain system (3.3.1).

There are at least several variations of either of these two concepts. We may want to reach or capture within a stipulated time; we may wish to reach not only the boundary of a target but also its core or interior, i.e., *penetrate* the target. We may want to reach a target in the close vicinity of some object without touching this object, so called *soft reaching*. The corresponding idea is *soft capture*, etc.

The Cartesian target was introduced in Section 1.2 and its transformation to the state space was discussed in Section 1.4. We explained there why the target must refer to all joint coordinates and velocities rather than those of the gripper only. Note also that, as much as each point in $\Delta \subset R^N$ instantaneously represents both configuration and movement of all joints, there are many such configurations in a target and their particular selection, that is, choice of the target, depends upon the objective attempted. For instance, a configuration quite different from that needed when the load is minimal may be needed for lifting a heavy load, in spite of the gripper being in the same position in both of the above cases. There is also a difference between a *stipulated target* as discussed above, either stationary or moving, for which the inverse kinematics is done a priori *off-line*, and a moving target with unpredictable path—the so-called *chasing rabbit* case—with the unpredictable path calculated on-line. In our study of the modular objectives, we assume that we know a closed and bounded stationary target set $\mathfrak{T}$ in Δ. Other interpretations will be given later while referring to complex objectives.

Definition 4.1.1. Given $(\bar{x}^0, t_0) \in \Delta \times R$, $\mathfrak{T} \subset \Delta$ a motion of $\mathscr{K}(\bar{x}^0, t_0)$ reaches $\mathfrak{T}$ if and only if (see Fig. 4.1)

$$k(\bar{u}, \bar{x}^0, t_0, R_0^+) \cap (\mathfrak{T} \times R) \neq \emptyset. \qquad (4.1.1)$$

It means there is some (unspecified) constant $T_\mathscr{R} > 0$ such that

$$k(\bar{u}, \bar{x}^0, t_0, t_0 + T_\mathscr{R}) \in \mathfrak{T}. \qquad (4.1.2)$$

Denote $\overset{\circ}{\mathfrak{T}}$ the interior of $\mathfrak{T}$.

Definition 4.1.2. Given $(\bar{x}^0, t_0) \in \Delta \times R$ and $\mathfrak{T} \subset \Delta$, a motion of $\mathscr{K}(\bar{x}^0, t_0)$ *penetrates* $\mathfrak{T}$, if and only if

$$k(\bar{u}, \bar{x}^0, t_0, R^+) \cap (\overset{\circ}{\mathfrak{T}} \times R) \neq \emptyset. \qquad (4.1.3)$$

It also means that there is some (unspecified) $T_P > 0$ such that

$$k(\bar{u}, \bar{x}^0, t_0, t_0 + T_P) \in \overset{\circ}{\mathfrak{T}}. \qquad (4.1.4)$$

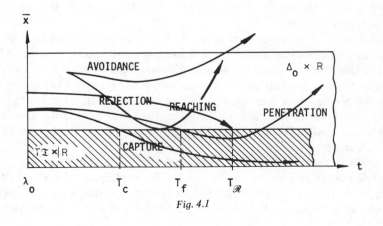

Fig. 4.1

4.1 REACHING AND RELATED OBJECTIVES

The time intervals $T_{\mathscr{R}}$, T_{P} may be prescribed (see Section 4.3). Although we have a separate chapter on avoidance, it is convenient to introduce the property here in a sort of preliminary and limited way.

Definition 4.1.3. Given $(\bar{x}^0, t_0) \in \Delta \times R$, $\mathfrak{T} \subset \Delta$, a motion $k(\bar{u}, \bar{x}^0, t_0, \cdot) \in \mathscr{K}(\bar{x}^0, t_0)$ *avoids* $\mathfrak{T}$, if and only if

$$k(\bar{u}, \bar{x}^0, t_0, R_0^+) \cap (\mathfrak{T} \times R) = \varnothing. \quad (4.1.5)$$

This implies that there is no $T_{\mathscr{R}} < \infty$ and possibly that $T_{\mathscr{R}} = \infty$ for reaching. Finally, we have

Definition 4.1.4. Given $(\bar{x}^0, t_0) \in \Delta \times R$, $\mathfrak{T} \subset \Delta$, a motion $k(\bar{u}, \bar{x}^0, t_0, \cdot) \in \mathscr{K}(\bar{x}^0, t_0)$ is *rejected* from $\mathfrak{T}$ if and only if

$$k(\bar{u}, \bar{x}^0, t_0, R^+) \cap (\mathring{\mathfrak{T}} \times R) = \varnothing. \quad (4.1.6)$$

This implies no $T_{\mathrm{P}} < \infty$, but possibly $T_{\mathrm{P}} = \infty$.

Later-introduced real-time avoidance and rejection, stipulated time avoidance, and ultimate avoidance (see Chapter 5) will slightly complicate the clear picture appearing in the above definitions and in Fig. 4.1. But these combined objectives will nevertheless be referring to the above four modular cases.

Reaching and penetration represent some Q's at some $(\bar{x}^0, t_0) \in \Delta \times R$, while avoidance and rejection represent the corresponding $\mathcal{Q}$'s again at some $(\bar{x}^0, t_0) \in \Delta \times R$ (see Section 3.3). The definitions of controllability and strong controllability for Q have been given in Section 3.3 and apply here to all above Q's and $\mathcal{Q}$'s, together with corresponding notation. Regions of controllability for reaching and penetration are Δ_r, Δ_p and corresponding regions of strong controllability are $\Delta_{\mathscr{R}}$, Δ_{P}. The region of controllability for avoidance is $\Delta_{\mathbf{R}}$ and rejection $\Delta_{\mathbf{p}}$, with corresponding strong regions Δ_r, Δ_p.

Observe that penetration implies reaching, whence we have

Property 4.1.1. $\Delta_p \subset \Delta_r$, $\Delta_{\mathrm{P}} \subset \Delta_{\mathscr{R}}$.

Similarly, avoidance implies rejection, whence

Property 4.1.2. $\Delta_r \subset \Delta_p$, $\Delta_{\mathbf{R}} \subset \Delta_{\mathbf{p}}$.

The same inclusions apply to intersections of those regions with a specified subset $\tilde{\Delta} \subset \Delta$.

Example 4.1.1. Consider a single-DOF elastically suspended overhead gripper subject to gravity forces and control only (see Fig. 4.2) with the controller polluted by noise. We attempt the soft-reaching of a target point at

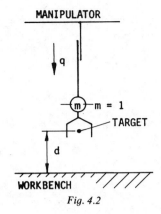

Fig. 4.2

the distance $d > 0$ above the workbench. The reduced motion equation is $\ddot{q} = uw - g$, where $w(t)$ is the noise parameter and g the constant of gravity. In state form, we have

$$\dot{x}_1 = x_2, \qquad \dot{x}_2 = uw - g, \tag{4.1.7}$$

with $\Delta: x_1 \geq 0$, $x_2 \in R$, and $\mathfrak{T} = \{(d, 0)\}$ (see Fig. 4.3).

The control action uw balances the gravitation and reduces the acceleration $\dot{x}_2$ and the velocity $\dot{x}_1$ to zero at the rest position $\mathfrak{T}$. The control is constrained by $u(t) \in [a_1, a_2]$, $a_2 > a_1 > 0$, and subject to the disturbing noise $w(t) \in [b_1, b_2]$, $b_2 > b_1 > g > 0$. Equation (4.1.7) is integrable:

$$x_2^2 - 2(uw - g)x_1 = \text{const.} \tag{4.1.8}$$

The gripper operator achieves his objective by maximizing the control $u(t) \equiv a_2$ against all w, that is, against $w(t) \equiv b_1$, which generates the family of trajectories

$$x_2^2 - 2(a_2 b_1 - g)x_1 = \text{const.} \tag{4.1.9}$$

sketched in Fig. 4.2. The trajectory that reaches $\mathfrak{T}$ without switching the control $u \equiv a_2$ is thus

$$x_1 = [x_2^2/2(a_2 b_1 - g)] + d.$$

Motions from all other paths must switch control. On the other hand, crashing into the workbench in spite of $u \equiv a_2$ occurs at $(0, 0)$ along $x_1 = x_2^2/(2(a_2 b_1 - g))$ which means that the strongly winning ability of $u \equiv a_2$ ends before the line

$$x_1 = \begin{cases} x_2^2/2(a_2 b_1 - g), & x_2 < 0, \\ 0, & x_2 \geq 0, \end{cases} \tag{4.1.10}$$

which thus determines $\partial \Delta_{\mathscr{R}} = B_{\mathscr{R}}$, making $\Delta_{\mathscr{R}}$ a well-determined open set.

4.1 REACHING AND RELATED OBJECTIVES

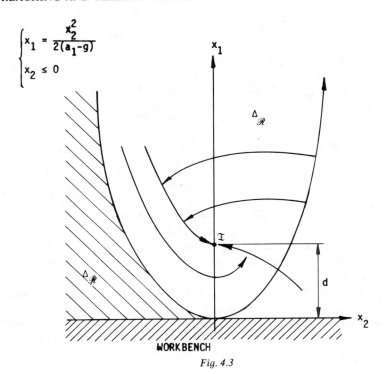

Fig. 4.3

Indeed, it is obvious that (4.1.10) is a semipermeable line S (see Section 3.3) that coincides with $\partial \Delta_{\mathscr{R}}$ and is closest to $\mathfrak{T}$, as by definition $S \not\subset \Delta_{\mathscr{R}}$. This makes (4.1.10) an obvious $B_{\mathscr{R}}$. Note that $\Delta - \bar{\Delta}_{\mathscr{R}}$ is filled up with $\Delta_{\mathscr{I}}$ (crashing motions) which makes $B_{\mathscr{R}}$ the barrier $\mathfrak{B}$. ∎

One of the halfway properties which lead to the combined manipulator objectives mentioned before is *maneuverability* (see Section 3.1), which in turn needs an interface of reaching with another basic concept which may be described in practical terms as "ability to return." It is a property that is dual to controllability for reaching, namely, *reachability* (see again Section 3.1).

Let us elaborate on the concept adjusting it to our present needs.

Definition 4.1.5. A point $\bar{x} \in \Delta$ is *reachable* from a set $\Delta_0 \times R$ if and only if there is $\mathscr{P}(\cdot)$ and $T_{\mathfrak{R}} > 0$ such that given $(\bar{x}^0, t_0) \in \Delta_0 \times R$ there is a motion of $\mathscr{K}(\bar{x}^0, t_0)$ for which $k(\bar{u}, \bar{x}^0, t_0, t_0 + T_{\mathfrak{R}}) = \bar{x}$. The set $\Delta_{\mathfrak{R}}(\Delta_0)$ of all such points $\bar{x}$ in Δ is called a *reachable set* from $\Delta_0 \times R$ (see Fig. 4.4).

Note that Δ_0 may consist of a singleton $\{\bar{x}^0\}$; then the reachable set is $\Delta_{\mathfrak{R}}(\bar{x}^0)$. The controllability for reaching told us *where the system came from* (region $\Delta_{\mathscr{R}}$), while the reachability tells us *where it may go* (set $\Delta_{\mathfrak{R}}$).

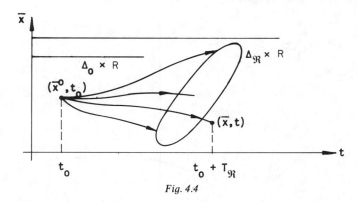

Fig. 4.4

It is a simple matter to see that we must have $\Delta_0 \subset \Delta_\Re$ and $\mathfrak{T} \subset \Delta_\mathscr{R}$ trivially. Then if $\Delta_0 \cap \mathfrak{T} \neq \emptyset$, we conclude that $\Delta_\Re \cap \Delta_\mathscr{R} \neq \emptyset$, and expect that every point $\bar{x}$ from the latter intersection is controllable to and reachable from every other point $\bar{y}$ of this intersection. If, moreover, there exists a control program $\mathscr{P}(\cdot)$ which generates motions that can transfer $\bar{x}$ into $\bar{y}$ and back again while remaining in $\Delta_\Re \cap \Delta_\mathscr{R} \triangleq \Delta_M$, we call Δ_M a *maneuverable set*. It follows that $\Delta_M \times R$ is positively invariant under $\mathscr{P}(\cdot)$.

Let us now qualify the region of strong controllability for reaching $\mathfrak{T}$ by denoting it $\Delta_\mathscr{R}(\mathfrak{T})$, in particular $\Delta_\mathscr{R}(\bar{x}^r)$. Gayek and Vincent [1] proved that Δ_M is connected and that if $G \triangleq \Delta_\Re(\bar{x}^0) \cap \Delta_\mathscr{R}(\bar{x}^0)$, then $G = \Delta_\Re(G) \cap \Delta_\mathscr{R}(G)$ and G is maneuverable unless it is a singleton. Obviously G may be either the target $\mathfrak{T}$ or a starting set Δ_0, depending on the case at hand.

Quite obviously, in most cases of manipulator work we are interested in that part of $\Delta_\mathscr{R}$ which does belong to Δ_M, that is, that part that is contained in the set Δ_M rather than in the entire $\Delta_\mathscr{R}$. Ideally, we would want the whole work space to be maneuverable: $\Delta_M = \Delta$, and in fact we may be prepared to narrow Δ in order to achieve it.

Let us now turn to the question of how to find $\Delta_\mathscr{R}$. In Example 4.1.1 we did it almost by inspection from the character of the motions. In general circumstances such a result may not be possible.

The method which we suggested already in Section 3.3 is to produce a suitable candidate for the semibarrier $B_\mathscr{R}$ by selecting the semipermeable surface that is closest to the target $\mathfrak{T}$, and verifying it against sufficient conditions for such a semibarrier (see Theorem 3.3.1). In practical terms we may attempt to pass some S in a suitable neighborhood of $\mathfrak{T}$ and see whether there is no other S closer to $\mathfrak{T}$. A slightly modified method will be presented later in this section by using the barrier for penetration to serve as an underestimate of $B_\mathscr{R}$.

Alternatively, having the S closest to $\mathfrak{T}$ as a suitable candidate for $B_\mathscr{R}$, instead of checking it against Theorem 3.3.1 we may consider the correspond-

4.1 REACHING AND RELATED OBJECTIVES

ing interior of such an S as a candidate for $\Delta_\mathcal{R}$ and check it against the sufficient conditions for strong controllability for reaching which are presented below.

Going further, we may not need to determine $\Delta_\mathcal{R}$. Indeed, in many practical situations we are given the set Δ_0 of initial states wherefrom we want the gripper to reach the target, and we may not be interested in whether or not the same can be done from any other set. This applies particularly when Δ_0 is our candidate for a maneuverable set. Thus, we may have a candidate set Δ_0 to check against the said sufficient conditions for strong controllability, and then perhaps reach a conclusion about its maneuverability as well.

The sufficient conditions, in the Liapunov sense, have been introduced by Blaquière Gerard, and Leitmann [1] and proved in more practical terms by Sticht, Vincent, and Schultz [1]. It seems instructive to mention the latter before generalizing.

The system considered is an autonomous version of (1.5.3″), that is,

$$\dot{\bar{x}} = \bar{f}(\bar{x}, \bar{u}, \bar{w}) \qquad (4.1.11)$$

and the control program sought is a single-valued $\bar{P}(\bar{x})$. The target is specified as the set $\mathfrak{T} = \{\bar{x} \in \Delta \,|\, \mathfrak{T}(\bar{x}) \leq 0\}$, where $\mathfrak{T}(\cdot): \Delta \to R$ is a C^1 function with nonzero gradient for all $\bar{x}$ in a neighborhood of $\partial \mathfrak{T}$. We also set up $\Delta_0 = \{x \in \Delta \,|\, g(\bar{x}) \leq 0\}$, where $g: \Delta \to R$ is a C^1 function with nonzero gradient such that $\partial \Delta_0$ is semipermeable (see Fig. 4.5). When some preassigned sets $\tilde{\mathfrak{T}}, \hat{\Delta}_0$ do not fit into the above specification, our arguments will remain valid if we take $\mathfrak{T}(\cdot), g(\cdot)$ such that the core $\mathfrak{T} \subset \tilde{\mathfrak{T}}$ and the envelope $\Delta_0 \supset \hat{\Delta}_0$ are suitably close to give satisfactory estimates.

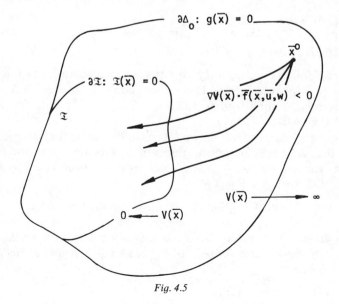

Fig. 4.5

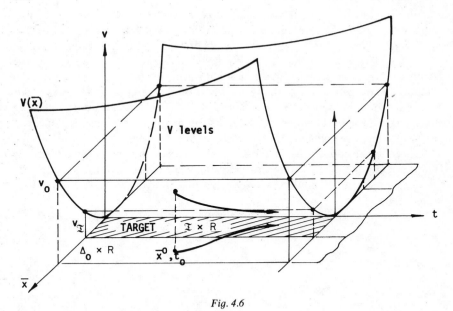

Fig. 4.6

Theorem 4.1.1. *The system (4.1.11) is strongly controllable on $\overset{\circ}{\Delta}_0$ for reaching $\mathfrak{T}$ if there is a C^1 function $V(\cdot): \Delta_0 \to R$ and a program $\bar{P}(\bar{x})$ such that*

(i) $V(\bar{x}) > 0, \bar{x} \notin \mathfrak{T}$;
(ii) $V(\bar{x}) \to 0, \bar{x} \to \partial \mathfrak{T}$;
(iii) $\nabla V(\bar{x}) \cdot \bar{f}(\bar{x}, \bar{u}, \bar{w}) < 0, \forall \bar{w} \in W$; (4.1.12)
(iv) $\partial \Delta_0$ is an envelope of levels $V(\bar{x}) = $ const.

The proof follows immediately from the fact that conditions (i)–(iv) satisfy Theorem 4.1.2 (proved below). The authors allow the case $\Delta_0 = \Delta = R^N$ but then require additionally that $V(\bar{x}) \to \infty$ as $|\bar{x} - \bar{x}^{\mathfrak{T}}| \to \infty$ for all $\bar{x}^{\mathfrak{T}} = \{\bar{x} \in R^N \mid \mathfrak{T}(\bar{x}) = 0\}$.

After the above introduction, we return to our general system (3.3.1) and suppose that $\mathfrak{T}$ and Δ_0 are given subsets of Δ, $\mathfrak{T}$ being closed and bounded.

We designate $C\mathfrak{T} \triangleq \Delta_0 - \mathfrak{T}$ and consider a function $V(\cdot): D \to R$, where open $D \supset \overline{C\mathfrak{T}}$, with (see Fig. 4.6)

$$v_{\mathfrak{T}} = \inf V(\bar{x}) \mid \bar{x} \in \partial \mathfrak{T}, \qquad v_0 = \sup V(\bar{x}) \mid \bar{x} \in \partial \Delta_0. \quad (4.1.13)$$

Theorem 4.1.2. *The system (3.3.1) is strongly controllable on $\Delta_0 \times R$ for reaching $\mathfrak{T}$ if there is a program $\mathscr{P}(\cdot)$ defined on $D \times R$ and a function $V(\cdot): D \to R$ such that*

(i) $V(\bar{x}) > v_{\mathfrak{T}}$, for $\bar{x} \notin \mathfrak{T}$;

4.1 REACHING AND RELATED OBJECTIVES

(ii) $V(\bar{x}_1) \leq v_0$, for $\bar{x} \in C\mathfrak{T}$;
(iii) for all $\bar{u} \in \mathscr{P}(\bar{x}, t)$ there is a constant $T_{\mathscr{R}}(\bar{x}^0, t_0, k(\cdot)) > 0$ such that

$$V(k(\bar{u}, \bar{x}^0, t_0, t_0 + T_{\mathscr{R}})) - V(\bar{x}^0) \leq -(v_0 - v_{\mathfrak{T}}) \quad (4.1.14)$$

for all $k(\bar{u}, \bar{x}^0, t_0, \cdot) \in \mathscr{K}(\bar{x}^0, t_0)$, $(\bar{x}^0, t_0) \in C\mathfrak{T} \times R$.

Proof. Motions from $\bar{x}^0 \in \mathfrak{T}$ reach trivially; thus let $\bar{x}^0 \notin \mathfrak{T}$. By (iii), $V(k(t_0 + T_{\mathscr{R}})) + [v_0 - V(\bar{x}^0, t_0)] \leq v_{\mathfrak{T}}$. By (ii), there is a $\kappa(\bar{x}^0) \geq 0$ such that the above becomes $V(k(t_0 + T_{\mathscr{R}})) + \kappa \leq v_{\mathfrak{T}}$, which by (i) means reaching $\mathfrak{T}$, and this completes the proof.

Remark 4.1.2. Given some constants $v_{\mathfrak{T}}, v_0 > 0$, Theorem 4.1.2 is true if we define $\mathfrak{T}$ and Δ_0 by

$$\partial \mathfrak{T}: V(\bar{x}) = v_{\mathfrak{T}}, \qquad \partial \Delta_0: V(\bar{x}) = v_0. \quad (4.1.15)$$

Note that the function $V(\cdot)$ does not have to be smooth, it may in fact even be discontinuous. On the other hand, the condition (4.1.14) for an arbitrary function like that is difficult to handle, unless it serves as a verification for some known function obvious from somewhere else, like the total energy. The "handier" sufficient conditions were given by Skowronski [21]:

Theorem 4.1.3. *The system (3.3.1) is strongly controllable on Δ_0 for reaching $\mathfrak{T}$ if there is $\mathscr{P}(\cdot)$ defined on $C\mathfrak{T} \times R$ and a C^1 function $V(\cdot): D \to R$ such that*

(i) $V(\bar{x}) > v_{\mathfrak{T}}$ for $\bar{x} \notin \mathfrak{T}$;
(ii) $V(\bar{x}) \leq v_0$ for $\bar{x} \in C\mathfrak{T}$;
(iii) for all $\bar{u} \in \mathscr{P}(\bar{x}, t)$, there is $T_{\mathscr{R}}(\bar{x}^0, t_0, k(\cdot)) > 0$, yielding

$$\int_{t_0}^{t_0 + T_{\mathscr{R}}} \nabla V(\bar{x}) \cdot \bar{f}(\bar{x}, t, \bar{u}, \bar{w}) \, dt \leq -(v_0 - v_{\mathfrak{T}}) \quad (4.1.16)$$

for all $\bar{w}(t) \in W$.

Proof. Observe that conditions (i)–(iii) imply those of Theorem 4.1.2, but we prefer to give a separate proof. Again assume the nontrivial case of $\bar{x}^0 \notin \mathfrak{T}$, and suppose some motion from $C\mathfrak{T} \times R$ avoids reaching $\mathfrak{T}$. Taking $V(\cdot)$ along such a motion, that is, calculating the flow of $V(t)$, (i) implies

$$V(k(\bar{u}, \bar{x}^0, t_0, t)) = V(\bar{x}^0) + \int_{t_0}^{t} \dot{V}(\bar{x}) \, d\tau > v_{\mathfrak{T}} \quad (4.1.17)$$

for all $t \in R_0^+$. Since $V(\bar{x}^0) \leq v_0$, (4.1.17) yields

$$\int_{t_0}^{t_f} \dot{V}(\bar{x}(t)) \, dt > -(v_0 - v_{\mathfrak{T}}), \quad (4.1.18)$$

which contradicts (iii), proving our hypothesis.

Corollary 4.1.3. *Suppose there is a C^1 function $V(\cdot): D \to R$ satisfying conditions (i) and (ii) of Theorem 4.1.3. Then in order for the system (3.3.1) to be strongly controllable on Δ_0 for reaching $\mathfrak{T}$ it is necessary and sufficient to have $\mathscr{P}(\bar{x}, t)$ such that for all $\bar{u} \in \mathscr{P}(\bar{x}, t)$ and all $(\bar{x}^0, t_0) \in C\mathfrak{T} \times R$, there is a constant $T_{\mathscr{R}} > 0$ yielding*

$$\int_{t_0}^{t_0 + T_{\mathscr{R}}} [\nabla V(\bar{x}) \cdot \bar{f}(\bar{x}, t, \bar{u}, \bar{w})] \, dt \leq -(v_0 - v_{\mathfrak{T}}) \quad (4.1.16')$$

for all $\bar{w}(t) \in W$.

Proof. Sufficiency follows from Theorem 4.1.3. Consider necessity. Indeed, if not (4.1.16'), then we have (4.1.18), wherefrom there is $(\bar{x}^0, t_0) \in \overline{C\mathfrak{T}} \times R$ yielding $V(\bar{x}^0) = v_0$ such that

$$V(t_f) = v(\bar{x}^0) + \int_{t_0}^{t_f} \dot{V}(\bar{x}(t)) \, dt > v_{\mathfrak{T}}, \quad (4.1.17')$$

which contradicts strong controllability on Δ_0 for reaching $\mathfrak{T}$ and thus completes the proof.

Remark 4.1.2 applies here as well. If assumption (4.1.15) does not fit the required $\hat{\mathfrak{T}}, \hat{\Delta}_0$, we can always use estimates taking V levels which define $\mathfrak{T}, \Delta_0$ such that $\mathfrak{T} \subset \hat{\mathfrak{T}}, \Delta_0 \supset \hat{\Delta}_0$.

In some circumstances discussed later, we may also need the following conditions which are more demanding but have the advantage of providing a control program almost immediately.

Theorem 4.1.4. *Theorem 4.1.3 holds if condition (iii) is replaced by*

(iii') *for all $\bar{u} \in \mathscr{P}(\bar{x}, t)$, there is a*

$$\nabla V(\bar{x}) \cdot \bar{f}(\bar{x}, t, \bar{u}, \bar{w}) < 0 \quad (4.1.19)$$

for all $\bar{w}(t) \in W$.

Proof. Consider a motion from $C\mathfrak{T} \times R$ avoiding $\mathfrak{T}$ and thus by (i) yielding (4.1.17) for all $t \in R_0^+$. By (4.1.19), $\dot{V}(\bar{x}(t)) < 0$, generating outflux $V(t)$ on the left-hand side of (4.1.17). Hence there is a $T_{\mathscr{R}} > 0$ large enough for $t = t_0 + T_{\mathscr{R}}$ to imply $V(k(t)) = v_{\mathfrak{T}}$ contradicting (4.1.17) and thus completing the proof.

Denote $\mathscr{L}(\bar{x}, t, \bar{u}, \bar{w}) \triangleq \nabla V(\bar{x}) \cdot \bar{f}(\bar{x}, t, \bar{u}, \bar{w})$. Theorem 4.1.4 has the following practical corollary.

Corollary 4.1.4. *Given $(\bar{x}, t) \in C\mathfrak{T} \times R$, if there is a pair $\tilde{\bar{u}}(\cdot), \tilde{\bar{w}}$ such that*

$$\mathscr{L}(\bar{x}, t, \tilde{\bar{u}}, \tilde{\bar{w}}) = \min_{\bar{u}} \max_{\bar{w}} \mathscr{L}(\bar{x}, t, \bar{u}, \bar{w}) < 0, \quad (4.1.20)$$

then condition (4.1.19) is met with $\tilde{\bar{u}} \in \mathscr{P}(\bar{x}, t)$.

4.1 REACHING AND RELATED OBJECTIVES

The proof follows by the same argument as that presented in the proof of Corollary 3.4.1, and here again we may conclude that a suitable control program $\mathscr{P}(\cdot)$ may be deduced from (4.1.20), either directly or indirectly using the L–G controller (3.4.6), (3.4.7) with slight adaptation due to the difference between (4.1.20) and (3.4.2). We shall illustrate the technique through examples (see Example 4.1.2 below).

To obtain maneuverability on Δ_0 using the above, we would have to cover Δ_0 with targets $\mathfrak{T}$ corresponding to a *sequence of reaching*, each such job requiring some Liapunov function and a control program. Then $\Delta_0 = \cup \mathfrak{T}$ is made a reachable set from every point in $\Delta_0 \times R$, which satisfies the definition of Δ_M. In our manipulator case, we may have the advantage of the single reference frame $\mathscr{H}$ with its levels, but still a *local Liapunov function* in terms of $H(\bar{x})$ *would have to be specified for each job*, depending very much on where in $\mathscr{H}$ the target $\mathfrak{T}$ is located.

For instance, it must be obvious to the reader that if the target is somewhere *surrounding* a Dirichlet stable equilibrium and the motion starts above it, dissipation of energy is needed and we may choose $V(\bar{x}) = \tilde{H}(\bar{x})$ directly as indicated, with stabilization below some energy level. On the other hand, when the target is at a *threshold*, the motion starts from the cup below and so we need accumulation of energy. Whence, in order to satisfy the conditions of Theorem 4.4.1, we have to take $V(\bar{x}) \equiv -\hat{H}(\bar{x})$, where, as a consequence of the worst-case-design philosophy, we invert the role of $\bar{w}$, that is, $\hat{H}(\bar{x}) = \max_{\bar{w}} H(\bar{x}, \bar{w})$ (see Section 3.4). Then, however, a different controller must be used because Corollary 4.1.4 now gives not only an inverted inequality (4.1.20) but also exchanged roles for $\bar{u}, \bar{w}$. Now u is the maximizer while w tries to minimize and we have

$$\max_{\bar{u}} \min_{\bar{w}} f_0(\bar{x}, t, \bar{u}, \bar{w}) > 0. \qquad (4.1.20')$$

One must then remember also that the equilibria are shifted to the other side of their noise band, namely, in (3.4.11) we have $\max Q_i^P = 0$ instead of $\min Q_i^P = 0$. See also the discussion on barriers in Section 3.4 and (3.4.15).

Still in a different case, when the target does not fill up any energy level set Δ_C entirely but stays on part of it only or on part of its boundary H_C, the motion may reach such a level with $V(\bar{x}) \equiv -\hat{H}(\bar{x}), \tilde{H}(\bar{x})$ for up or down reaching, respectively; but then *it still must travel along this level* to reach the actual target using perhaps an altogether different Liapunov function. This function may in fact depend very much on where on the level the motion starts and where the target is, and it may have nothing to do with energy.

For instance, suppose that the motion has reached a local $\mathfrak{T}$ at level h_C in the cup $\Delta_H^{k=1}$ in Fig. 4.7, cutting $\partial \mathfrak{T}$ at the point A below the threshold while our terminal target is at the point B on the other side of this level. We have two options, either switch to a controller calculated from the zero power $\tilde{H}(\bar{x}) = 0$ and move the arm along $\partial \mathfrak{T}$ until it hits B or take a boat and sail it across the

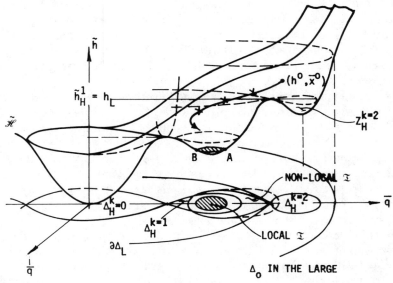

Fig. 4.7

lake $\mathfrak{T}$ from A to B. We do the latter by shifting the origin of coordinates to B and specifying a Liapunov function with levels centered at B, which attracts us there as to the target. Such function may be, for instance, the function $V = x_1 \cdot x_2 \cdots x_{N-2}$, which with N even would be positive within a certain region of attraction to B enclosing A, with simple partial derivatives $\partial V/\partial x_i = x_1,\ldots,x_{i-1},x_{i+1},\ldots,x_{N-2}$ likely to produce negative inner products with projections of f.

When successive targets like the above must be reached, it may be possible to work out sufficient conditions for the composite objective, but in practical terms it is of no advantage to search for a single Liapunov function covering the whole work space. It is often much easier to compose $V(\cdot)$ of a sequence of branches $V_\sigma(\cdot)$, $\sigma = 1,\ldots,l < \infty$, each defined on a subset D_σ, the subsets not necessarily disjoint. We shall illustrate this later on some composite objectives other than successive reaching. On the other hand, even assuming continuous differentiability of a particular $V_\sigma(\cdot)$, it is usually rather difficult to secure an overall smoothness of the resulting global function in order to comply with the sufficient conditions. This, however, may not be necessary if the union of D_σ's fills up Δ_0, since then we can use the relaxed conditions of Theorem 4.1.2 (see Section 5.1).

Considering (3.3.1) as representative of either (1.3.24″) or (1.5.16″), we refer to *definite* $\mathscr{H}$ and *specified* power characteristics $f_0(\bar{x}, t, \tilde{\bar{u}}, \tilde{\bar{w}})$.

The location of $\mathfrak{T}$ and Δ_0 under the surface $\tilde{\mathscr{H}}$ affects the satisfacting of conditions (i) and (ii) of Theorems 4.1.2 and 4.1.3 in an obvious way. When the

4.1 REACHING AND RELATED OBJECTIVES

set Δ_0 is restricted to a local cup Δ_H^k or if $\mathfrak{T}$ encloses all the equilibria and Δ_0 covers at least the lower energy part of $C_\Delta \Delta_L$, these conditions are obviously satisfied by taking $V(\bar{x}) \equiv \tilde{H}(\bar{x})$, with $\mathscr{L}(\bar{x}, t, \bar{u}, \bar{w}) = f_0(\bar{x}, t, \bar{u}, \bar{w})$, and monotone dissipativeness $f_0(\bar{x}, t, \bar{u}, \bar{w}) < 0$ for all $\bar{w}$ may be used to satisfy (4.1.19) instead of (4.1.16) in order to produce the fastest result and to give the control program based on (4.1.20) (see Fig. 4.7). Obviously, the reader realizes that the monotone dissipativeness is stronger than necessary. Indeed, if we do not care what happens to the energy flow after $\mathfrak{T}$ has been reached, that is, if we do not want to stabilize the system below $\tilde{H}(\bar{x}) = \text{const} = h_{\mathfrak{T}}$, corresponding to $\partial \mathfrak{T}$, then only a single drop in energy suffices to reach the local $\mathfrak{T}$ and it does not matter whether this drop is produced by a monotone decrease. In fact, on some time intervals, increases are welcome if useful as long as the integral in (4.1.16) is below the required estimate. This is obviously secured by a suitable outflux of energy satisfying the dissipation inequality (2.3.19) and thus, with the energy flow dissipative on $C\mathfrak{T} \times R: f_0(\bar{x}, t, \bar{u}, \bar{w}) \leq 0$, not necessarily the monotone dissipative $f_0(\bar{x}, t, \bar{u}, \bar{w}) < 0$. On the other hand, as seen from Theorem 4.1.4 and Corollary 4.1.4, monotone dissipativeness may not only be the fastest way to travel to local and in-the-large targets but also may provide an effective controller for the task.

Envisage now, as already mentioned, that the system reached the local target $\mathfrak{T}$ at the point A (see Fig. 4.7) but the assignment is such that it must get to B on the other side of the energy level which specifies $\partial \mathfrak{T}$. The condition $f_0(\bar{x}, t_0, \bar{u}, \bar{w}) < 0$, wherefrom we calculated the controller to hit $\partial \mathfrak{T}$, did not give us the location of A on $\partial \mathfrak{T}$. How do we learn about it in order to control the system to B? The answer is simple: We do not learn this. However, no matter where in $\partial \mathfrak{T}$ the arm is, all we do is *switch to the controller calculated from* $f_0(\bar{x}, t, \bar{u}, \bar{w}) = 0$, which will push the arm along the constant-energy = zero-power contour $\partial \mathfrak{T}$ as long as it does not hit B. This step *does not follow from Theorem 4.1.3 formally*, but it is a logical extension to the case for which $v_0 = v_{\mathfrak{T}}$, that is, that for which we stay on the same energy level.

Now suppose that the target is nonlocal, reaching across thresholds but not enclosing all the cups; like, for instance, in the case of the manipulator working on a job that requires several rotations of links and thus forces the motions to cross several unstable equilibria. If the Δ_0 is such that the highest energy summit in its interior is lower than its boundary, we still have condition (ii) satisfied for the choice of $V(\bar{x}) \equiv \tilde{H}(\bar{x})$ and the whole route to $\mathfrak{T}$ may be achieved with the outflux of energy satisfying (4.1.16) but allowing some threshold climbing, with temporary accumulation of energy, see the motion indicated in Fig. 4.7 with crosses.

When Δ_0 is such that its boundary is lower in energy than some of the summits inside Δ_0, then for at least part of the traveling to $\mathfrak{T}$ we must invert the inequalities of (i), (ii) and consequently that of (iii) by taking a branch $V_\sigma(\bar{x}) \equiv -\hat{H}(\bar{x})$. This allows for the required climb on the threshold, thus making the latter an intermediate target and the itinerary to $\mathfrak{T}$ composed of at least two reachings. The reader realizes here that, should we not have Theorem 4.1.2 allowing corners in $V(\cdot)$ along motions, we could have trouble in sowing the particular pieces of $V_\sigma(t)$ along the route. For illustration of the case, the reader may imagine the arm starting from the cup $\Delta_H^{k=2}$ in Fig. 4.7 and attempting the nonlocal target single dashed in this figure. The most energetically economic route for the motion is toward the threshold $\tilde{h}_H^1$ representing the mountain pass to get to the target. This requires an influx of energy if the motion started anywhere below the threshold.

In successive reachings, the estimates $v_0 - v_\mathfrak{T} = h_0 - h_\mathfrak{T}$ of the outflux integral (4.1.16) depend on the energy-level difference during each step and are additive along the whole itinerary, as much as the energy flow is. The size of each intermediate energy flux may require a different branch of the control program concerned. Sowing the branches is no problem since $\mathscr{P}(\cdot)$ is expected to be set-valued anyway. To determine the branches we use (4.1.20) and perhaps the G–L design, or the positive counterpart of (4.1.20) for $V \equiv -\hat{H}$ (see (4.1.20′) and the discussion following it). The selection is best shown in the case studies and examples that follow.

Example 4.1.2. Let us study a single-DOF overhead manipulator, with the lumped mass of link and gripper $m = 1$ rotating at the end of a link (see Fig. 4.8). The system is subject to gravity, unknown but bounded external perturbation and control torque, which is expected to damp the motion to a target about a basic equilibrium. The motion equation takes the form

$$\ddot{q} + u + G(q) = w \qquad (4.1.21)$$

where $u \in \mathscr{P}(q, \dot{q})$ follows from a damping program, $G(\bar{q})$ represents mainly gravity force either in the full form $G = \omega^2 \sin q$, ω being the frequency of

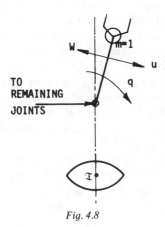

Fig. 4.8

4.1 REACHING AND RELATED OBJECTIVES

oscillations, or in the trunkated form $G \triangleq aq - bq^3$, and finally $w(t) \in W$ is the external perturbation unknown but with limited power, that is, there is $N < \infty$ such that

$$|w(t)\dot{q}(t)| \leq N, \qquad t \geq t_0, \tag{4.1.22}$$

which defines W. We let the target be given as

$$\mathfrak{T} = \{q, \dot{q} \mid \tfrac{1}{2}\dot{q}^2 + \tfrac{1}{2}aq^2 - \tfrac{1}{4}bq^4 \leq 1\}.$$

The state equations obtained from (4.1.21) are

$$\dot{x}_1 = x_2, \qquad \dot{x}_2 = -G(x_1) + w - u, \tag{4.1.23}$$

with the equilibria defined by [see (1.5.24)]

$$x_2^e = 0, \qquad G(x_1^e) = 0. \tag{4.1.24}$$

In particular, $G = \omega^2 \sin q$ gives $x_1^e = 0, \pm\pi, \pm 2\pi, \ldots$, etc. and $G = ax_1 - bx_1^3$ gives $x_1^e = 0, \sqrt{a/b}, -\sqrt{a/b}$. The total energy is $H(\bar{x}) = \tfrac{1}{2}x_2^2 + \int G(q)\,dx_1$, which in the case of $G = ax_1 - bx_1^3$ gives

$$H(\bar{x}) = \frac{x_2^2}{2} + \frac{ax_1^2}{2} - \frac{bx_1^4}{4},$$

with the energy levels sketched in Fig. 4.9, and representing our reference frame.

It seems reasonable to assume that both the disturbing force w and the damping control will not act when the system is at rest $x_2 \equiv 0$. Indeed, we shall have to design $\mathscr{P}(\bar{x})$ to attain this effect. Consequently we may also assume that the equilibria mentioned are those of the full system (4.1.23).

A natural candidate for Δ_0 to reach $\mathfrak{T}$ is $\Delta_H^{k=0}$, which encloses $\mathfrak{T}$ and stretches up to the first neighboring equilibrium $x_1^e = \sqrt{a/b}, x_2^e = 0$. Such Δ_0

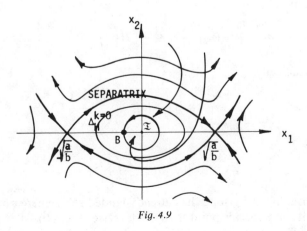

Fig. 4.9

accomodates the axiom of the potential energy restoration (2.2.12), (2.2.9) and thus is defined by

$$\Delta_0 : G(q)q = aq^2 - bq^4 \geq 0.$$

Then $\partial \Delta_0$ is one of the levels, namely, that passing through $x_1^c = \sqrt{a/b}$, $x_2^c = 0$. It is called the *conservative separatrix*:

$$\partial \Delta_0 : \frac{1}{2}x_2^2 + \frac{a}{2}x_1^2 - \frac{b}{4}x_1^4 = \frac{a^2}{4b}. \tag{4.1.25}$$

Since the *target is enclosed in the local cup*, we choose

$$V(\bar{x}) \equiv H(\bar{x}) = \frac{1}{2}x_2^2 + \frac{a}{2}x_1^2 - \frac{b}{4}x_1^4;$$

whence $v_0 = h_0 = a/4b$, $v_{\mathcal{I}} = \frac{1}{2}$.

Conditions (i), (ii) of Theorem 4.1.3 hold. Next, given $u \in \mathscr{P}(\bar{x})$, condition (4.1.19) becomes

$$\mathscr{L}(\bar{x}, t, \bar{u}, \bar{w}) = wx_2 - ux_2 < 0$$

for all w subject to (4.1.22). Corollary 4.1.4 leads to the maximizing assumption $|wx_2| = N$ and the minimizing choice $u = [lx_2|x_2| \pm N/x_2]$, for $x_2 \gtrless 0$, respectively, and for $l > 0$. Note that such $\mathscr{P}$ is an obvious physical choice. Then (4.1.19) is satisfied which makes Theorem 4.1.4 hold. Indeed, (4.1.23) becomes

$$\dot{x}_1 = x_2, \qquad \dot{x}_2 = -G(x_1) - lx_2|x_2|. \tag{4.1.26}$$

It is thus controlled by positive viscous quadratic damping with equilibria (4.1.24). We may obviously relax condition (4.1.19) and use Theorem 4.1.3, replacing it by

$$\int_{t_0}^{t_0 + T_{\mathcal{R}}} -lx_2^2|x_2|\, dt \leq -\left[H\left(\sqrt{\frac{a}{b}}, 0\right) - \frac{1}{2}\right], \tag{4.1.27}$$

or, more specifically,

$$\int_{t_0}^{t_0 + T_{\mathcal{R}}} -lx_2^2|x_2|\, dt \leq -\left(\frac{a^2}{4b} - \frac{1}{2}\right),$$

which is met if

$$\int_{t_0}^{t_0 + T_{\mathcal{R}}} x_2^2|x_2|\, dt \geq \frac{1}{2l}\left(\frac{a^2}{2b} - 1\right). \tag{4.1.28}$$

Now suppose that the gripper has already landed somewhere on $\partial \mathcal{I}$ and we want it to hit some specific point B on this level (see Fig. 4.9). All we have to do is to switch off the damping controller which pushes the gripper below the $\partial \mathcal{I}$

4.1 REACHING AND RELATED OBJECTIVES

level and switch on a new controller from the condition

$$\mathscr{L}(\bar{x}, t, \tilde{u}, \tilde{w}) = \pm N - ux_2 = 0, \qquad x_2 \gtrless 0, \qquad (4.1.29)$$

that is, $u = \pm N/x_2$ for $x_2 \gtrless 0$, meaning that the control force will be used to cover the uncertainty only, the rest being done by the system itself. It will circle $\partial \mathfrak{T}$ as long as is necessary to hit B no matter from what point the gripper started this step. It may not be the shortest path to B, but it is definitely the most energy-economic path.

Alternatively, we may consider B an origin of an angular coordinate θ along $\partial \mathfrak{T}$, form the Liapunov function $V(\theta) = \theta$, and by the same argument presented in Example 3.3.3 secure strong controllability for reaching B. ∎

Exercises

4.1.1 Solve the soft-reaching problem of Example 4.1.1 but extend it to the case of reaching under water with the water resistance (drag force) acting upon the manipulator:

$$\dot{x}_1 = x_2, \qquad \dot{x}_2 = u - g - wx_2^2. \qquad (4.1.30)$$

[*Hint*: Note that (4.1.30) without the drag $-wx_2^2$ is integrable, which suggests the Liapunov function.]

4.1.2 Let the origin of R^2 be the source of a potential with intensity inversely proportional to the square of distance: $\mathscr{V}(q_1, q_2) = k/(q_1^2 + q_2^2)$, and consider a representing point with a given initial kinetic energy $T = \frac{1}{2}[(\dot{q}_1^0)^2 + (\dot{q}_2^0)^2]$. Ignoring all other forces, find a control program under which the point is charged with the full potential $\mathscr{V}_{\max}$, that is, the point reaches the source subject to the repelling force $w \in [-1, 1]$. [*Hint*: $V = (\mathscr{V}_{\max} - k/(q_1^2 + q_2^2)) + \dot{q}_1^2 + \dot{q}_2^2$.]

4.1.3 Consider the system $\ddot{q} + wq = u, q \in R$, with the program $\mathscr{P}$ defined by $u = \pm k$ for $\dot{q} \gtrless 0$, respectively, and with $w = \text{const} \in [1, 2]$. By completing the square, show that the motions consist of elipses centered at $q = \mp k/w$, $\dot{q} = 0$ for $q \gtrless 0$, respectively. Then verify strong controllability for reaching $\mathfrak{T} = \{(0, 0)\}$.

4.1.4 Consider the RP manipulator of Section 1.1 and Example 3.2.3, and write Eqs. (1.1.27) as

$$(m_1 r_1^2 + m_2 q_2^2)\ddot{q}_1 + 2m_2 q_2 \dot{q}_1 \dot{q}_2 + \lambda_1 |\dot{q}_1|\dot{q}_1 + 9.81(m_1 r_1 + m_2 q_2)\cos q_1$$
$$- 9.81 m_1 r_1 + a q_1 + b q_1^3 = B_1 u_1 + w_1$$
$$m_2 \ddot{q}_2 - m_2 q_2 \dot{q}_1^2 + \lambda_2 \dot{q}_2 + 9.81 m_2 \sin q_1 = B_2 u_2 + w_2$$

with m_2, w_1, w_2 uncertain, $m_2 = \text{const} \in [1, 2]$, $w_i \in [-1, 1]$, $i = 1, 2$. Using the calculation in Example 3.2.3, sketch the surfaces $\mathscr{H}, \tilde{\mathscr{H}}$ and discuss the reaching of a neighborhood of basic equilibrium. Specify the feedback controller for doing so.

4.2 REACHING WITH AND WITHOUT PENETRATION

Behavior of the arm with respect to the boundary surface of a target is essential, particularly when uncertainties are involved in the manipulator dynamics and when the surface is either not-well-defined or difficult to model mathematically for the use in the robot software. We discussed the case of *soft reaching*. Let us mention now the control dynamical background for another two cases, *reaching with penetration* (see Definitions 4.1.1, 4.1.2) and without it, that is, *reaching with rejection* (see Definition 4.1.3).

Isaacs [1] calls our noise-robust penetration "capture with escape" and provides a number of case studies in which it is possible to determine sets of neutral behavior between a strong region for penetration and a region for avoidance. Our conditions for penetration are those of reaching $\mathring{\mathfrak{T}}$; thus there is only slight modification.

Again, we may either search for $\Delta_{\mathscr{R}}$ or a given $\Delta_0 \subset \Delta_{\mathscr{R}}$. Note that by definition $\mathring{\mathfrak{T}} \subset \Delta_{\mathscr{R}}$ and thus only $\Delta_0 \supset \mathring{\mathfrak{T}}$ is a proper candidate. Denote $C\mathfrak{T} \triangleq \Delta_0 - \mathring{\mathfrak{T}}$, D (open) $\supset C\mathring{\mathfrak{T}}$ and consider $V(\cdot): D \to R$ with

$$v_P = \sup V(\bar{x}) \,|\, \bar{x} \in \partial \Delta_0.$$

Theorem 4.2.1. *The system (3.3.1) is strongly controllable on $\Delta_0 \times R$ for penetration of $\mathfrak{T}$ if there is a program $\mathscr{P}(\cdot)$ on $D \times R$ and a C^1 function $V(\cdot): D \to R$ such that*

(i) $V(\bar{x}) \geq v_{\mathfrak{T}}$ *for* $\bar{x} \notin \mathring{\mathfrak{T}}$,
(ii) $V(\bar{x}) < v_P$ *for* $\bar{x} \in C\mathring{\mathfrak{T}}$,
(iii) *for all* $\bar{u} \in \mathscr{P}(\bar{x}, t)$, *there is* $T_P(x^0, t_0, k(\cdot)) > 0$ *yielding*

$$\int_{t_0}^{t_0 + T_P} [\nabla V(\bar{x}, t) \cdot \bar{f}(\bar{x}, t, \bar{u}, \bar{w})] \, dt \leq -(v_P - v_{\mathfrak{T}}) \tag{4.2.1}$$

for all $\bar{w}(t) \in W$.

Proof. The motions from $\bar{x}^0 \in \mathring{\mathfrak{T}}$ penetrate trivially; thus we take a motion from $\bar{x}^0 \notin C\mathring{\mathfrak{T}}$, $t_0 \in R$ and suppose that it avoids penetration. Then by (i),

$$V(k(\bar{u}, \bar{x}^0, t_0, t)) = V(\bar{x}^0) + \int_{t_0}^{t} \dot{V}(\bar{x}(\tau)) \, d\tau \geq v_{\mathfrak{T}}$$

for all $t \geq t_0$. Since $V(\bar{x}^0) < v_P$, the above gives

$$\int_{t_0}^{t} \dot{V}(\bar{x}(t)) \, dt > -(v_P - v_{\mathfrak{T}})$$

for all $t \geq t_0$, contradicting (iii), thus completing the proof.

Remark 4.1.2 applies to penetration without change except that v_0 is replaced by v_P. Theorem 4.1.2 holds for penetration (reaching $\mathring{\mathfrak{T}}$) with its

4.2 REACHING WITH AND WITHOUT PENETRATION

conditions (i), (ii) replaced by those of Theorem 4.2.1. Identical arguments as in the case of reaching lead to the following three statements.

Corollary 4.2.1. *Suppose there is $V(\cdot)$ satisfying conditions (i) and (ii) of Theorem 4.2.1. Then in order for system (3.3.1) to be strongly controllable for penetration of $\mathfrak{T}$, it is necessary and sufficient to have condition (iii) of Theorem 4.2.1 satisfied.*

Theorem 4.2.2. *Theorem 4.2.1 holds if its condition (iii) is replaced by*

(iii′)   *for each $\bar{u} \in \mathcal{P}(\bar{x}, t)$,*

$$\nabla V(\bar{x}, t) \cdot \bar{f}(\bar{x}, t, \bar{u}, \bar{w}) < 0 \qquad (4.2.2)$$

for all $\bar{w}(t) \in W$.

Corollary 4.2.2. *Given $(\bar{x}^0, t_0) \in C\overset{\circ}{\mathfrak{T}} \times R$, if there is a pair $(\tilde{\tilde{u}}, \tilde{w})$ such that (4.1.20) holds, then (4.2.2) is met with $\tilde{\tilde{u}} = \mathcal{P}(\bar{x}, t)$.*

The implementation of these theorems and corollaries is done in the same way as for reaching. Let us demonstrate it by an example.

Example 4.2.1. We return to the single-DOF manipulator of Example 4.1.2, but with a different target, namely $\mathfrak{T} = \Delta_H^{k=0}$, thus defined by

$$\mathfrak{T} \triangleq \{(x_1, x_2) \mid -\sqrt{a/b} \leq x_1 \leq \sqrt{a/b}, -a/\sqrt{2b} \leq x_2 \leq a/\sqrt{2b}\}$$

and bounded by the conservative separatrix (see Fig. 4.9). Our objective is the strong region for penetration located somewhere in $\Delta_{\mathcal{R}}$.

Note that the motion equations with the control program and the maximal noise inserted, that is, Eqs. (4.1.26), are integrable. Indeed, if we write (4.1.26) in the phase-space form (1.5.8), it becomes

$$\frac{dx_2}{dx_1} = \frac{-lx_2|x_2| - G(x_1)}{x_2}.$$

Since $|x_2| = x_2 \operatorname{sign} x_2$, we have

$$x_2 \frac{dx_2}{dx_1} + lx_2^2 \operatorname{sign} x_2 = -G(x_1)$$

or

$$\frac{dx_2^2}{dx_1} + 2lx_2^2 \operatorname{sign} x_2 = -2G(x_1), \qquad (4.2.3)$$

linear equation in x_2^2 with x_1 the independent variable. Thus

$$x_2(x_1) = \pm\sqrt{2} e^{\mp lx_1} \left(C - \int_{x_1^0}^{x_1} G(\xi) e^{\pm 2l\xi} d\xi \right)^{1/2} \qquad (4.2.4)$$

for $x_2 \gtrless 0$, respectively. The constant of integration C is to be determined for the first half-rotation of the manipulator by the initial conditions, and for the subsequent half by requiring the continuity of x_1 at $x_2 = 0$. In the case $G = \omega^2 \sin q$, (4.2.4) becomes

$$x_2^2 = Ce^{\mp 2lx_1} + \frac{2\omega^2}{1 + 4l^2}\cos x_1 \mp \frac{4\omega^2 l}{1 + 4l^2}\sin x_1,$$

or in the abbreviated case of Figs. 4.9 and 4.10, that is, with $G = ax_1 - bx_1^3$, it is

$$x_2^2 = Ce^{\mp 2lx_1} + \frac{-ax_1 + bx_1^3}{2l} + \frac{a - 3bx_1^2}{4l^2} + \frac{3bx_1}{4l^3} - \frac{3b}{8l^4}. \quad (4.2.5)$$

The region Δ_P may be obtained directly by determining its boundary $\partial\Delta_P$ in terms of the so-called *dissipative separatrix*, which corresponds in the damped, and for us, controlled case, to the conservative separatrix. This dissipative separatrix is a trajectory of (4.1.26) or one of the curves (4.2.4) that passes through the equilibrium $(\sqrt{a/b}, 0)$ as illustrated in Fig. 4.10. Substituting $x_2(\sqrt{a/b}) = 0$ into (4.2.4), by (4.1.24), we obtain $C = 0$, yielding

$$x_2(x_1) = \mp\sqrt{2}\,e^{\mp lx_1}\left(\int_{x_1^0}^{x_1} G(x_1)e^{\pm 2lx_1}\,dx_1\right)^{1/2} \quad (4.2.6)$$

for $x_1 \gtrless 0$, where $-$ corresponds to $<$ and $+$ to $>$. This curve is an obvious boundary $\partial\Delta_P$ and belongs to Δ_P, making it closed. Hence there is no semibarrier B_P for penetration, although the reader may perhaps note that $\partial\Delta_P$

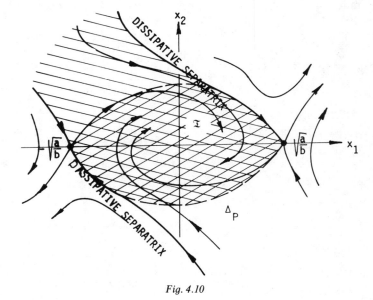

Fig. 4.10

4.2 REACHING WITH AND WITHOUT PENETRATION

is the semibarrier $B_{\not{P}}$, which is at the same time a boundary to the strong region of rejection which is open and covers $\Delta - \Delta_P$. Obviously, because we have no B_P, we also do not have a barrier between Δ_P and $\Delta_{\not{P}}$. ∎

Let us now look more closely at the case of reaching with rejection (see Definition 4.1.3).

Definition 4.2.1. The system (3.3.1) is strongly controllable at $(\bar{x}^0, t_0) \in \Delta \times R$ for *rejected reaching* of $\mathfrak{T}$ if and only if it is strongly controllable at this point for reaching $\mathfrak{T}$ and rejection from this target.

Denote $C\Delta_P \triangleq \Delta_{\mathscr{R}} - \Delta_P$. This is obviously the region of strong controllability for rejected reaching.

Observe then that any motion leaving $C\Delta_P$ must either miss $\mathfrak{T}$ or penetrate it. Hence, we have

Property 4.2.1. The region $C\Delta_P$ is positively strongly invariant under $\mathscr{P}$.

Then we also have

Property 4.2.2. $\mathfrak{T} \subset \Delta_P$ if and only if $C\Delta_P = \varnothing$.

Indeed, no motion from $C\Delta_P$ can reach $\mathfrak{T}$ without penetration, which proves necessity. Sufficiency is obvious as $\Delta_{\mathscr{R}} - \Delta_P = \varnothing \Leftrightarrow \Delta_{\mathscr{R}} = \Delta_P$ and we have $\mathfrak{T} \subset \Delta_{\mathscr{R}}$.

Now let $\partial^{\not{P}}\mathfrak{T} \triangleq \partial \mathfrak{T} \cap C\Delta_P$ be the rejecting part of $\partial \mathfrak{T}$ termed *nonusable*. Immediately we obtain the contrapositive to the above sufficiency:

Property 4.2.3. $C\Delta_P \neq \varnothing \Rightarrow \partial^{\not{P}}\mathfrak{T} \neq \varnothing$.

We conclude the obvious fact that with rejected reaching, not all of $\partial \mathfrak{T}$ is used for penetration. This also means that $\partial^{\not{P}}\mathfrak{T} \subset C\Delta_P$. On the other hand, $\overset{\circ}{\mathfrak{T}} \subset \Delta_P$, whence $\overset{\circ}{\mathfrak{T}} \cap C\Delta_P = \varnothing$ and we have

Property 4.2.4. There is no protruding of $\mathfrak{T}$ into $C\Delta_P$ except for $\partial \mathfrak{T}$ (see Fig. 4.11).

Thus we may have two cases: (i) nonprotruding: $\partial^{\not{P}}\mathfrak{T} = \varnothing$, that is, all of $\partial \mathfrak{T}$ is *usable for penetration*; (ii) protruding: $\partial^{\not{P}}\mathfrak{T} \neq \varnothing$, that is, some part of $\partial \mathfrak{T}$ is nonusable. In the first case, by Property 4.2.2, we have $C\Delta_P = \varnothing$, that is, there is no rejected reaching; whence all reaching motions penetrate. This case is illustrated in Example 4.2.1.

Let us consider now the protruding case (ii) with $\partial^{\not{P}}\mathfrak{T} \neq \varnothing$ (see Fig. 4.11). Suppose that Δ_P is closed, that is, $C\Delta_P$ is "open below", then $\partial^{\not{P}}\mathfrak{T} \subset \widehat{C\Delta_P}$ and there may be a point internal to both $\overset{\circ}{\mathfrak{T}}$ and $C\Delta_P$, which contradicts the obvious $\overset{\circ}{\mathfrak{T}} \subset \Delta_P$. Hence we have

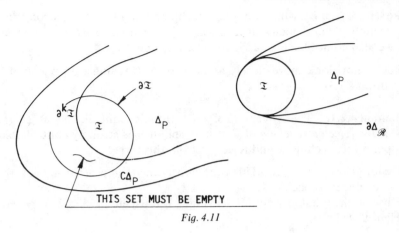

Fig. 4.11

Property 4.2.5. $C\Delta_P \neq \emptyset$ implies that Δ_P is open.

Granted that Δ_P is open and that only $\partial \mathfrak{T}$ can have common points with $C\Delta_P$, it follows that only $\partial \mathfrak{T}$ of the entire $\mathfrak{T}$ can protrude into the closed $C\Delta_P$. Hence

Property 4.2.6. $C\Delta_P \neq \emptyset$ implies $\partial^F \mathfrak{T} \subset \partial \Delta_P$.

In turn, Definition 4.2.1 implies that all motions from $C\Delta_P$ terminate in $\partial^F \mathfrak{T}$ and thus, by the above, we have

Property 4.2.7. Motions from closed $C\Delta_P \neq \emptyset$ terminate in $\partial \Delta_P$ and through this set in $\partial^F \mathfrak{T}$.

This means that $\partial \Delta_P$ is strongly positively invariant under $\mathscr{P}(\bar{x}, t)$ of Definition 4.2.1. Obviously $\partial \Delta_P$ is the semipermeable surface that is closest to $\mathfrak{T}$ and thus an obvious B_P candidate. The latter fact is used to obtain an algorithm for $\partial \Delta_P = B_P$ by retrograde integration of the state equations with maximal noise and minimal controller substituted (see Example 4.2.1), provided the system is autonomous (see Grantham [1] and Grantham and Vincent [1]). Grantham and Vincent introduce necessary conditions which, given $\mathscr{P}$ and w, generate trajectories that fill up $\partial \Delta_P = B_P$. The conditions correspond to (3.5.11), with the cost $\mathcal{V}$ specified as the time of reaching $\mathfrak{T}$.

The following brief example illustrates the case geometrically.

Example 4.2.2. Consider a planar manipulator arm with the gripper operating against the resistance of a stream of water (oil, air) with velocity v_2 allowed by the radius of the circular vectogram shown in Fig. 4.12 (adapted from Isaacs [1]). The controller's choice is indicated by the cone of velocity vectors

4.3 STIPULATED REACHING TIME. MANEUVERING

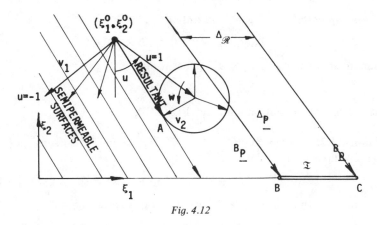

Fig. 4.12

v_1 placed at the initial position (ξ_1^0, ξ_2^0) of the gripper. We assume $v_2 < v_1$, and for some constant c, $v_2 \leq c < 1$ (see Isaacs [1]). The resultant velocity of the gripper is to be the vector sum of a choice from each vectogram. Its components may be written analytically as

$$\dot{\xi}_1 = v_1 u + v_2 \sin w, \qquad \dot{\xi}_2 = -1 + v_2 \cos w,$$

where $u \in [-1, 1]$ and w are the angles indicated in Fig. 4.12. The target is also shown in Fig. 4.12 between points B and C. In order to penetrate, the controller uses his best $u = 1$ against the angle w of v_2. The resultant motion $(\xi_1^0, \xi_2^0) \to A$ lies in a semipermeable line determined by (ξ_1^0, ξ_2^0). Varying these initial conditions, one obtains a family of such surfaces. Some of them evidently miss $\mathfrak{T}$, two of them hit $\mathfrak{T}$ at the boundary points B and C of $\partial \mathfrak{T}$. These two form B_P. It is obvious that such a B_P is strongly positively invariant under the chosen u. Observe also that Δ_P is open and all of $\partial \mathfrak{T}$ is usable, that is, there is no protruding of $\mathfrak{T}$ except for $\partial \mathfrak{T}$ consisting of points B and C. ∎

4.3 STIPULATED REACHING TIME. MANEUVERING

Maneuvering in the work space, in particular, *sequential reachings or penetrations*, needs a prespecified time for handling particular operations. So far the intervals $T_\mathcal{R}$ and T_P, before reaching or penetration, have been made irrelevant as long as they were finite. This is rather unrealistic. We must have them bounded (below, above, or both) within *prescribed bounds* $T_{\mathcal{R},P}^-$, $T_{\mathcal{R},P}^+$, or quite often we want the values $T_\mathcal{R}$, T_P to be *prescribed* exactly. Modifying Definition 4.1.1, we have accordingly

Definition 4.3.1. Given $(\bar{x}^0, t_0)$, $T_\mathcal{R}^+ > 0$, a motion from $\mathcal{K}(\bar{x}^0, t_0)$ *reaches* $\mathfrak{T}$

before the stipulated time $T_{\mathscr{R}}^+$, if and only if

$$k(\bar{u}, \bar{x}^0, t_0, [t_0, t_0 + T_{\mathscr{R}}^+]) \cap (\mathfrak{T} \times R) \neq \emptyset. \qquad (4.3.1)$$

This means that there is a $T_{\mathscr{R}} \leq T_{\mathscr{R}}^+$ generating

$$k(\bar{u}, \bar{x}^0, t_0, t_0 + T_{\mathscr{R}}) \in \mathfrak{T}, \qquad (4.3.2)$$

for the motion of Definition 4.3.1 (see Fig. 4.13).

Definition 4.3.2. Given $(\bar{x}^0, t_0) \in \Delta \times R$, $T_{\mathscr{R}}^-$, a motion from $\mathscr{K}(\bar{x}^0, t_0)$ *reaches* $\mathfrak{T}$ *ultimately* if and only if

$$k(\bar{u}, \bar{x}^0, t_0, [t_0 + T_{\mathscr{R}}^-, \infty)) \cap (\mathfrak{T} \times R) \neq \emptyset. \qquad (4.3.3)$$

This means that there is a $T_{\mathscr{R}} \geq T_{\mathscr{R}}^-$ yielding (4.3.2) (see Fig. 4.13).

Now if we want the *stipulated instant reaching*, say, at $t_0 + T_{\mathscr{R}}^*$, we let $T_{\mathscr{R}}^+ = T_{\mathscr{R}}^- = T_{\mathscr{R}}^*$ in (4.1.2).

The above two definitions are immediately adjustable for penetration by changing $\mathfrak{T}$, $T_{\mathscr{R}}^{\pm}$, $T_{\mathscr{R}}^*$ into $\mathring{\mathfrak{T}}$, $T_{\mathrm{P}}^{\pm}$, T_{P}^*. For the regions of controllability and strong controllability, we use the notation Δ_{r+}, $\Delta_{\mathscr{R}+}$ for stipulated time reaching, Δ_{r-}, $\Delta_{\mathscr{R}-}$ for ultimate reaching, and similarly $\Delta_{p\pm}$, $\Delta_{P\pm}$ for penetration. For a specified-instant reaching/penetration, we use the notation $\Delta_{\mathscr{R}T}$, Δ_{PT}, respectively.

It is perhaps worth noting here that the strong regions depend upon the corresponding time constants, and generally those with smaller constants (earlier reaching) are enclosed in those corresponding to larger constants (later reaching). Then, in turn, every strong region belongs to its corresponding weak region.

Obviously, if we replace the search for the regions by setting up candidates Δ_0, the above serves only as an indicative estimate and excludes candidates

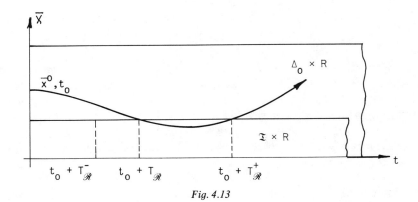

Fig. 4.13

4.3 STIPULATED REACHING TIME. MANEUVERING

that are incompatible with a complex objective composed of elementary reachings or penetrations.

Sufficient conditions are obtainable by slight adjustment of the theorems for reaching and penetration. It should suffice to show this for Theorem 4.1.3:

Theorem 4.3.1. *The system (3.3.1) is strongly controllable on Δ_0 for reaching before $T_\mathcal{R}^+$ if the conditions of Theorem 4.1.3 hold for $T_\mathcal{R}(\bar{x}^0, t_0, k(\cdot)) \leq T_\mathcal{R}^+$.*

Proof. The proof of Theorem 4.1.3 is adjusted as follows: Given $(\bar{x}^0, t_0) \in C\mathfrak{T} \times R$, suppose some motion from $\mathcal{K}(\bar{x}^0, t_0)$ avoids $\mathfrak{T}$ during $T_\mathcal{R}^+$. Then (4.1.17) holds for all $t \in [t_0, t_0 + T_\mathcal{R}^+]$. Hence we have

$$\int_{t_0}^{t_0+T_\mathcal{R}^+} \dot{V}(\bar{x}(t), t)\, dt > -(v_0 - v_\mathfrak{T}), \tag{4.3.4}$$

contradicting (iii) of Theorem 4.1.3 and thus completing the proof.

Theorem 4.3.2. *The system (3.3.1) is strongly controllable on Δ_0 for ultimate reaching if, given $T_\mathcal{R}^-$, the conditions of Theorem 4.1.3 hold for $T_\mathcal{R}(\bar{x}^0, t_0, k(\cdot)) \geq T_\mathcal{R}^-$.*

The proof is the same as for Theorem 4.3.1 but with t_0 in (4.1.17) and (4.1.18) adjusted to become $t_0' = t_0 + T_\mathcal{R}^-$.

Corollaries 4.1.3 and 4.1.4 are immediately adjustable, and so are the methods defining $\mathcal{P}(\cdot)$.

By referring now to (1.3.15″) and (1.3.24″) or (1.5.16‴) and a definite $\mathcal{H}$ (either $\tilde{\mathcal{H}}$ or $\hat{\mathcal{H}}$), we may repeat our discussion from Section 4.1 verbatim except for the fact that now not only $\mathfrak{T}$ (or intermediate targets) and Δ_0 but also time are stipulated. All stipulated time reachings give an additional advantage on $\mathcal{H}$, which is particularly important in sequential reaching. They allow us to find, if $T_\mathcal{R}^*$ is given, or to estimate, if $T_\mathcal{R}^+$ and $T_\mathcal{R}^-$ are given, the energy flux either dissipated or accumulated during the time of reaching and thus give the average power the actuators must use during the corresponding motion. Indeed, since condition (4.1.16) is necessary as well as sufficient, with $T_\mathcal{R}^*$ given, we have

$$\int_{t_0}^{t_\mathcal{R}} f_0(\bar{x}, t, \bar{u}, \bar{w})\, dt \leq -(h_0 - h_\mathfrak{T}) \tag{4.3.5}$$

with the rate $(h_\mathfrak{T} - h_0)/T_\mathcal{R}^*$ estimating the power the actuators must use during the step measured by $T_\mathcal{R}^*$. Conversely, knowing the shape of $\tilde{\mathcal{H}}$ or $\hat{\mathcal{H}}$, we may estimate the size of the step in the energy-state space, when the flux of energy can be measured.

The expression (4.3.5) gives the work of energy changing (nonconservative) forces and may (and should) be optimized (minimized) during particular operational steps, as an important means to decrease the energy consumption during industrial exploitation of manipulators. It seems that this is a far more important feature than time optimization, which is pursued by many authors simply because it is easier to investigate. The method of parametric optimization may be used here quite successfully (see Klein [1] and Beletskii and Chudinov [1]). We shall return to this topic in the context of optimal reaching with capture in Section 4.4, but the reader may see that our discussion is immediately applicable now as well.

In Example 4.1.2 condition (4.1.16) gives

$$\int_{t_0}^{t_\mathscr{R}} (wx_2 - ux_2)\, dt \leq -(h_0 - h_\mathfrak{T}), \tag{4.3.6}$$

or, inserting the maximizing uncertainty and minimizing control program, we have

$$\int_{t_0}^{t_\mathscr{R}} lx_2^2|x_2|\, dt \geq (h_0 - h_\mathfrak{T}). \tag{4.3.7}$$

When the step in sequential reaching requires moving along the energy level, as mentioned in Section 4.1, we use the controller obtained from the control condition that $f_0(\bar{x}, t_0, \bar{u}, \bar{w}) = 0$, with suitable $\tilde{u}$ and $\tilde{w}$ substituted depending upon whether the reaching is on $\mathscr{H}$ or $\hat{\mathscr{H}}$.

Then, however, the flux during the corresponding $T_\mathscr{R}$ is zero and we have no information from the energy about the distance traveled by the system. As mentioned in Section 4.1, in such cases we use a different Liapunov function that is embedded in the hypersurface of such an energy level and that attracts motions to the target. Then (4.3.5) works again for the new function.

Example 4.3.1. Let us continue with the R manipulator of Example 4.1.2 and refine the model by taking the gravity force with a better approximation, that is, $G(q) = \omega^2 \sin q \cong ax_1 - bx_1^3 + cx_1^5$, which gives five equilibria (see Fig. 4.14),

$$x_1^e = 0, \quad \left[\frac{b \pm \sqrt{b^2 - 4ac}}{2c}\right], \quad -\left[\frac{b \pm \sqrt{b^2 - 4ac}}{2c}\right], \tag{4.3.8}$$

allowing the gripper to rotate more times. The energy is then $H(\bar{x}) = \tfrac{1}{2}x_2^2 + \tfrac{1}{2}ax_1^2 - \tfrac{1}{4}bx_1^4 + \tfrac{1}{6}cx_1^6$, and the threshold is obtained by substituting the first unstable equilibrium into $H(\bar{x})$. Specifying a, b, c will facilitate calculations: We let $\omega^2 = 1$, then $a = 1$, $b = \tfrac{1}{6}$, $c = \tfrac{1}{120}$ which makes the

$$x_1^e = (\tfrac{1}{6} + \sqrt{\tfrac{1}{36} - \tfrac{4}{120}})/\tfrac{1}{60} = 89 \quad \text{and} \quad h_H = 618.10^6 > 0.$$

Then *we plan the following itinerary for the gripper*: Start at $\bar{x}^0$ on the level $h_1 = 300.10^6$ energy units (see Fig. 4.14) at time $t_0 = 0$ and reach $\mathfrak{T}^1$, which is the level h_H, before $T_{\mathscr{R}1}^+ = 10$ time units. Then reach $\mathfrak{T}^2$, which is the equi-

4.3 STIPULATED REACHING TIME. MANEUVERING

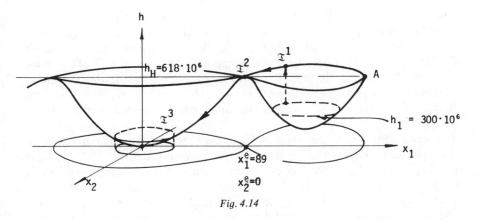

Fig. 4.14

librium $x_1^e = 89$ at the h_H level, before $T_{\mathcal{R}2}^+ = 2T_{\mathcal{R}1}^+$, that is, at the time $t_{\mathcal{R}2} = 0 + T_{\mathcal{R}} \leq 2T_{\mathcal{R}1}^+$. Then descend to $\mathfrak{T}^3$, which is the terminal target about $(0,0)$ specified by $x_1^2 + x_2^2 \leq 1$, before $T_{\mathcal{R}3}^+ = 3T_{\mathcal{R}1}^+$, that is, for some $t_{\mathcal{R}3} = 0 + T_{\mathcal{R}} \leq 3T_{\mathcal{R}1}^+$.

Between $\bar{x}^0$ and $\mathfrak{T}^1$ we follow the $V(\bar{x}(t)) = -H(\bar{x}(t))$ energy-state motion whence we need the energy influx

$$\int_{t_0=0}^{t_0+T_{\mathcal{R}1}^+=10} \max_{\bar{u}} \min_{\bar{w}} f_0(\bar{x}, t, \bar{u}, \bar{w}) \, dt \geq h_0 - h_{\mathfrak{T}}). \tag{4.3.9}$$

Observe that while the minimizing u and the maximizing w have been $u = [lx_2|x_2| \pm N/x_2]$, $x_2 \gtrless 0$, and $wx_2 = \pm N$, $x_2 \gtrless 0$, respectively, now the *maximizing* u and *minimizing* w are $u = [lx_2|x_2| \mp N/x_2]$, $x_2 \gtrless 0$, and $wx_2 = \mp N$, $x_2 \gtrless 0$, respectively. Thus (4.3.9) becomes

$$\int_0^{10} -lx_2^2|x_2| \, dt \geq (h_0 - h_{\mathfrak{T}}). \tag{4.3.10}$$

Obviously, we now have $h_0 - h_{\mathfrak{T}} < 0$. Compare the above with (4.3.7) calculated for the dissipative step during $T_{\mathcal{R}}$ where $h_0 - h_{\mathfrak{T}} > 0$. Hence, with the input from the actuator determined by $u = lx_2|x_2| \mp N/x_2$, $x_2 \gtrless 0$ during some $T_{\mathcal{R}} \leq T_{\mathcal{R}1}^+$ we achieve $\mathfrak{T}$ on level h_H. Here we switch the controller to the "conservative" $u = \pm N/x_2$, $x_2 \gtrless 0$ calculated from (4.1.29), and follow h_H until we reach $\mathfrak{T}^2$. The problem is, how do we know that we can do this at an instant $t_{\mathcal{R}} \leq t_0 + 2T_{\mathcal{R}1}^+$? The energy flux with the above controller substituted is

$$\int_{t_0}^{t_{\mathcal{R}}} 0 \, dt = 0$$

and offers no information. Remember, however, that we are on the constant energy level which is a known conservative trajectory, in particular, separatrix (4.1.25), and we can calculate the length of any definite arc of it between two given points. One of these points is $\mathfrak{T}^2$ the other is unknown, but it cannot be further than the distance to the point A opposite $\mathfrak{T}^2$ (see Fig. 4.14). Thus if we subdivide the distance between $\mathfrak{T}^2$ and A by $T_{\mathscr{R}1}^+$, we obtain the upper estimate of travelling speed along the separatrix. Since the state velocity is measured by the vector $(\dot{x}_1, \dot{x}_2) = (x_2, f_2)$ we can calculate the forces needed for reaching $\mathfrak{T}^2$ in time.

There is also a more general way. Take the Liapunov function of Theorem 4.3.1 as a distance from $\mathfrak{T}^2$ on the plane $0x_1x_2$ with the origin at $\mathfrak{T}^2$ and use (4.3.5) in terms of the "flux" of such a function, obviously being prepared to readjust the controller to the required rate of change of the function.

Having arrived at $\mathfrak{T}^2$ in time, the rest is a simple matter of descending down the basic cup as already shown in Example 4.1.2, with the same controller but with a stipulated $t_{\mathscr{R}} \leq 3T_{\mathscr{R}1}^+$, which requires a dissipative flux and Theorem 4.3.1 instead of Theorem 4.1.2. ∎

Note that the above example does not impede the generality of the conclusions. The technique is just the same for any n-DOF manipulator, except for one aspect. In this example, the uncertainty was in the external perturbation only and did not affect $\mathscr{H}$. When it does affect $\mathscr{H}$, we have to decide where we are to place the intermediate targets $\mathfrak{T}^j$, $j = 0, 1, \ldots$: on either $\tilde{\mathscr{H}}$ or $\hat{\mathscr{H}}$. Once this is defined, we use the min–max or max–min controller, depending upon how the next target is placed. For instance, if we had the case that pertains at step 2 in the example, we would use $u = \pm N/x_2$, $x_2 \gtrless 0$, if referenced to $\tilde{\mathscr{H}}$ and $u = \mp N/x_2$, $x_2 \gtrless 0$, if referenced to $\hat{H}$.

Exercises and Comments

4.3.1 Consider the system of Example 4.1.2 with the work steps discussed in Example 4.3.1 and the energy flux estimates (4.3.6). Using the Pontriagin principle of Section 3.6 with $f^V(\bar{x}, t, \bar{u}, \bar{w}) \equiv f_0 = wx_2 - ux_2$, the reader is asked to make an attempt on his own to optimize the energy consumption during the successive reachings concerned.

4.3.2 Make a similar analysis of successive reachings for the RP manipulator of Section 1.1 specified in Exercise 4.1.4. Use both the min–max and the L–G controllers swapping it to max–min each time you change the sign of the energy flux needed on the step concerned.

4.4 REACHING WITH CAPTURE. OPTIMAL CAPTURE

As previously mentioned, capture is required when the operation to be done asks for a definite real-time contact between the arm and the target. In state

4.4 REACHING WITH CAPTURE. OPTIMAL CAPTURE

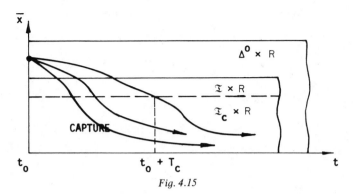

Fig. 4.15

coordinates redefined as relative to the target, the capture of such a target in the basic Liapunov cup covers the case of planned path tracking (see Section 4.6).

We may require that the motions be enclosed in $\mathfrak{T}$ either ultimately (*capture*) or at least for some time (*rendezvous*), preferably with the intervals preassigned. Controllability for capture in Liapunov terms was introduced by Vincent and Skowronski [1]. Obviously, capture implies reaching; hence there is no need to consider reaching with capture unless you have an opponent who can deny the second objective. Strong controllability for reaching (called collision) with capture was developed by Skowronski [19, 21] and Skowronski and Vincent [1]. The study gives necessary and sufficient conditions for strong controllability for capture, including reaching, but if the necessary conditions cannot be satisfied, offers the flexible alternative that reaching can be achieved, but, due to the opposing uncertainties, capture cannot. In particular, the uncertain perturbations may deny capture (*reaching without capture*) or they may be capable of driving the reference point out of $\mathfrak{T}$ in finite time and keeping it out (*reaching with escape*) no matter what the controller does. Such an escape leaves us with rendezvous for the time duration, which becomes shorter as the opposition becomes stronger.

The discussion in this section applies equally well to states $\bar{x}$ represented either by joint variables or by variables relative to the target.

Definitions 4.4.1. Given $(\bar{x}^0, t_0) \in \Delta \times R$ and some motion of the corresponding class $\mathscr{K}(\bar{x}^0, t_0)$, we have (see Fig. 4.15) *capture* if and only if there is a $T_C, 0 \leq T_C < \infty$, such that

$$k(\bar{u}, \bar{x}^0, t_0, [t_0 + T_C, \infty)) \subset \mathfrak{T} \times R, \qquad (4.4.1)$$

and we have *rendezvous* if and only if there is a $T_Z, 0 \leq T_Z < \infty$, such that (see Fig. 4.16)

$$k(\bar{u}, \bar{x}^0, t_0, [t_0, t_0 + T_Z]) \subset \mathfrak{T} \times R. \qquad (4.4.2)$$

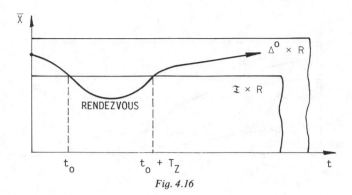

Fig. 4.16

We also have *capture after* T_C and *rendezvous during* T_Z if and only if these constants are preassigned (see Section 3.1).

Controllability for capture implies controllability for reaching and we have $\Delta_c \subset \Delta_r$, where Δ_c is the region of controllability for capture. For a system subject to uncertainty, our objective is combined to include reaching and capture.

Definition 4.4.2. The system (3.3.1) is strongly controllable at $(\bar{x}^0, t_0) \in \Delta \times R$ for *reaching with capture in* $\mathfrak{T}$, if and only if there is a $\mathscr{P}(\cdot)$ and a $T_C \geq T_\mathscr{R}$ such that for each $k(\cdot) \in \mathscr{K}(\bar{x}^0, t_0)$ we have (4.4.1). The region for strong controllability for reaching with capture is denoted by Δ_C, and we immediately have (see Fig. 4.17)

$$\Delta_c \subset \Delta_r, \qquad \Delta_C \subset \Delta_\mathscr{R}. \tag{4.4.3}$$

We may conclude that $\Delta_C \cap \mathfrak{T} \neq \varnothing$ but not necessarily that $\mathfrak{T} \subset \Delta_C$. Hence, if the system is strongly controllable for reaching with capture in $\mathfrak{T}$, then there is a strongly winning $\mathscr{P}(\cdot)$ such that for all $w(t)$ the motions are in $\mathfrak{T}$ after some time interval. However, it does not follow that the manipulator

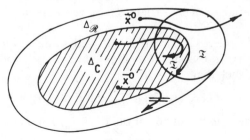

Fig. 4.17

4.4 REACHING WITH CAPTURE. OPTIMAL CAPTURE

state can be maintained in an arbitrary subset of $\mathfrak{T}$. It does not follow either that the state can be maintained even in a given subset of $\mathfrak{T} \cap \Delta_C$, since for a certain point of this set it may be possible that the state passes through the point but that there is a $w(\cdot)$ such that no $\mathscr{P}(\cdot)$ is able to return the state to a neighborhood of the point. However, it does follow that there is at least one nonempty subset of $\mathfrak{T}$ located in $\mathfrak{T} \cap \Delta_C$ which is positively strongly invariant under $\mathscr{P}(\cdot)$. Any such subset is called a *capturing subtarget* $\mathfrak{T}_C$ in $\mathfrak{T}$, and there may exist an indefinite number of them. Thus, in practical terms we refer to a given *candidate set* $\widetilde{\mathfrak{T}}_C$ which is required to be such a subtarget. Note, however, that for fixed T_C the choice of $\mathfrak{T}_C$ affects the size of Δ_C, as is shown later in this section [see (4.4.10)].

We can get $\widetilde{\mathfrak{T}}_C$ as a real life proposition or find it via some necessary conditions. The first is usually the case when a given path is to be followed. For the second, observe that the largest possible $\mathfrak{T}_C$ in $\mathfrak{T}$ is determined by the boundary $\partial \Delta_C$ intersecting $\mathfrak{T}$ (see Fig. 4.17). Hence we can try the method we used previously for $\partial \Delta_\mathscr{R}$ and search for the capture-referenced semipermeable surface passing closest to $\Delta_C \cap \mathfrak{T}$. To determine such a surface, we obviously use the condition (3.3.4). The problem of finding the semipermeable candidate may be simpler now than for reaching if we remember (4.4.3), which means that $\partial \Delta_\mathscr{R}$ estimates $\partial \Delta_C$, provided that we know $\partial \Delta_\mathscr{R}$ at all.

For instance, consider the system

$$\dot{x}_2 = \cos u + x_1^2 + x_2^2 - w, \qquad \dot{x}_2 = \sin u, \qquad (4.4.4)$$

with $(x_1, x_2) \in \Delta = R^2$, $u \in R$, $w \in [0, 1] \subset R$ and target $\mathfrak{T} = \{\bar{x} \in R^2 \,|\, x_1^2 + x_2^2 \leq 2\}$. Taking $\bar{n} = \nabla \rho(\bar{x})$ with ρ meaning the distance from the origin, we have from (3.3.4)

$$\frac{\partial \rho(\bar{x})}{\partial x_1}(\cos u + x_1^2 + x_2^2 - w) + \frac{\partial \rho(\bar{x})}{\partial x_2} \sin u = 0, \qquad (4.4.5)$$

which cannot be satisfied if there is $w(\cdot)$ such that $x_1^2 + x_2^2 - w > 1$ for every point of $\partial \widetilde{\mathfrak{T}}_C$. However, if $x_1^2 + x_2^2 \leq 1$, then for any $w(t) \in [0, 1]$ some u can be found which satisfies (4.4.5). Thus an apparently good candidate for $\mathfrak{T}_C$ is specified by

$$\partial \widetilde{\mathfrak{T}}_C : x_1^2 + x_2^2 = 1. \qquad (4.4.6)$$

Whether Δ_0 and $\widetilde{\mathfrak{T}}_C$ are candidates for a region and a subtarget obtained through one of the mentioned methods or they are just desired sets, we need sufficient conditions to verify them. Let $C\widetilde{\mathfrak{T}}_C \triangleq \Delta_0 - \widetilde{\mathfrak{T}}_0$ and introduce the function $V(\cdot): D \to R$, $D(\text{open}) \supset \overline{C\widetilde{\mathfrak{T}}_C}$ such that

$$v'_\mathfrak{T} \triangleq \inf V(\bar{x}) \,|\, \bar{x} \in \partial \widetilde{\mathfrak{T}}_C, \qquad v'_0 \triangleq \inf V(\bar{x}) \,|\, \bar{x} \in \partial \Delta_0. \qquad (4.4.7)$$

Note that generally v'_0 and $v'_\mathfrak{T}$ differ from v_0 and $v_\mathfrak{T}$ as defined for reaching. The following is proved in Skowronski [19, 21]:

Theorem 4.4.1. *The system (3.3.1) is strongly controllable on $\Delta_0 \times R$ for reaching with capture in $\tilde{\mathfrak{T}}_C$ if there is a $\mathscr{P}(\cdot)$ defined on $D \times R$ and a C^1 function $V(\cdot)$: $D \to R$ such that*

  (i) $V(\bar{x}) \geq v'_{\mathfrak{T}}$ *for* $\bar{x} \notin \tilde{\mathfrak{T}}_C$;
  (ii) $V(\bar{x}) \leq v'_0$ *for* $\bar{x} \in C\tilde{\mathfrak{T}}_C$;
  (iii) *for all $\bar{u} \in \mathscr{P}(\bar{x}, t)$ there is a $c > 0$ such that*

$$\nabla V(\bar{x}) \cdot \bar{f}(\bar{x}, t, \bar{u}, \bar{w}) \leq -c \tag{4.4.8}$$

for all $\bar{w}(t) \in W$.

Proof. We show first that Δ_0 is positively strongly invariant under $\mathscr{P}(\cdot)$. Suppose not: Then $(\bar{x}^0, t_0) \in \Delta_0 \times R$ generates $k(\cdot) \in \mathscr{K}(\bar{x}^0, t_0)$ such that for some $t_1 \geq t_0$, we have $k(\bar{u}, \bar{x}^0, t_0, t_1) = \bar{x}^1 \in \partial\Delta_0$. Then by (ii), $V(\bar{x}^1) \geq V(\bar{x}^0)$ contradicting (4.4.8) and proving the invariance postulated.

Now take any $k(\cdot) \in \mathscr{K}(\bar{x}^0, t_0), (\bar{x}^0, t_0) \in \overline{C\tilde{\mathfrak{T}}_C} \times R$. From (4.4.8), by integrating, one obtains the estimate

$$t - t_0 \leq c^{-1}(V(\bar{x}^0) - V(\bar{x})) \tag{4.4.9}$$

for the interval of time spent in $\overline{C\tilde{\mathfrak{T}}_C}$. From (i) and (ii) we have $V(\bar{x}) - v'_{\mathfrak{T}} \geq 0$, $V(\bar{x}^0) - v'_0 \leq 0$, or $V(\bar{x}^0, t_0) - V(\bar{x}) \leq v'_0 - v'_{\mathfrak{T}}$; whence

$$t \leq t_0 + c^{-1}(v'_0 - v'_{\mathfrak{T}}). \tag{4.4.10}$$

Letting $T_C = c^{-1}(v'_0 - v'_{\mathfrak{T}})$, we conclude that for $t > t_0 + T_C$ the motion must leave $\overline{C\tilde{\mathfrak{T}}_C}$. Since it cannot leave the strongly invariant Δ_0, it must enter $\tilde{\mathfrak{T}}_C$. A return to $\overline{C\tilde{\mathfrak{T}}_C}$ is not possible since then (iii) and (i) would be contradictory. Since we have the above for any motion from any point in $C\tilde{\mathfrak{T}}_C \times R$, the theorem is proved.

Remark 4.4.1. Note that if $\Delta_0, \tilde{\mathfrak{T}}_C$ are defined by some V levels as

$$\partial\Delta_0: V(\bar{x}) = v'_0, \qquad \partial\tilde{\mathfrak{T}}_C: V(\bar{x}) = v'_{\mathfrak{T}}, \tag{4.4.11}$$

conditions (i) and (ii) hold automatically.

Theorem 4.4.2. *Consider a C^1 function $V(\cdot)$: $D \to R$ and constants $v'_0 > v'_{\mathfrak{T}} > 0$ such that (4.4.11) holds. Then in order for (3.3.1) to be strongly controllable on $\Delta_0 \times R$ for reaching with capture in $\tilde{\mathfrak{T}}_C$ it is necessary and sufficient that there be a $\mathscr{P}(\cdot)$ and a constant $c > 0$ such that for all $\bar{u} \in \mathscr{P}(\bar{x}, t)$,*

$$\nabla V(\bar{x}) \cdot \bar{f}(\bar{x}, t, \bar{u}, \bar{w}) \leq -c \tag{4.4.8}$$

for all $\bar{w}(t) \in W$.

Proof of sufficiency follows from Theorem 4.4.1. To show the necessity, suppose along some $k(\cdot) \in \mathscr{K}(\bar{x}^0, t_0)$, $(x^0, t_0) \in C\tilde{\mathfrak{T}}_C \times R$ for each $c > 0$, we have $\dot{V}(\bar{x}(t)) > -c$. Then no reaching is possible.

4.4 REACHING WITH CAPTURE. OPTIMAL CAPTURE

Corollary 4.4.1. *Given* $(\bar{x}^0, t_0) \in C\tilde{\mathfrak{T}}_0 \times R$, *if there is a pair* $\tilde{\tilde{u}}, \tilde{w}$ *such that*

$$\mathscr{L}(\bar{x}, t, \tilde{\tilde{u}}, \tilde{w}) + c = \min_{\bar{u}} \max_{\bar{w}} \mathscr{L}(\bar{x}, t, \bar{u}, \bar{w}) + c \leq 0, \quad (4.4.8')$$

condition (4.4.8) is met with $\tilde{\tilde{u}} = \mathscr{P}(\bar{x}, t)$, *and one can deduce the control program from* (4.4.8').

The proof follows from Theorem 4.4.1 by the same argument as for Corollary 3.4.1. Similarly, as in Section 3.4, we may either use (4.4.8') directly to produce the control program $\mathscr{P}(\cdot)$ or indirectly using the L-G controller (3.4.7). Examples of both cases have been given in Section 3.4.

Remark 4.4.2. Note that given the bound c of the rate of change of $V(\cdot)$ of Theorem 4.4.2, we may find

$$T_C = c^{-1}(v_0' - v_{\mathfrak{T}}'),$$

or, alternatively, given T_C, the constant c may be found.

By definition, there is no such thing as an objective composed of a sequence of captures, but one of the obvious applications of the study on reaching with capture is the objective composed of *sequential reachings* (see Section 4.3) *terminating with capture*. Then, if the reachings must be in *stipulated time*, so must be the terminal capture, which in turn *requires stipulated* $\mathfrak{T}_C$ (see Fig. 4.18). Also in the case of $\bar{x}$ represented by variables relative to the target, that is, in path tracking, both the paths $\mathfrak{T}_C$ and T_C are stipulated. Obviously there are other situations in which arbitrary $\mathfrak{T}_C$, T_C cannot be allowed.

Definition 4.4.3. The system (3.3.1) is strongly controllable at $(\bar{x}^0, t_0) \in \Delta \times R$ for *reaching with capture in* $\mathfrak{T}_C$ *after* T_C if and only if, given $\mathfrak{T}_C$, T_C, there is a $\mathscr{P}(\cdot)$ such that for each $k(\cdot) \in \mathscr{K}(\bar{x}^0, t_0)$,

$$k(\bar{u}, \bar{x}^0, t_0, [t_0 + T_C, \infty)) \subset \mathfrak{T}_C \times R. \quad (4.4.12)$$

Let us denote the corresponding region by Δ_{CT}, with an obvious conclusion

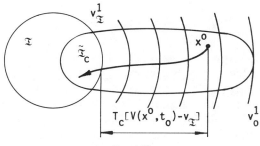

Fig. 4.18

that $\Delta_{CT} \subset \Delta_C$ with possibly, but not necessarily, the converse enclosure holding. The latter depends on how the stipulated T_C, $\mathfrak{T}_C$ relate to their indefinite counterparts. The reader is advised to return to Section 4.3 for comparison.

Theorem 4.4.3. *The system (3.3.1) is strongly controllable on $\Delta_0 \times R$ for reaching with capture in $\mathfrak{T}_C$ after T_C if there is a $\mathscr{P}(\cdot)$ defined on $C\mathfrak{T}_C \triangleq \Delta_0 - \mathfrak{T}_C$ and a C^1 function $V(\cdot): D \to R$, $D(\text{open}) \supset \overline{C\mathfrak{T}_C}$, such that*

(i) $V(\bar{x}) \geq v'_{\mathfrak{T}}$ for $\bar{x} \notin \mathfrak{T}_C$,
(ii) $V(\bar{x}) \leq v'_0$ for $\bar{x} \in C\mathfrak{T}_C$,
(iii) for all $\bar{u} \in \mathscr{P}(\bar{x}, t)$

$$\nabla V(\bar{x}) \cdot \bar{f}(\bar{x}, t, \bar{u}, \bar{w}) \leq -(v'_0 - v'_{\mathfrak{T}})/T_C \qquad (4.4.13)$$

for all $\bar{w}(t) \in W$.

The constants v'_0, $v'_{\mathfrak{T}}$ are defined as in (4.4.7). The proof follows from Theorem 4.4.1 if we specify that

$$c = (v'_0 - v'_{\mathfrak{T}})/T_C \qquad (4.4.14)$$

in condition (iii). From Theorem 4.4.2 we have

Corollary 4.4.2. *Suppose Δ_0 and $\mathfrak{T}_C$ are defined by (4.4.11). Then the system (3.3.1) is strongly controllable on such Δ_0 for reaching with capture of such $\mathfrak{T}_C$ after stipulated T_C, if and only if Theorem 4.4.2 holds with c in (4.4.8) as specified by (4.4.14).*

Sufficiency follows from Theorem 4.4.3. To show necessity, observe that if not (iii), then for some $\tilde{u}, \tilde{w}$ we have

$$\dot{V}(t) > -T_C^{-1}(v'_0 - v'_{\mathfrak{T}}).$$

Integrating this along an arbitrary motion from anywhere in $C\mathfrak{T}_C$,

$$V(\bar{x}) - V(\bar{x}^0) > -\frac{v'_0 - v'_{\mathfrak{T}}}{T_C}(t - t_0)$$

or

$$V(\bar{x}^0) - V(\bar{x}) < \frac{v'_0 - v'_{\mathfrak{T}}}{T_C}(t - t_0),$$

which means that the drop in the value of V is not sufficient to cross $\partial \mathfrak{T}_C$ for all $t - t_0 \geq T_C$, which contradicts capture.

The constant c in (4.4.14) determines the upper bound of the dissipation rate along the motions concerned, and given $v'_0 - v'_{\mathfrak{T}} = h_0 - h_{\mathfrak{T}} \triangleq \delta h$ and the time interval T_C, c may be calculated. In turn, knowing this *speed of dissipation* and T_C, we may find δh for the objective involved (see Fig. 4.14). By the same argument as before, we have

4.4 REACHING WITH CAPTURE. OPTIMAL CAPTURE

Corollary 4.4.3. *Given* $(x^0, t_0) \in \Delta_0 \times R$, *if there is a pair* $(\tilde{\tilde{u}}, \tilde{w})$ *such that*

$$\mathscr{L}(\bar{x}, t, \tilde{\tilde{u}}, \tilde{w}) = \min_{\bar{u}} \max_{\bar{w}} \mathscr{L}(\bar{x}, t, \bar{u}, \bar{w}) + (v'_0 + v_{\bar{\mathfrak{x}}})/T_C \leq 0, \qquad (4.4.15)$$

then condition (iii) of Theorem 4.4.3 is met with $\tilde{\tilde{u}} \in \tilde{\mathscr{P}}(\bar{x}, t)$; *that is, it may be possible to deduce a winning control program from (4.4.15)*.

For a brief example, let us consider the system

$$\dot{x}_1 = -x_1 - ux_2^2 + w, \qquad \dot{x}_2 = x_2(u - x_1^2)$$

with $u, w \in [-1, 1] \subset R$, and find $P(x_1, x_2)$ that generates strong controllability for capture in $\mathfrak{T}_C : x_1^2 + x_2^2 \leq 1$ from $\Delta_0 : x_1^2 + x_2^2 < 6$, during a stipulated T_C beginning at $t_0 = 0$. After a brief calculation, Corollary 4.4.3 with the designed function $V(\bar{x}) = x_1^2 + x_2^2$ gives

$$u = [(-5/T_C) + 2x_1^2(1 + x_2^2)]/2x_2^2(1 - x_1)$$

for $1 < x_1^2 + x_2^2 < 6$, which specifies $C\mathfrak{T}_C$.

This can easily be generalized. To do so we need to introduce the following property from vector calculus.

Property 4.4.1. *Let* $\bar{y}, \bar{z}$ *be two real* n *vectors and* $\bar{y} \neq 0$. *Given any scalar* α, *we have* $\bar{y} \cdot \bar{z} = \alpha$ *if and only if there is a real skew symmetric* $n \times n$ *matrix* $\mathfrak{C}$ *such that*

$$\bar{z} = [(\alpha/\|\bar{y}\|^2)I + \mathfrak{C}]\bar{y}, \qquad (4.4.16)$$

where I *is the unit matrix.*

The above gives an explicit but nonunique solution to the implicit algebraic equation

$$\bar{z} \cdot \bar{y} = \alpha, \qquad \bar{y} \neq 0.$$

When $\bar{y} = 0$, α must be zero for a solution to exist, and z is arbitrary in this case. Obviously, $\mathfrak{C}$ is not unique so it produces a class of $\bar{z}$. Liu and Leake [2] proved that this class is an equivalence class. We use this fact to obtain the controller from (4.4.16). Indeed, by (4.4.15) and (4.4.16),

$$\bar{f}(\bar{x}, t, \tilde{\tilde{u}}, \tilde{w}) = \left(\frac{v'_0 - v'_{\bar{\mathfrak{x}}}}{T_C \|V(\bar{x})\|^2} I + \mathfrak{C}\right) \nabla V(\bar{x}), \qquad (4.4.17)$$

which gives the control condition to calculate $\bar{u}$, when $\bar{f}(\cdot)$ is specified. In view of the arbitrariness of $\mathfrak{C}$, we will have a class of controls which satisfies (4.4.17). Such a class is narrowed by other accompanying conditions.

As an alternative approach, we still have the L–G controller (3.4.7) for the case of separate control $\bar{u}$, which covers our systems. Cvetkovič and Vukobratovič [1] give a very good application of such a controller to the six-DOF anthropomorphic manipulator.

From Corollary 4.4.2 it follows that, given $V(\cdot)$, we may also determine Δ_0, $\mathfrak{T}_C$ to within the accuracy of the V levels. Indeed, note that (4.4.13) is a field condition. Then by Corollary 4.4.4, this condition defines a set

$$C\mathfrak{T}_C|_V = \{\bar{x} \in \Delta \,|\, \mathscr{L}(\bar{x}, t, \tilde{\bar{u}}, \tilde{\bar{w}}) + (v'_0 - v'_{\mathfrak{T}})/T_C \le 0\}, \quad (4.4.18)$$

which, for given $V(\cdot)$, produces a strongly controllable set for collision with capture. If we can find the largest simply connected set so defined for the case study concerned, we have a V-related region of strong controllability. Obviously, a better choice of $V(\cdot)$ may produce a still larger set (see Lewandowska [1] and Mansour [1]). However, when we have to choose energy or its function for physical reasons, then the approximation by the largest $C\mathfrak{T}_C|_H$ is satisfactory. For instance, in Example 4.2.1 (see Fig. 4.10) we quite obviously have $\Delta_P \cup \partial\Delta_P = \Delta_C$ with the dissipative separatrix bounding the region of monotone dissipation, that is, the region specified by (4.4.18), while penetration in $\mathfrak{T}$ means in fact capture in $\mathfrak{T}_C = \mathring{\mathfrak{T}}$. In this case the region Δ_C is well defined by the dissipative separatrixes and no matter what the choice of $V(\cdot)$ it would not be any larger.

Now referring directly to [(1.3.15″) or (1.3.24″)] or (1.5.16″), we may specify the above obtained conditions further, but much in the same vein as the discussion given in Section 3.4. Let us assume here that $\bar{x}$ represents the joint variables (not relative variables). As a rule, when controlling the system through a sequence of reachings (see Sections 4.1 and 4.3) and capture, the capture is the last intermediate objective. Thus we may look at it as a subject of *local studies* on the surface $\mathscr{H}$ (either $\tilde{\mathscr{H}}$ or $\hat{\mathscr{H}}$) either within the energy cups Δ_H^k or in-the-large. In the latter case the capture in $\mathfrak{T}$ expressed in terms of an energy level $h_B > h_L$ does not differ at all from stabilization under that level and the discussion in Section 3.4 applies directly without change [see Proposition 3.4.1 and formulas (3.4.12)–(3.4.17)]. It is left to the reader to see that this is a direct application of Theorems 4.4.1–4.4.3 and their corollaries.

The discussion in Section 4.1 on reaching remains valid here. Namely, for capture in a target about a Dirichlet stable equilibrium we use $V(\bar{x}) \equiv \tilde{H}(\bar{x})$ and about a threshold we use $V(\bar{x}) = -\hat{H}(\bar{x})$, with $f_0(\bar{x}, t, \bar{u}, \bar{w}) < 0$ and $f_0(\bar{x}, t, \bar{u}, \bar{w}) > 0$, respectively. The only difference is that now for capture we *must use monotone dissipation* of energy along the whole route and the rate of this dissipation is specified by (4.4.14). This, however, does not differ much from reaching, since for local reaching, we also recommend monotone dissipation, which obviously gives the quickest results. Now, the time steps in getting to the final target remain the same as discussed in Section 4.3 with only the last step $T_{\mathscr{R}}$ replaced by our present T_C. In the above sense, Example 4.3.1 illustrates reaching with capture as well, and there is no need for an extra illustration if the reader replaces the last step by capture and checks that the proposed controller $u = lx_2|x_2| \pm N/x_2$ does in fact satisfy (4.4.15) and

4.4 REACHING WITH CAPTURE. OPTIMAL CAPTURE

(4.4.17), while the Δ_C can be found from (4.4.18) bounded by the dissipative separatrixes (4.2.6).

Let us return for a short moment to the energy flux estimates as discussed in Section 4.3 [see (4.3.5)]. Time instantaneously, we have from Corollary 4.4.2 the power

$$\min_{\bar{u}} \max_{\bar{w}} f_0(\bar{x}, t, \bar{u}, \bar{w}) \leq -(h_0 - h_{\mathfrak{x}})/T_C$$

for outflux (dissipation) and

$$\max_{\bar{u}} \min_{\bar{w}} f_0(\bar{x}, t, \bar{u}, \bar{w}) \geq (h_0 - h_{\mathfrak{x}})/T_C$$

for influx (accumulation), as measuring the rate of energy use by the *nonpotential forces*. Then the work of such forces is given by

$$W_{\not{P}}(\bar{x}, t) = \int_{t_0}^{t_0 + T_C} f_0(\bar{x}, t, \tilde{\bar{u}}, \tilde{\bar{w}}) \, dt \qquad (4.3.5')$$

with $\tilde{\bar{u}}, \tilde{\bar{w}}$ acting as a pair that generates min–max or max–min pair depending upon the sign of f_0, and calculated from the above conditions on power. Then the *quality criterion* for the motion is the ratio

$$\Re = \frac{1}{h_0 - h_{\mathfrak{x}}} \int_{t_0}^{t_0 + T_C} f_0(\bar{x}, t, \tilde{\bar{u}}, \tilde{\bar{w}}) \, dt. \qquad (4.3.5'')$$

Suppose now that we deal with input forces that are uncertain and bounded but for which the bound is unknown. When the selecting state equation (1.5.3'') can be written in the form (3.4.21), which is the case for our specification (1.3.15'')/(1.3.24'') or (1.5.16'') when the actuator transmission function is linear in $\bar{u}$: $\bar{Q}^F(\bar{q}, \dot{\bar{q}}, \bar{u}) = B(\bar{q}, \dot{\bar{q}})\bar{u}$, we may use the adaptive control as described in Section 3.4 without any change to our strong controllability for reaching with capture. The adaptive laws (3.4.24') and (3.4.27) and the controller (3.4.23') and (3.4.28) apply directly and Propositions 3.4.2 and 3.4.3 can be rewritten with "stabilization" replaced by "strong controllability for reaching with capture." A number of examples will illustrate the adaptive procedure in Section 4.6 as well as in Chapter 5.

Suppose now that we want reaching with capture to be joined with some optimization objective which requires, say, that we minimize a cost assumed in the form (3.5.14):

$$\mathcal{V} = \int_{t_0}^{t_0 + T_C} f_0^V(\bar{x}, t, \bar{u}, \bar{w}) \, dt \qquad (4.4.19)$$

where $f_0^V(\cdot)$ is the characteristic of the cost flow $\mathcal{V}(\mathcal{P}, \bar{x}^0, t_0, k, \cdot)$, $t \to \mathcal{V}(t) \in R$ or the cost-flow rate (see Section 3.5) on the cost-surface Σ (see Fig. 3.16). The characteristic is assumed to be positive.

The opposition loses its bet against capture, but does its worst for us and is assumed to be attempting to maximize (4.4.19). Definition 3.5.2 implies

Definition 4.4.4. System (3.3.1) is strongly optimally controllable at $(\bar{x}^0, t_0) \in \Delta \times R$ for reaching with capture in $\mathfrak{T}_C^*$ after T_C if and only if there are $\mathscr{P}(\cdot)$'s such that Definition 4.4.3 holds and among such $\mathscr{P}(\cdot)$'s we may find $\mathscr{P}^*(\cdot)$ which together with some $\bar{w}^*(\cdot)$ generates the saddle condition (3.5.4).

The corresponding region is denoted Δ_C^* and we have $\Delta_C^* \subset \Delta_C$. Then since the corresponding $\mathfrak{T}_C^*$ must satisfy $\mathfrak{T}_C^* \subset \Delta_C^* \cap \mathfrak{T}$, it may differ from $\mathfrak{T}_C$. Using our usual argument we stipulate a $\mathfrak{T}_C^*$ that satisfies the necessary conditions, but this time from the cost semipermeability, that is, (3.5.4), passing $\partial \mathfrak{T}^*$ through the cost-semipermeable surface closest to Δ_C^*. It also justifies a candidate Δ_0 for Δ_C^* itself. Then the pair $\mathfrak{T}_C^*$, Δ_0 must be confirmed by some sufficient conditions.

The sufficient conditions for optimization used below have been introduced by Stalford and Leitmann [1]. On the same ground, the optimal capture was investigated by Skowronski [19], which we apply here. Let $C\mathfrak{T}_C^* \triangleq \Delta_0 - \mathfrak{T}_C^*$ and consider a function $V(\cdot): D \to R$, D(open) $\supset \overline{C\mathfrak{T}_C^*}$, with v_0^*, $v_{\mathfrak{T}}^*$ defined as v_0', $v_{\mathfrak{T}}'$ in (4.4.7).

Theorem 4.4.4. *The system (3.3.1) is strongly optimally controllable on Δ_0 for reaching with capture in $\mathfrak{T}_C^*$ if there is a $\mathscr{P}(\cdot)$ defined on D and a C^1 function $V(\cdot): D \to R$ such that*

  (i) $V(\bar{x}) \geq v_{\mathfrak{T}}^*$ for $\bar{x} \notin \mathfrak{T}_C^*$,
  (ii) $V(\bar{x}) \leq v_0^*$ for $\bar{x} \in C\mathfrak{T}_C^*$,
  (iii) for all $u^* \in \mathscr{P}_*(\bar{x}, t)$,

$$\nabla V(\bar{x}) \cdot \bar{f}(\bar{x}, t, \bar{u}^*, \bar{w}) \leq -f_0^V(\bar{x}, t, \bar{u}^*, \bar{w}) \quad (4.4.20)$$

for all $\bar{w} \in W$ and with

$$0 \leq v_0^* - v_{\mathfrak{T}}^* \leq T_C \cdot f_0^V(\bar{x}, t, \bar{u}^*, \bar{w}), \quad (4.4.21)$$

  (iv) *there is a $\bar{w}^*(\cdot)$ such that*

$$\nabla V(\bar{x}) \cdot \bar{f}(\bar{x}, t, \bar{u}, \bar{w}^*) \geq -f_0^V(\bar{x}, t, \bar{u}, \bar{w}^*) \quad (4.4.22)$$

for all $\bar{u} \in \mathscr{P}(\bar{x}, t)$ and all $\mathscr{P}(\cdot)$ winning capture.

Proof. Since $0 < f_0^V(\bar{x}, t, \bar{u}, \bar{w}) < \infty$, conditions (i)–(iii) satisfy Theorem 4.4.3, yielding Definition 4.4.3. Thus, we only have to prove optimality. From (iii) and (iv) one obtains, respectively,

$$\begin{aligned} f_0^V(\bar{x}, t, \bar{u}^*, \bar{w}) + \nabla V(\bar{x}) \cdot \bar{f}(\bar{x}, t, \bar{u}^*, \bar{w}) &= -l(t), \\ f_0^V(\bar{x}, t, \bar{u}, \bar{w}^*) + \nabla V(\bar{x}) \cdot \bar{f}(\bar{x}, t, \bar{u}, \bar{w}^*) &= l(t) \end{aligned} \quad (4.4.23)$$

4.4 REACHING WITH CAPTURE. OPTIMAL CAPTURE

for some $l(t) > 0$ and all $t \geq t_0$. Integrating both equations (4.4.23) successively between t_0 and $t_C = t_0 + T_C$, we obtain

$$\int_{t_0}^{t_C} f_0^V(\bar{x}, t, \bar{u}^*, \bar{w}) \, dt = V(\bar{x}^0) - V(\bar{x}^C) - \int_{t_0}^{t_C} l(t) \, dt,$$
$$\int_{t_0}^{t_C} f_0^V(\bar{x}, t, \bar{u}, \bar{w}^*) \, dt = V(\bar{x}^0) - V(\bar{x}^C) + \int_{t_0}^{t_C} l(t) \, dt, \qquad (4.4.24)$$

where $\bar{x}^C = k(u, \bar{x}^0, t_0, t_C)$. On the other hand, from both (iii) and (iv), we have

$$f_0^V(\bar{x}, t, \bar{u}^*, \bar{w}^*) + \nabla V(\bar{x}) \cdot \bar{f}(x, t, u^*, w^*) = 0, \qquad (4.4.25)$$

and integrating, we have

$$\int_{t_0}^{t_C} f_0^V(\bar{x}, t, \bar{u}^*, \bar{w}^*) \, dt = V(\bar{x}^0) - V(\bar{x}^C). \qquad (4.4.26)$$

Comparing (4.4.24) and (4.4.25) with (4.4.26) gives the optimality defined by the saddle condition (3.5.4), thus completing the proof.

Observe that, given the Liapunov function, condition (4.4.21) estimates the size of Δ_C^*. Observe further that from Remark 4.4.2 and (4.4.14),

$$c \leq f_0(\bar{x}, t, \bar{u}^*, \bar{w}), \qquad \forall \bar{w} \in W, \qquad (4.2.27)$$

which generated (4.4.21) and yields

$$T_C = \frac{1}{f_0^V}(v_0^* - v_\mathrm{T}^*). \qquad (4.2.28)$$

Thus, either T_C may be calculated from given $V(\cdot)$, or Δ_C^* may be established with stipulated T_C, which is our present case. On the other hand, the necessary conditions for optimality (3.5.11) produce the control program $\mathcal{P}^*(\cdot)$ the same way that Corollary 4.4.1 produced $\mathcal{P}$, winning capture. If such $\mathcal{P}^*(\cdot)$ satisfies Theorem 4.4.4, it must also satisfy Theorem 4.4.3.

As mentioned in Section 3.5, the problem simplifies considerably when the cost is time: $f_0^V(\bar{x}, t, \bar{u}, \bar{w}) \equiv 1$, that is, for the so-called *time optimization*. Unfortunately, for a robotic task, this type of optimization does not seem to be that essential. The reader may recall our comments from Section 4.3.

First, the stipulated time needed on almost every assembly line as a rule prevents time minimization. Several attempts have been made to use pointwise time minimization (see Barnard [1], Spong, Thorp, and Kheradpir [1], Krogh [1], and Barmish, Thomas, and Liu [1]) in spite of the fact that Barmish and Feuer [1] showed that it can destabilize the system. In general, however, the application of time-optimization routines is limited to cases of our capture or ultimate reaching type, where there are no constraints upon the time interval except at the lower end and where it may be to our benefit to make the time shorter.

Further, there are other objectives, such as, for instance, energy saving (minimization of the energy influx needed), which seem to have priority in robotic work. Here, besides direct flux minimization via Pontriagin, confirmed by some sufficiency conditions, we may use the proposed routine of parametric optimization (see Klein [1]): We make the joint coordinates dependent not only upon t but also upon some other parameters such as initial and terminal gripper position coordinates and mechanical and geometrical parameters within the characteristics. Analyzing the energy flux $f_0^V(\bar{x}, t, \bar{u}, \bar{w}) \equiv f_0(\bar{x}, t, \bar{u}, \bar{w})$ this way, we may attempt to state the path along which the motion of our manipulator is most energy-economical.

With conditions for both the types of optimal objectives at our disposal, we may illustrate as well their use on two corresponding examples.

Example 4.4.1. Consider the R manipulator of Example 4.1.2 (see Fig. 4.8), continued in Examples 4.2.1 and 4.3.1 with the target $\mathfrak{T}$ of Example 4.1.2. Suppose that we have passed the steps of sequential reaching described in Example 4.3.1 (see Fig. 4.14) and that we are now at the intermediate threshold target $\mathfrak{T}_2$ looking down toward $\mathfrak{T}_3 = \mathfrak{T}$ of Example 4.1.2 to attain capture there, but with the aim of *minimizing* T_C. As mentioned already in this section, capture is attained with the same min–max controller as for reaching, namely, $u = lx_2|x_2| \pm N/x_2$, calculated from the control condition $f_0 = x_2 u - wx_2 < 0$, $x_2 \neq 0$, securing monotone dissipation of energy, suggestive of the quickest descent down the cup (route *I* in Fig. 4.19). Indeed, for the time optimization $f_0^V \equiv 1$ and with $h_0^* = h_H$ and $h_{\mathfrak{T}}$ given, we have T_C of route *I* estimated by (4.4.21) as $T_C \geq h_0^* - h_{\mathfrak{T}}^*$. Then, to minimize it, we let $V(\cdot)$ of Theorem 4.4.4 be $V(\bar{x}) = H(\bar{x})$ of our manipulator. Conditions (i) and (ii),

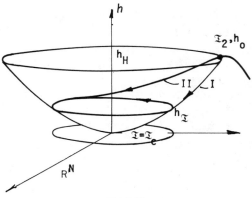

Fig. 4.19

4.4 REACHING WITH CAPTURE. OPTIMAL CAPTURE

are obviously satisfied. The left-hand side in (4.4.20) is the power with the minimizing $u^* = lx_2|x_2| \pm (N+1)/x_2$ plugged in: $\nabla H(\bar{x}) \cdot \bar{f}(\bar{x}, t, \bar{u}^*, \bar{w}) = wx_2 - lx_2^2|x_2| \mp N$, $x_2 \gtrless 0$. We want the above to be ≤ -1 for all w, that is,

$$wx_2 + 1 - lx_2^2|x_2| \mp (N+1) \leq 0, \qquad x_2 \gtrless 0,$$

which in view of $|wx_2| < N$ obviously holds, satisfying (iii).

Last, we have to check (4.4.22), with just the capturing $u = lx_2|x_2| \pm N/x_2$, but the maximizing $w^*x_2 = \pm N$, $x_2 \gtrless 0$:

$$\pm N + 1 - lx_2^2|x_2| \mp N = 0, \qquad x_2 \gtrless 0,$$

which satisfies (iv). Note that the "improved" u^* is also capture-winning by Theorem 4.4.2, with $c = 1$. ∎

Consider now the energy usage and take the work (4.3.5′) as the cost $\mathcal{V}$ to be optimized

$$\mathcal{V} = \pm \int_{t_0}^{t_0 + T_C} f_0(\bar{x}, t, \bar{u}, \bar{w}) \, dt, \qquad (4.4.29)$$

for $f_0 \gtrless 0$ respectively, with $\bar{u} \in \mathscr{P}(\bar{x})$ a winning program for strong controllability for reaching with capture. Theorem 4.4.4 holds and the mentioned program $\mathscr{P}^*(\cdot)$ selected from among the above winning $\mathscr{P}(\cdot)$'s secures the best quality criterion $\Re$ of (4.3.5″), at the passage between h_0, $h_\mathfrak{T}$ concerned.

Example 4.4.2. Let us now illustrate the case in terms of the R manipulator time-optimized in Example 4.4.1 along the same itinerary in the state space. The min–max controller $u = lx_2|x_2| \pm N/x_2$, $x_2 \gtrless 0$, calculated from $f_0 = x_2 u - wx_2 < 0$, $x_2 \neq 0$, secures monotone dissipation, quickest descent (see Example 4.4.1), and at the same time minimizes the work (4.4.29), which immediately follows from our general comments above and Theorem 4.4.4. The situation will require switching the sign of u if we want to capture the arm in a target about a threshold, or in fact in any target (see Example 4.3.1).

Exercise

Consider the RP manipulator oriented on a target as shown in Fig. 4.20, with the rotation joint fixed under the convenient angle and the prismatic joint controlled to push the gripper into the target, subject to a friction force $\pm F$ constant for $|\dot{q}| \neq 0$. The other forces reduce to gravity $G(q) = \partial V/\partial q$ only. The target $\mathfrak{T} = [a, b]$ is on the q axis, and the motion equation may be written as

$$\ddot{q} \pm F + G(q) = u, \qquad \dot{q} \neq 0,$$

where $u \in [-K, K]$. Find the control program and regions of strong

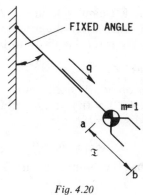

Fig. 4.20

controllability for reaching with capture. Taking the cost

$$\mathscr{V} = \int_{t_0}^{t_\mathscr{R}} |u|\, dt,$$

specify the optimal program that minimizes the effort.

4.5 REACHING WITHOUT CAPTURE. STIPULATED HANDLING TIME

Let us now look at the case in which the system fails capture.

Definition 4.5.1. The system (3.3.1) is strongly controllable at $(x^0, t_0) \in \Delta \times R$ for *reaching without capture in* $\mathfrak{T}$ if and only if it is strongly controllable there for reaching and for each $\mathscr{P}(\cdot)$ that wins reaching there is a $w(\cdot)$ that denies capture, as specified by Definition 4.4.1.

The corresponding strong region is denoted by Δ_C. By Definition 4.4.2, it is the complement of Δ_C to $\Delta_\mathscr{R}$:

$$\Delta_\mathscr{R} = \Delta_C \cup \Delta_\mathscr{C}. \qquad (4.5.1)$$

Hence $\Delta_\mathscr{C} = \varnothing$ if and only if $\mathfrak{T} = \mathfrak{T}_C$, and $\Delta_C = \Delta_\mathscr{R}$ if and only if $\mathfrak{T}_C = \varnothing$. The noncapturing part of $\mathfrak{T}$, denoted by $\mathfrak{T}_\mathscr{C} \triangleq \mathfrak{T} - \mathfrak{T}_C$, is obviously enclosed in Δ_C.

It is a consequence of choosing the boundary of $\mathfrak{T}_\mathscr{C}$ on the semipermeable surface (3.3.4) closest to Δ_C that $\mathfrak{T}_C$ must be assumed open. Consequently $\mathfrak{T}_\mathscr{C}$ is closed.

Recall now that due to $\Delta_C \subset \Delta_\mathscr{R}$, we have an option of estimating Δ_C by $\Delta_\mathscr{R}$. Indeed, for $\bar{\mathfrak{T}}_C = \mathfrak{T}$ we must have $\bar{\Delta}_C = \Delta_\mathscr{R}$. Hence if it came to the worst, the methods for defining $\Delta_\mathscr{R}$ may do for $\dot{\Delta}_C$ as well. The case was discussed in Section 4.4.

4.5 REACHING WITHOUT CAPTURE. STIPULATED HANDLING TIME

In general, however, all $\bar{\mathfrak{T}}_C \subsetneq \mathfrak{T}$, whence $\Delta_C \subsetneq \Delta_{\mathscr{R}}$, and further refining is of interest. Here the converse of the above estimation argument may work. Observe that choosing $\mathfrak{T} = \bar{\mathfrak{T}}_C$ produces the same effect, namely, *any region of strong controllability for reaching $\bar{\mathfrak{T}}_C$ is $\bar{\Delta}_C$* or *any region of strong controllability for penetration in $\mathfrak{T}_C$ is $\mathring{\Delta}_C$*, provided that $C\Delta_P = B_P = \partial\Delta_{\mathscr{R}}$ (see Fig. 4.21), that is, $\partial\Delta_C$ is the semibarrier B_C.

With the above, obviously motions from the closed $\Delta_{\mathscr{C}}$ tend to the closed $\mathfrak{T}_{\mathscr{C}}$ and avoid $\Delta_C \supset \mathfrak{T}_C$. Recall that (4.4.8) has been shown to be a necessary and sufficient condition, thus the contradiction of it yields sufficiency for strong controllability for reaching without capture as defined in Definition 4.4.5, referring to $\mathfrak{T}_{\mathscr{C}} = \mathfrak{T} - \mathfrak{T}_C$, with stipulated target $\mathfrak{T}$. Hence there is no need for a separate sufficient condition. We simply make sure that (4.4.8) does not hold for the circumstances described by Theorem 4.4.2.

When reaching without capture, the arm has a contact with $\mathfrak{T}$. How long does it last and is it of any use to us? The answer is found by investigating the rendezvous specified in Definition 4.4.1, together with reaching.

Definition 4.5.2. The system (3.3.1) is strongly controllable at $(\bar{x}^0, t_0) \in \Delta \times R$ for reaching with rendezvous in $\mathfrak{T}$ if and only if there is a $\mathscr{P}(\cdot)$ and a constant $T_Z > 0$ such that for each $k(\cdot) \in \mathscr{K}(\bar{x}^0, t_0)$ and some $T_{\mathscr{R}} \leq T_Z$ we have

$$k(\bar{u}, \bar{x}^0, t_0, [t_{\mathscr{R}}, t_{\mathscr{R}} + T_Z]) \subset \mathfrak{T} \times R, \tag{4.5.2}$$

where $t_{\mathscr{R}} = t_0 + T_{\mathscr{R}}$.

Clearly $t_{\mathscr{R}}$ need not be unique. From Definition 4.5.2 it follows that there is a proper subset $\mathfrak{T}_Z$ of $\mathfrak{T}$ consisting of points from which the rendezvous takes place. Moreover, the region of strong controllability for reaching $\mathfrak{T}_Z$ is the *region of strong controllability for reaching with rendezvous* denoted by Δ_Z. Given $\mathfrak{T}_Z$, the latter statement indicates the method of determining Δ_Z.

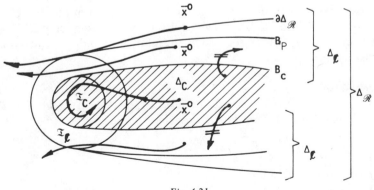

Fig. 4.21

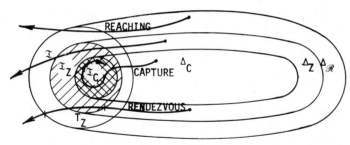

Fig. 4.22

Clearly Δ_Z is related to $\mathfrak{T}_Z$. Conversely, $\mathfrak{T}_Z$ is a proper subset of $\Delta_Z \cap \mathfrak{T}$ (see Fig. 4.22). If $\mathfrak{T}_Z = \Delta_Z \cap \mathfrak{T}$, we might have had $T_Z = 0$ yielding reaching *without* rendezvous, that is, with rejection.

As much as rendezvous is a temporary capture, we may say that capture is a permanent rendezvous: $T_Z \to \infty$. Consequently, capture implies rendezvous, whence

$$\mathfrak{T}_C \subset \mathfrak{T}_Z, \qquad \Delta_C \subset \Delta_Z. \tag{4.5.3}$$

Thus selected $\mathfrak{T}_C, \Delta_C$ may serve as lower estimates. Moreover, $\mathfrak{T}_Z$ is not unique and $\mathfrak{T}$ estimates their union from above. Similarly, $\Delta_{\mathscr{R}}$ estimates from above the corresponding Δ_Z.

As before, we stipulate candidates $\tilde{\mathfrak{T}}_Z, \Delta_0$ for $\mathfrak{T}_Z$ and a subset of Δ_Z, respectively, and need sufficient conditions to verify them. Designate $C\tilde{\mathfrak{T}}_Z \triangleq \Delta_0 - \tilde{\mathfrak{T}}_Z, C'\tilde{\mathfrak{T}}_Z = \mathfrak{T} - \tilde{\mathfrak{T}}_Z$ (see Fig. 4.23), and consider a function $V(\cdot)$: $D \to R$ with D(open) $\supset (C\tilde{\mathfrak{T}}_Z \cup C'\tilde{\mathfrak{T}}_Z)$ (dashed in Fig. 4.23) and with

$$\begin{aligned} v_Z &= \inf V(\bar{x}) \,|\, \bar{x} \in \partial\Delta_0, \\ v_{\mathfrak{T}Z} &= \sup V(\bar{x}) \,|\, \bar{x} \in \partial\tilde{\mathfrak{T}}_Z, \\ v_{\mathfrak{T}} &= \inf V(\bar{x}) \,|\, \bar{x} \in \partial\mathfrak{T}. \end{aligned} \tag{4.5.4}$$

Similarly, as with capture, we attempt to force the motions of $\mathscr{K}(\bar{x}^0, t_0)$, $(x^0, t_0) \in \Delta_0 \times R$ into $\tilde{\mathfrak{T}}_Z$ but now require only that they stay in $\mathfrak{T}$ for no shorter time than T_Z. This is secured by the following conditions introduced in Skowronski [29].

Theorem 4.5.1. *The system (3.3.1) is strongly controllable on $\Delta_0 \times R$ for reaching with rendezvous in $\mathfrak{T} \supset \tilde{\mathfrak{T}}_Z$ if, given $\tilde{\mathfrak{T}}_Z$, there is $\mathscr{P}(\cdot)$ defined on $D \times R$ and a C^1 function $V(\cdot)$ defined on D such that*

(i) $V(\bar{x}) > v_{\mathfrak{T}Z}$ *for* $\bar{x} \notin \tilde{\mathfrak{T}}_Z, t \in R$;
(ii) $V(\bar{x}) < v_{\mathfrak{T}}$ *for* $\bar{x} \in C'\tilde{\mathfrak{T}}_Z, t \in R$;
(iii) $V(\bar{x}) \le v_Z$ *for* $\bar{x} \in C\tilde{\mathfrak{T}}_Z, t \in R$;
(iv) *for all* $(\bar{x}, t) \in C\tilde{\mathfrak{T}}_Z \times R, \bar{u} \in \mathscr{P}(\bar{x}, t)$, *there is a constant*

4.5 REACHING WITHOUT CAPTURE. STIPULATED HANDLING TIME

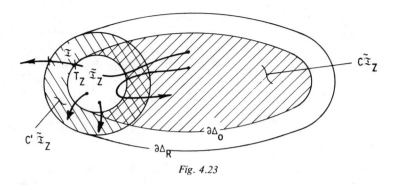

Fig. 4.23

$T_{\mathscr{R}}(\bar{x}^0, t_0, k(\cdot)) > 0$ such that for $t_{\mathscr{R}} = t_0 + T_{\mathscr{R}}$,

$$\int_{t_0}^{t_{\mathscr{R}}} [\nabla V(\bar{x}) \cdot \bar{f}(\bar{x}, t, \bar{u}, \bar{w})] \, dt \leq -(v_Z - v_{\mathfrak{X}Z}) \quad (4.5.5)$$

for all $w(t) \in W$;

(v) for all $(\bar{x}, t) \in C'\tilde{\mathfrak{X}}_Z \times R$, $u \in \mathscr{P}(\bar{x}, t)$, there is a constant $c_Z(\bar{x}^0, t_0, k(\cdot)) > 0$ such that

$$\nabla V(\bar{x}) \cdot \bar{f}(\bar{x}, t, \bar{u}, \bar{w}) \leq c_Z \quad \forall w \in W. \quad (4.5.6)$$

Proof. By Theorem 4.4.1, conditions (i), (iii), and (iv) imply strong controllability on Δ_0 for reaching $\tilde{\mathfrak{X}}_Z$. It remains to prove that no motion from $\tilde{\mathfrak{X}}_Z$ may leave $\mathfrak{X}$ earlier than after $t_0 + T_Z$. Motions which do not leave $\mathfrak{X}$ (captured) satisfy this postulate trivially, hence we consider any $k(\cdot) \in \mathscr{K}(\bar{x}^0, t_0), (\bar{x}^0, t_0) \in \tilde{\mathfrak{X}}_Z \times R$ which crosses $\partial \mathfrak{X}$ at some $t_2 \geq t_0 : k(\bar{u}, \bar{x}^0, t_0, t_2) = \bar{x}^2 \in \partial \mathfrak{X}$. To cross $\partial \mathfrak{X}$ such a motion must pass over $\partial \tilde{\mathfrak{X}}_Z$ at some t_1, $t_0 \leq t_1 \leq t_2$, that is, $k(\bar{u}, \bar{x}^0, t_0, t) = \bar{x}^1 \in \partial \tilde{\mathfrak{X}}_Z$. Now, if the Liapunov derivative $\dot{V}(t)$ of (v) were nonpositive, the passage to $\partial \mathfrak{X}$ would contradict (ii). So there may not be a crossing $\partial \mathfrak{X}$ unless $\dot{V}(t) > 0$. Then integrating (v),

$$t_2 - t_1 \geq (1/c_Z)[V(\bar{x}^2) - V(\bar{x}^1)]. \quad (4.5.7)$$

Letting $T_Z(\bar{x}^0, t_0) = (1/c_Z)(v_{\mathfrak{X}} - v_{\mathfrak{X}Z})$, (4.5.4) implies $t_2 - t_1 \geq T_Z$ for the duration in $C'\tilde{\mathfrak{X}}_Z$, which complete the proof.

Observe that T_Z depends upon c_Z which is the upper bound of the rates of change of $V(t)$ along a motion. We have

$$c_Z(\bar{x}^0, t_0, k(\cdot)) = (1/T_Z)[v_{\mathfrak{X}Z} - V(x^0)] \quad (4.5.8)$$

with $v_{\mathfrak{X}Z} - V(x^0) \geq 0$ by (ii). Given $\tilde{\mathfrak{X}}_Z$, it is easy to envisage a design with c_Z made suitable for the required duration T_Z. Alternatively, given c_Z, the candidate $\tilde{\mathfrak{X}}_Z$ may be adjusted to fit T_Z. Indeed,

$$T_Z c_Z = v_{\mathfrak{X}Z} - V(x^0) \geq 0 \quad (4.5.9)$$

provides the condition for such an adjustment. For capture, $T_Z \to \infty$, we have $\dot{V}(t) < 0$ implying that c_Z must be zero for sufficiently large T_Z [see (4.5.8)]. For stipulated time reaching without rendezvous, $T_Z \to 0$, and (4.5.8) yields $c_Z \to \infty$, wherefrom Theorem 4.5.1 reduces to Theorem 4.4.1.

Suppose now that T_Z is stipulated.

Definition 4.5.3. The system (3.3.1) is strongly controllable at $(\bar{x}^0, t_0)$ for *reaching with rendezvous in $\mathfrak{T}$ during T_Z* if and only if Definition 4.5.2 holds for stipulated T_Z.

The corresponding $\mathfrak{T}_{ZT}$ will depend upon T_Z and obviously may differ from $\mathfrak{T}_Z$ even for the same T_Z. Moreover, we have $\mathfrak{T}_{ZT}(T'_Z) \subset \mathfrak{T}_{ZT}(T''_Z)$ for any two T'_Z, T''_Z such that $T'_Z \geq T''_Z$. The corresponding region is denoted by Δ_{ZT}. This is the same as the region of strong controllability for reaching $\mathfrak{T}_{ZT}$, which gives the method for its determination. Obviously, Δ_{ZT} is related to $\mathfrak{T}_{ZT}$ and thus also to $T_Z: \Delta_{ZT}(T'_Z) \subset \Delta_{ZT}(T''_Z)$ for $T'_Z \geq T''_Z$. Hence, capture with its $T_Z \to \infty$ gives the lower estimate for all $\mathfrak{T}_{ZT}, \Delta_{ZT}: \mathfrak{T}_{ZT} \supset \mathfrak{T}_C, \Delta_{ZT} \supset \Delta_C$, for any T_Z.

Theorem 4.5.2. *The system (3.3.1) is strongly controllable on Δ_0 for reaching with rendezvous in $\mathfrak{T}$ during T_Z if, given $\tilde{\mathfrak{T}}_{ZT}$, the conditions of Theorem 4.5.1 hold with (v) replaced by*

(v') *for all $(x, t) \in C'\tilde{\mathfrak{T}}_Z \times R$, $\bar{u} \in \mathscr{P}(\bar{x}, t)$, we have*

$$\nabla V(\bar{x}) \cdot f(\bar{x}, t, \bar{u}, \bar{w}) \leq (v_{\mathfrak{T}} - v_{\mathfrak{T}Z})/T_Z \qquad (4.5.10)$$

for all $w(t) \in W$.

The proof follows from Theorem 4.5.1 by choosing c_Z as

$$c_Z = (v_{\mathfrak{T}} - v_{\mathfrak{T}Z})/T_Z. \qquad (4.5.11)$$

The min–max and L–G controllers are obtainable by the same methods as for reaching and capture, that is, from conditions (4.5.5) and (4.5.10) through corresponding corollaries that the reader may easily derive by the same arguments as for Corollaries 4.1.4 and 4.4.3.

Exercise and Comments

The same discussion as made for sequential reaching in Section 4.3 on the energy surface H can be made for the present rendezvous case using Theorems 4.5.1 and 4.5.2. The reader is invited to do so, including the reinterpretation of the assignment in Example 4.3.1. The time of such sequential rendezvous may not be optimized, but the corresponding work of the nonpotential forces [see (4.3.5')] may. Using (4.4.29), calculate the best quality criteria for each rendezvous step investigated above.

4.6 PLANNED PATH TRACKING

As we did for all targets before, we assume that a target in the form of a planned path had been transformed off-line into the joint variables $q_{m1},\ldots,q_{mn}$, possibly with specified planned velocities $\dot{q}_{m1},\ldots,\dot{q}_{mn}$ (see Section 1.4). Hence the planned *target path* is given in terms of the vector $\bar{x}_m(t) = (\bar{q}_m(t), \dot{\bar{q}}_m(t))^T$ moving in Δ and thus producing a curve $\bar{x}_m(R_0^+)$. This may obviously be augmented into $\Delta \times R$. Note, however, that if it was given in $\Delta \times R$ as a unique curve, its image in Δ would not necessarily be unique.

Let us briefly review the method of *relative coordinates* mentioned several times before (see, in particular, Section 3.1). This method is based on redefining the state from the joint-coordinate representation into the coordinates relative to the target path. Then we apply to the new system all the tools previously obtained for attaining capture and the derived notions, including optimal capture.

First let us consider the system without uncertainty (1.5.3′) specified by either (1.3.24′) or (1.5.16″) in Section 1.5. Substituting $\bar{x}_m(t) = (\bar{q}_m(t), \dot{\bar{q}}_m(t))^T$, we may calculate the open loop controller $\bar{u}_m(t)$ of the (3.1.1) type. Granted this, we may in turn put this controller into (1.5.3′) or the particular representations and, given the same initial conditions as the target path starting point $\bar{x}_m^0 = (\bar{q}_m^0, \dot{\bar{q}}_m^0)^T$, we obtain the motion $\bar{x}(t) = k(\bar{u}_m, \bar{x}_m^0, t_0, t)$, $t \geq t_0$, in $\Delta \times R$, called *target motion*, which is briefly denoted by $\bar{k}_m(t)$. All of the above is subject to the proviso that the given $x_m(t)$ was compatible with a solution to the equation concerned.

Without uncertainty present, $k_m(t)$ should be identical with the path (or very close to it anyway!). Note here that $\bar{k}_m(t)$ is a given, well-defined function of time and that it does not depend upon any other solutions of (1.5.3′).

We now set up the *relative state vector* $\bar{z}(t) \triangleq \bar{x}(t) - \bar{k}_m(t)$, $t \geq t_0$, implying immediately $\dot{\bar{z}}(t) = \dot{\bar{x}}(t) - \dot{\bar{k}}_m(t)$, which transforms (1.5.3′) into the relative state equation

$$\dot{\bar{z}} = \bar{f}(\bar{z} + \bar{k}_m(t), t, \bar{u}) - \bar{f}(\bar{k}_m(t), t, \bar{u}_m). \tag{4.6.1}$$

The reader is asked to recall the opening discussion on stability in Section 3.2. The second term in the right-hand side of (4.6.1) is a given well-defined function of t and may be considered a given perturbation of the first term. Hence, even if we had the autonomous (1.5.3) instead of (1.5.3′), it would still become nonautonomous after adding this perturbation. However, adding it has no additional effect on the already nonautonomous (1.5.3′), particularly in terms of (1.3.15) where we also have an additive term $\bar{Q}^R(\bar{q}, \dot{\bar{q}}, t)$ as a perturbation. Denote $\bar{f}_m(t) \triangleq \bar{f}(\bar{k}_m(t), t, \bar{u}_m)$. By definition, $\bar{f}_m(t)$ is bounded the same way as $\bar{f}(\cdot)$ was.

Obviously, with all the above, (4.6.1) generates the same type of motions $z(\bar{u}, \bar{z}^0, t_0, t)$, $t \geq t_0$, as (1.5.3′) did, and the *zero motion* $\bar{z}(t) \equiv 0$ *now represents the target path.* This means that all the tools developed for asymptotic stabilization in Section 3.2 apply to target-path tracking, with $\bar{x}$ in joint variables replaced now by the relative variable $\bar{z}$ (see Fig. 4.24). The same obviously applies to controllability for capture in $\mathfrak{T}_C = \{\bar{z}(t) \equiv 0\}$, as was discussed in Section 3.1. Moreover, since substituting a given $\bar{w}$ instead of keeping it uncertain makes it a particular case of our results in Sections 3.3–3.5 and Chapter 4, all the methods and objectives discussed so far are directly applicable to path tracking via (4.6.1). Clearly, capture during the stipulated (real-time) T_C in the singleton $\{\bar{z} \equiv 0\}$ gives a better result than the asymptotic stabilization of Section 3.2. We stipulate some Δ_0, possibly equal to $\mathfrak{T}_C = \{\bar{z} \equiv 0\}$, since we can always set up the arm almost on the target path. Then the study is absolutely local and capture during T_C means security that after T_C the system will not deviate from the target path, while rendezvous secures the same for a desired period of time.

Observe that $\bar{f}_m(t)$ balances the right-hand side of (4.6.1) to zero uniformly in t at the target path, that is, for $\bar{z}(t) \equiv 0$, provided $\bar{u} = \bar{u}_m$ at that point. This makes our control program $\bar{u}_m$-dependent (see the opening of Section 3.1). On the other hand, the programs obtained either from the corollaries or as the L–G controllers are specified for nonzero velocities, that is now for $\dot{\bar{q}} - \dot{\bar{q}}_m \neq 0$. Thus the *above specification* $\bar{u} \triangleq \bar{u}_m$ *for* $\dot{\bar{q}} = \dot{\bar{q}}_m = 0$ *gives an extra defining condition.* Note that $\bar{z} \equiv 0$ is the only relative equilibrium in the new system (4.6.1). Referring to (1.3.24′), we denote $\bar{q}_z(t) \triangleq (\bar{q}(t) - \bar{q}_m(t))$, $\dot{\bar{q}}_z(t) \triangleq (\dot{\bar{q}}(t) - \dot{\bar{q}}_m(t))$, and write (4.6.1) as

$$\ddot{\bar{q}}_z + \bar{\Gamma}(\bar{q}_z + \bar{q}_m, \dot{\bar{q}}_z + \dot{\bar{q}}_m) + \bar{D}(\bar{q}_z + \bar{q}_m, \dot{\bar{q}}_z + \dot{\bar{q}}_m) + \Pi(\bar{q}_z + \bar{q}_m)$$
$$= \bar{F}(\bar{q}_z + \bar{q}_m, \dot{\bar{q}}_z + \dot{\bar{q}}_m, \bar{u}) + \bar{R}(\bar{q}_z, \dot{\bar{q}}_z, \bar{q}_m, \dot{\bar{q}}_m, t, \bar{u}_m). \quad (4.6.2)$$

Letting $V(\bar{z}) \equiv H(\bar{z})$, we have all the conditions for stabilization and controllability for capture satisfied by the shape of the corresponding $\mathcal{H}$, except for the requirement of monotone dissipativeness of the energy flow,

$$\dot{H}(\bar{z}) = \bar{F}(\bar{q}_z + \bar{q}_m, \dot{\bar{q}}_z + \dot{\bar{q}}_m, \bar{u})\dot{\bar{q}} - \bar{R}(\bar{q}_z, \dot{\bar{q}}_z, \bar{q}_m, \dot{\bar{q}}_m, t, \bar{u}_m)\dot{\bar{q}}_z \leq -c, \quad (4.6.3)$$

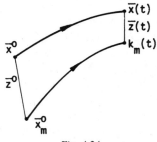

Fig. 4.24

4.6 PLANNED PATH TRACKING

which now becomes the control condition to give the program for $\bar{u}$. Obviously, if we want stipulated time capture of the target path, we specify c as the rate of approach, but if we start at

$$\mathfrak{T}_C = \{\bar{z} \equiv 0\}, \qquad (4.6.4)$$

we simply need $\dot{H}(\bar{z})$ negative to keep $\bar{z} \equiv 0$ positively invariant.

Example 4.6.1. We return to the gripper motion of Example 4.1.2, but ignore the uncertainty:

$$\ddot{q} + u + aq - bq^3 = 0, \qquad (4.6.5)$$

or in state form

$$\dot{x}_1 = x_2, \qquad \dot{x}_2 = -ax_1 + bx_1^3 - u. \qquad (4.6.6)$$

The objective is to follow the closed target path somewhere in Δ_H and given by

$$\frac{1}{2}x_{m2}^2 + \frac{a}{2}x_{m1}^2 - \frac{b}{4}x_{m1}^4 = h_C \leq \frac{a^2}{4b} \qquad (4.6.7)$$

[see (4.1.25)], which immediately secures the required compatibility with solutions to (4.6.5) (see Fig. 4.25). Now we substitute (4.6.7) into (4.6.6) to obtain

$$\dot{x}_{m1} = x_{m2}, \qquad \dot{x}_{m2} = -ax_{m1} + bx_{m1}^3 - u_m, \qquad (4.6.8)$$

where $u_m(t) \equiv 0$, to produce (4.6.7) as a solution. The relative coordinates are

$$z_i = x_i - x_{mi}, \qquad i = 1, 2. \qquad (4.6.9)$$

Substituting $x_i = z_i + x_{mi}, i = 1, 2$, into (4.6.6), in view of (4.6.8) we obtain

$$\dot{z}_1 = z_2, \qquad \dot{z}_2 = -az_1 + bz_1^3 + 3b(z_1 x_{m1}^2 + z_1^2 x_{m1}) - u. \qquad (4.6.10)$$

In the relative coordinate system, we want the capture in the stipulated $\mathfrak{T}_C$: $z_1^2 + z_2^2 < \eta$, with η suitably small for the purpose concerned. We can thus

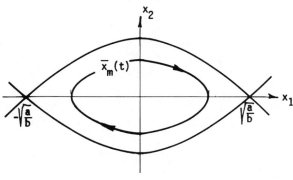

Fig. 4.25

use Theorem 4.4.1 or, if we wish it to be done in stipulated time, Theorem 4.4.3. By letting

$$V(\bar{z}) = \frac{1}{2}z_2^2 + \frac{a}{2}z_1^2 - \frac{b}{4}z_1^4,$$

which is an obvious choice, we have conditions (i) and (ii) satisfied as long as our initial differences from the target path $z_1^0, z_2^0 \in \Delta_0$ do not make Δ_0 overflow the basic cup: $V(z_1, z_2) \leq a^2/4b$. Then we have to verify (iii) either for Theorem 4.4.1, that is, (4.4.8), or for Theorem 4.4.3, that is, (4.4.13). Hence, we need the control program $P(z_1, z_2)$ that secures $f_0(z_1, z_2, u) \leq -c$, that is,

$$3b(z_1 x_{m1}^2 + z_1^2 x_{m1})z_2 - uz_2 + c \leq 0. \qquad (4.6.11)$$

Thus we let

$$u = 3b(z_1 x_{m1}^2 + z_1^2 x_{m1}) + c/z_2, \qquad (4.6.12)$$

with c specified by (4.4.14) if we want a stipulated T_C; otherwise we only let c be suitably small. The program works on Δ_0 given by $z_1 \leq \sqrt{a/b}$, $z_2 \leq a/4b$ (see Example 4.1.2).

Let us now suppose that the target path is not such a curve which may be approximated satisfactorily by an immediately verifiable solution (4.6.7). Suppose, in particular, that the curve is a closed contour given by

$$x_{m2}^2 + ax_{m1}^2 = 2h_C. \qquad (4.6.13)$$

The target equation (4.6.8) is then replaced by

$$\dot{x}_{m1} = x_{m2}, \qquad \dot{x}_{m2} = -ax_{m1} - u_m \qquad (4.6.14)$$

with $u_m \not\equiv 0$ but compensating for the difference from (4.6.8), that is, $u_m = bx_{m1}^3$. The rest of the calculations remain the same. Observe that the controller u_m will then be canceled when deriving (4.6.10). ∎

By taking the target path in the above example in the implicit form (4.6.7) instead of parametrically in t, we have avoided the inconvenience of having $\ddot{\bar{q}}$ appear in the program for $\bar{u}_m$. Such an approach may work many times, but is not always applicable, particularly for $N > 2$.

Generally, $P(\bar{z}, t)$ calculated from (4.6.3) will include $\bar{u}_m(t)$ and, through that open-loop controller (3.1.1), all velocities *and* accelerations of the path: $\dot{\bar{q}}_m, \ddot{\bar{q}}_m$. There is an efficient method other than that shown in Example 4.6.1 which helps to get rid of the accelerations. For lack of room here, we must however refer the interested reader to the work by Voroneckaia and Fomin [1] and Aksenov and Fomin [1].

Since we operate at the bottom of the local cup, linearization of (4.6.1) might be appropriate. Indeed, if one puts the first term on the right-hand side of

4.6 PLANNED PATH TRACKING

(4.6.1) into a Taylor series about $\bar{x}_m(t)$ and then takes the linear part of the result with the second term subtracted, it leads to

$$\dot{\bar{z}} = \nabla_x \bar{f}|_m \bar{z} + \nabla_u \bar{f}|_m (\bar{u} - \bar{u}_m), \qquad (4.6.15)$$

where $\nabla_x \bar{f}|_m$, $\nabla_u \bar{f}|_m$ are Jacobian matrices of $\bar{f}$ evaluated at $\bar{x}_m(t)$ and $\bar{u}_m(t)$, respectively. Lee and Chung [1] developed this approach, also inserting adaptive signals in the obtained controller. The difficulty they encountered was that the state and control Jacobian matrices needed to produce the controller were very difficult to determine because of their complexity. Lee and Chung used some parameter identification techniques in order to obtain the data. Trying to calculate the program from (4.6.3), we face the same problem, or worse. The adaptive identification technique is discussed in Chapter 6.

For the uncertain system (3.3.1) an essential modification must be made since such a system does not produce a unique target motion. First, if we calculate the u_m controller from (1.5.3''), we do not get a unique answer. It becomes w-dependent. Second, even if this were not the case (say because of the min–max or a similar technique), we still have a class $\mathscr{K}_m(\bar{x}_m^0, t_0)$ of motions $k_m(t)$ instead of a single target motion at the beginning of the target path. So which one do we subtract?

The answer is that we do not subtract. Instead we consider the product system

$$\dot{\bar{x}}_m = f(\bar{x}_m, t, \bar{u}_m(\bar{w}, t), \bar{w}), \qquad \dot{\bar{x}} = f(\bar{x}, t, \bar{u}, \bar{w}) \qquad (4.6.16)$$

in $\Delta^2 \triangleq \Delta \times \Delta$ of the product state space R^{2N} with the product state vector $\bar{X}_m(t) \triangleq (\bar{x}(t), \bar{x}_m(t))^T$ and investigate the strong controllability for reaching with capture, possibly in a stipulated time interval T_C, in the required subtarget $\mathfrak{T}_C = \{\bar{X}_m \in \Delta^2 \,|\, \|\bar{x} - \bar{x}_m\| < \eta\}$, with $\eta > 0$ a stipulated suitably small number as before (see Fig. 4.26).

As an alternative to capture, if we wish, we may go for some other suitable objective, say for asymptotic stabilization about the diagonal $\mathscr{M} = \{\bar{X}_m \,|\, \bar{x} = \bar{x}_m\}$ on the one hand or for rendezvous in stipulated time in $\mathfrak{T}_C$ on the other. The case (4.6.16) is a particular version of model following (the target equation is the model to be followed) which we develop in Section 4.7, so the description of details is left to that section. Note that in the latter interpretation, the controller is adaptive and also the system is made adaptive. We did not indicate this in (4.6.16) but that is easily added, after the procedure is learned in Section 4.7.

The significant advantage of the above proposed *product space method* over others known to the author is that the coordinates $q_{m1}, \ldots, q_{mn}$, appearing in $\bar{X}_m$ are independent, or may be made so, of the joint variables $q_1, \ldots, q_n$. Also there is no need to have the product space built up of two vectors of the same

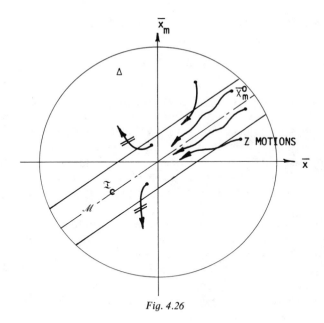

Fig. 4.26

dimensions, that is, $\bar{q}_m(t)$ could be far less dimensional than $\bar{q}$. These two facts allow us to *take Cartesian coordinates of the planned path directly as the target coordinates in $\bar{q}_m$*, thus avoiding the difficult inverse kinematic calculations. In fact, for this reason alone, one may like to use the product space method rather than the previously mentioned relative coordinate approach, even when there is no uncertainty.

An alternative method of path tracking by an uncertain system was introduced by Corless, Leitmann, and Ryan [1]. The uncertain system, taken in a slightly more particular form than (1.5.3″), is

$$\dot{\bar{x}} = \bar{f}(\bar{x}, t) + \delta\bar{f}(\bar{x}, t, \bar{r}) + [B + \delta B(\bar{x}, t, \bar{r})]\bar{u}, \quad (4.6.17)$$

with $B, \delta B$ obviously being $N \times n$ matrices. The target path $\bar{x}_m(t)$ is required to be an absolutely continuous desired state motion—thus it is not an arbitrary curve.

The following assumptions are imposed on (4.6.17):

A1. Linearizability and stabilizability. There exists an $N \times N$ matrix A and a continuous function $\bar{g}(\cdot): R^N \times R \to R^n$ bounded in $\bar{x}$ and measurable in t such that

$$\bar{f}(\bar{x}, t) = A\bar{x} + B\bar{g}(\bar{x}, t) \quad (4.6.18)$$

with A, B being stable matrices (see Example 3.2.2).

4.6 PLANNED PATH TRACKING

A2. Matching of uncertainties. There are continuous functions of the same type as $g(\cdot)$ of A1, $\bar{h}(\cdot): R^N \times R \times \mathfrak{R} \to R^n$ and $E(\cdot): R^N \times R \times \mathfrak{R} \to R^{n \times n}$ and a continuous function $\beta(\cdot): R^N \times R \to R^+$ such that

$$\delta \bar{f}(\bar{x}, t, \bar{r}) = B\bar{h}(\bar{x}, t, \bar{r}),$$

$$\delta B(\bar{x}, t, \bar{r}) = BE(\bar{x}, t, \bar{r}), \quad (4.6.19)$$

$$\|E(\bar{x}, t, \bar{r})\| \le \beta(\bar{x}, t) < 1.$$

A3. Feasibility of $\bar{x}_m(\cdot)$. There is a function $\bar{\theta}^f(\cdot): R \to R^n$, Lebesque integrable on bounded intervals, such that

$$\dot{\bar{x}} = A\bar{x} + B\bar{\theta}^f(t). \quad (4.6.20)$$

It is seen that for any A, $g(\cdot)$ satisfying A1 there is some $n \times N$ matrix K such that $\tilde{A} = A + BK$ is stable. Recalling that $\bar{z} = \bar{x} - \bar{x}_m$, the proposed feedback controllers are given by

$$\bar{P}(\bar{x}, t) = K\bar{z} - \bar{g}(\bar{x}, t) + \bar{\theta}^f(t) + \bar{P}_\varepsilon(\bar{x}, t). \quad (4.6.21)$$

Here $\bar{P}_\varepsilon(\cdot)$ is a function that is continuous, bounded in $\bar{x}$, and measurable in t, and such that $\|\bar{P}_\varepsilon(\bar{x}, t)\| < \rho(\bar{x}, t)$, defined by

$$\bar{P}_\varepsilon(\bar{x}, t) = -\rho(\bar{x}, t)\|\eta(\bar{x}, t)\|^{-1}\eta(\bar{x}, t), \quad \|\eta(\bar{x}, t)\| > \varepsilon \quad (4.6.22)$$

for any $\varepsilon > 0$, with $\eta(\bar{x}, t) = B^T P \bar{z} \rho(\bar{x}, t)$. In the latter P is a solution of the Liapunov matrix equation in $\tilde{A}$ (see Example 3.2.2) while

$$\rho(\bar{x}, t) \ge \rho^0(\bar{x}, t) \triangleq \max_{\bar{r}}(1 - \|E\|)^{-1}\|h + E[K\bar{z} - \bar{g}(\bar{x}, t) + \bar{\theta}^f(t)]\|,$$

for any E, h satisfying A2 and any θ^f satisfying A3.

As a particular case satisfying the above, the authors give

$$\bar{P}_\varepsilon(\bar{x}, t) = \begin{cases} \rho^0(\bar{x}, t)\|\eta(\bar{x}, t)\|^{-1}\eta(x, t), & \text{for } \|\eta(\bar{x}, t)\| > \varepsilon \\ -\rho^0(\bar{x}, t)\varepsilon^{-1}\eta(\bar{x}, t), & \text{for } \|\eta(\bar{x}, t)\| \le \varepsilon. \end{cases} \quad (4.6.23)$$

Under such controllers it is proved that $\bar{x}(t)$ tracks $\bar{x}_m(t)$ in the following sense:

(a) given $\delta \ge 0$ there is $d(\delta) \ge 0$ such that $\|\bar{z}^0\| < \delta$ implies $\|\bar{z}(t)\| \le d$ for all $t \ge t_0$ along any motion of (4.6.17);

(b) there is a neighborhood $\mathscr{B}_0$ of the origin in R^N such that $\bar{z}^0 \in \mathscr{B}_0$ implies $\bar{z}(t) \in \mathscr{B}$, $t \ge t_0$, with $\mathscr{B}$ a stipulated bounded set, along motions of (4.6.17);

(c) for each $\delta \ge 0$ there is $T \ge 0$ such that $\|\bar{z}^0\| \le \delta$ implies $\bar{z}(t) \in \mathscr{B}$, $t \ge t_0 + T$, along motions of (4.6.17).

Example 4.6.2. Corless, Leitmann and Ryan [1] offer the following illustration of their method. Consider the two-DOF horizontal manipulator

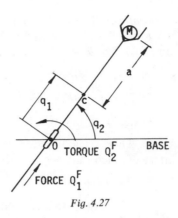

Fig. 4.27

shown in Fig. 4.27. The arm can be rotated about the vertical axis and extended, as shown. The joint coordinates q_1, q_2 specify the position of the center of mass C and the load is an uncertain mass M bounded between the constants $\underline{M}$ and $\bar{M}$, while m is the mass of the arm. The motion equations are

$$(M + m)\ddot{q}_1 - Q^P(q_1, M)\dot{q}_2^2 = Q_1^F,$$
$$I(q_1, M)\ddot{q}_2 + 2Q^P(q_1, M)\dot{q}_1\dot{q}_2 = Q_2^F, \quad (4.6.24)$$

where $Q^P = mq_1 + M(q_1 + a)$, $I = J_1 + J_2 + mq_1^2 + M(q_1 + a)^2$, with J_1, J_2 moments of inertia about 0 and C, respectively. The equations (4.6.24) become in state form

$$\dot{x}_1 = x_3,$$
$$\dot{x}_2 = x_4,$$
$$\dot{x}_3 = (m + M)^{-1} Q^P(x_1, M) x_4^2 + (m + M)^{-1} Q_1^F,$$
$$\dot{x}_4 = -[I(x_1, M)]^{-1} 2Q^P(x_1, M) x_3 x_4 + [I(x_1, M)]^{-1} Q_2^F.$$

Assuming the nominal load mass $M_0 = \frac{1}{2}(\underline{M} + \bar{M})$ and letting $\bar{u} = (u_1, u_2)^T = ((m + M_0)^{-1} Q_1^F, [I(x_1, M^0)]^{-1} Q_2^F)^T$, we obtain

$$\dot{x}_1 = x_3,$$
$$\dot{x}_2 = x_4,$$
$$\dot{x}_3 = (m + M)^{-1} Q^P(x_1, M) x_4^2 + (m + M)^{-1}(m + M_0) u_1, \quad (4.6.25)$$
$$\dot{x}_4 = -[I(x_1, M)]^{-1} 2Q^P(x_1, M) x_3 x_4 + [I(x_1, M)]^{-1} I(x_1, M_0) u_2.$$

Moreover, let us specify $\bar{x}_m = ((s - a), \theta, \dot{s}, \dot{\theta})^T$ in polar coordinates $s(t), \theta(t)$. In

4.6 PLANNED PATH TRACKING

terms of (4.6.17) the above means

$$f_1 = x_3; \quad f_2 = x_4; \quad f_3 = (m + M_0)^{-1} Q^P(x_1, M_0) x_4^2;$$
$$f_4 = -[I(x_1, M_0)]^{-1} 2 Q^P(x_1, M_0) x_3 x_4;$$
$$(\delta f)_1 = 0; \quad (\delta f)_2 = 0; \quad (\delta f)_3 = (m + M)^{-1}(m + M_0)^{-1}(M - M_0) a m x_4^2;$$
$$(\delta f)_4 = [I(x_1, M)]^{-1}[I(x_1, M_0)]^{-1} 2(M - M_0)(a m x_1 - J_1 - J_2)(x_1 + a) x_3 x_4;$$

$$B = \begin{bmatrix} 0^{2 \times 2} \\ I^{2 \times 2} \end{bmatrix}; \quad 0^{2 \times 2} = \begin{bmatrix} 0 & 0 \\ 0 & 0 \end{bmatrix}; \quad I^{2 \times 2} = \begin{bmatrix} 1 & 0 \\ 0 & 1 \end{bmatrix};$$

$$\delta B = \begin{bmatrix} 0^{2 \times 2} \\ E_1(\bar{x}, t, M) & 0 \\ 0 & E_2(\bar{x}, t, M) \end{bmatrix};$$

$$E_1 = -(m + M)^{-1}(M - M_0); \quad E_2 = -[I(x_1, M)]^{-1}(M - M_0)(x_1 + a)^2.$$

To demonstrate that assumption A1 holds, let

$$A = \begin{bmatrix} 0^{2 \times 2} & I^{2 \times 2} \\ 0^{2 \times 2} & 0^{2 \times 2} \end{bmatrix}, \quad \bar{g}(x, t) = (f_3, f_4)^T.$$

Assumption A2 is shown to hold by taking

$$h = ((\delta f)_3, (\delta f)_4)^T, \quad E = \begin{bmatrix} E_1 & 0 \\ 0 & E_2 \end{bmatrix},$$

for which it is readily seen that for all arguments

$$\|E(\bar{x}, t, M)\| \le \beta = \tfrac{1}{2} [\underline{M}(J_1 + J_2 + m a^2)$$
$$+ m(J_1 + J_2)]^{-1} (\bar{M} - \underline{M})(J_1 + J_2 + m a^2).$$

Thus if $(\bar{M} - \underline{M})$ is sufficiently small, then $\beta < 1$ and A2 is satisfied. In assumption A3 we let $\theta^f = (\ddot{s}, \ddot{\theta})^T$ almost everywhere. Now we select $K = [-k^2 I^{2 \times 2}$ and $2k I^{2 \times 2}]$, $k < 0$, wherefrom

$$\tilde{A} = A + BK = \begin{bmatrix} 0^{2 \times 2} & I^{2 \times 2} \\ -k^2 I^{2 \times 2} & 2k I^{2 \times 2} \end{bmatrix}$$

is stable. Choosing

$$Q = \begin{bmatrix} k^2 I^{2 \times 2} & 0^{2 \times 2} \\ 0^{2 \times 2} & I^{2 \times 2} \end{bmatrix},$$

and solving the corresponding Liapunov matrix equation give

$$P = \frac{1}{2} \begin{bmatrix} -3k I^{2 \times 2} & I^{2 \times 2} \\ I^{2 \times 2} & -k^{-1} I^{2 \times 2} \end{bmatrix}.$$

We may now apply (4.6.21) with

$$\rho(\bar{x}, t) = (1 - \beta)^{-1}[\rho_1(\bar{x}, t) + \beta\|K\bar{z}\| + \gamma], \qquad \gamma \geq 0,$$

$$\rho_1(\bar{x}, t) = \tfrac{1}{2}(\bar{M} - \underline{M})[(m + \underline{M})^{-1}|\ddot{s}(t) - (x_1 + a)x_4^2|$$

$$+ [J_1 + J_2 + mx_1^2 + \underline{M}(x_1 + a)^2]^{-1}$$

$$\cdot|\ddot{\theta}(t)(x_1 + a)^2 + 2(x_1 + a)x_3 x_4|].$$

Given the Cartesian planned path as (see Fig. 4.28)

$$\xi_i(t) = \begin{cases} \xi_i^0, & t < t_0 \\ \xi_i^0 + \mu_i(t - t_0)^3, & t \in [t_0, t_0 + \tfrac{1}{4}T] \\ \xi_i^0 + \mu_1[(t - t_0)^3 - 2(t - t_0 - \tfrac{1}{4}T)^3], & t \in [t_0 + \tfrac{1}{4}T, t_0 + \tfrac{3}{4}T] \\ \xi_i^0 + \mu_i[(t - t_0)^3 - 2(t - t_0 - \tfrac{1}{4}T)^3 + 2(t - t_0 - \tfrac{3}{4}T)^3], \\ \qquad\qquad t \in [t_0 + \tfrac{3}{4}T, t_0 + T] \\ \xi_i^f, & t \geq t_0 + T, \end{cases}$$

where $\mu_i = 16(\xi_i^f - \xi_i^0)/3T^3$, $i = 1, 2$, and the numerical data $T = 5$ sec, $(\xi_1^0, \xi_2^0) = (2, 0)$ m, $(\xi_1^f, \xi_2^f) = (1, 2)$ m, $\varepsilon = 0$, $\gamma = 0.01$, $k = -1$, $m = 100$ kg, $\bar{M} = 100$ kg, $\underline{M} = 0$, $J_1 = J_2 = 100$ kg m^2, $a = 1$ m, we have the tracking shown in Fig. 4.28. ∎

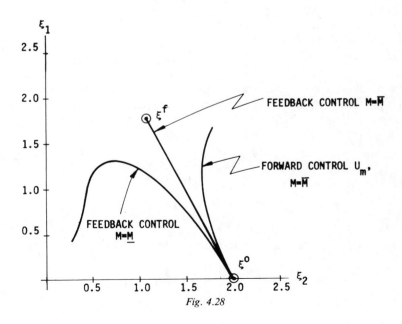

Fig. 4.28

4.7 MODEL REFERENCE ADAPTIVE CONTROL

As mentioned several times already, high-speed and high-accuracy demands on manipulator work require robustness of control against variations of system parameters and cross coupling. This is achieved by adapting the control program $\mathscr{P}(\bar{x}, \bar{\lambda}, t)$, that is, by so-called signal synthesis adaptation. This adaptation may, however, not suffice in situations which call for adapting the system structure (*self-organization*) to cover a wide range of changes in parameter values. This is the case when path following is replaced by *dynamic model following*, which has recently begun to dominate research in design and control of manipulators. Perhaps a most spectacular if not most frequent application of model following is for the case in which the manipulator works in a "hot cell" and must be controlled remotely. A typical case here is the so-called master–slave manipulator of relatively common use in many laboratories throughout the world.

Model following is implemented by using mainly the model reference adaptive control (MRAC) technique (see Dubovsky and DesForges [1], Stoten [1], Balestrino, DeMaria, and Sciavicco [1], and Tomizuka and Horowitz [1]). This technique seems to work for manipulators as it has worked for a number of other control applications in the past (see Winsor and Roy [1] and Donaldson and Leondes [1]) such as autopilots in the sixties (see Narendra and Tripathi [1]), particularly VTOL. There are, however, some difficulties. As we know, the manipulator is a strongly nonlinear and multiply coupled system, very often having several equilibria, while the MRAC technique, ever since its introduction by Butchard and Shakcloth [1] in 1966, does not accommodate nonlinearity; for a very good review see Landau [2]. A nonlinear extension of MRAC is needed (see Choe and Nikiforuk [1]).

Moreover, for technical reasons the reference model should be as simple as possible (see Dubovsky and DesForges [1]). This may be done by assuming the model to be linear and killing the manipulator nonlinearity outright by a powerful controller (see Balestrino, DeMaria, and Sciavicco [1]. The disadvantage of such a method is that a linear model with its single equilibrium may be incompatible with the multiequilibria manipulator to the extent of rendering the tracking ineffective. Simply stated, one may not have a controller powerful enough to jump over all the energy thresholds involved in the itinerary. But even if we have enough power, using it in excess all the time is obviously penalizing in various terms, beginning with the economic. Both these shortcomings are avoided when using the representation of an arbitrary nonlinear model and manipulator as the product system in the Cartesian product of the corresponding state spaces and making the product motions convergent to a diagonal set in such a product space (see Skowronski [18, 26–28]). The diagonal contains the equilibria of both the model and the manipulator.

It seems useful to begin with a brief outline of the linear technique. Let the plant be described by the linear version of (1.5.3):

$$\dot{\bar{x}} = A(t)\bar{x} + B(t)\bar{u} \qquad (4.7.1)$$

where A and B are the system and input matrices of appropriate dimensions. We let the reference model be given as

$$\dot{\bar{x}}_m = A_m \bar{x}_m + B_m \bar{u}_m, \qquad (4.7.2)$$

where $\bar{x}_m(t) \in \Delta$ is the model state vector, $\bar{u}_m(t) \in U$ the model control vector, while A_m and B_m are constant matrices of the same dimensions as $A(t)$ and $B(t)$. The control $\bar{u}_m$ is assumed to have been selected already to secure some desired behavior of the model.

We denote the state "error" by $\bar{e}(t) \triangleq \bar{x}(t) - x_m(t)$ for all $t \geq t_0$ and the parameter "errors" by $A_e(t) \triangleq A(t) - A_m$, $B_e(t) \triangleq B(t) - B_m$, with $A(t), B(t)$ adjustable (see Fig. 4.29). Subtracting (4.7.2) from (4.7.1) gives the so called *error equation*

$$\dot{\bar{e}} = A_m \bar{e} + A_e(t)\bar{x} + B_e(t)\bar{u}. \qquad (4.7.3)$$

To concentrate on tracking rather than other fringe objectives, we assume that the control vector $\bar{u}$ has already been selected to secure some objective which is not incompatible with tracking. The plant system (4.7.1) follows the model, or more exactly in this linear version, the plant converges asymptotically to the model if

$$\bar{e}(t) \to 0, \qquad A_e(t) \to 0, \qquad B_e(t) \to 0 \quad \text{as} \quad t \to \infty, \qquad (4.7.4)$$

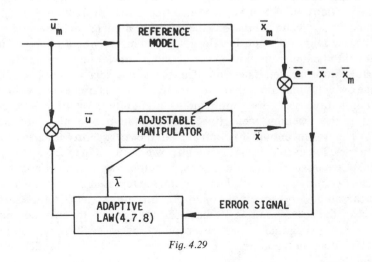

Fig. 4.29

4.7 MODEL REFERENCE ADAPTIVE CONTROL

which thus becomes our present tracking objective (see Fig. 4.30). What is needed is the uniform asymptotic stability of the zero error $\bar{e}(t) \equiv 0$, which by Theorem 3.2.3 is attained when we have a Liapunov function that is a square form and possesses a negative-definite derivative. The obvious candidate is the traditional square from

$$V(e, A_e, B_e) = \bar{e}^T P \bar{e} + \sum_{i=1}^{N} \sum_{j=1}^{N} a_{eij}^2 + \sum_{i=1}^{N} \sum_{k=1}^{n} b_{eik}^2, \qquad (4.7.5)$$

where $P = (p_{ij})$ is a positive definite symmetric matrix and a_{eij}, b_{eik} are components of $A_e(t)$, $B_e(t)$, respectively. The only condition to check is (3.2.4), that is, the negative derivative, the other two being obviously satisfied. Differentiating (4.7.5) and substituting (4.7.3), we have

$$\dot{V}(t) = (A_m \bar{e} + A_m \bar{x} + B_e \bar{u})^T P \bar{e} + \bar{e}^T P (A_m \bar{e} + A_e \bar{x} + B_e \bar{u})$$
$$+ 2 \sum_i \sum_j a_{eij} \cdot \dot{a}_{eij} + 2 \sum_i \sum_k b_{eik} \cdot \dot{b}_{eik},$$

or

$$\dot{V}(t) = \bar{e}^T (A_m^T P + P A_m) \bar{e} + \bar{x}^T A_e^T (P + P^T) \bar{e} + u^T B_e^T (P + P^T) \bar{e}$$
$$+ 2 \sum_i \sum_j a_{eij} \dot{a}_{eij} + 2 \sum_i \sum_k b_{eik} \dot{b}_{eik}. \qquad (4.7.6)$$

The negative-definiteness of the first term on the right-hand side of (4.7.6) follows immediately from the Liapunov matrix equation (see Example 3.2.2),

$$A_m^T P + P A_m = -Q, \qquad (4.7.7)$$

where Q is a positive-definite matrix. However, Q need not be determined if a stable reference model is chosen. Granted the above, it suffices to reduce the other terms of (4.7.6) to zero. Simple calculation shows that the latter happens

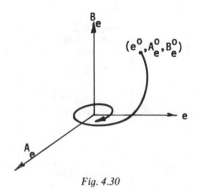

Fig. 4.30

if the *adaptive laws* hold

$$\dot{a}_{eij} = -\frac{1}{2} x_j \sum_{\sigma=1}^{N} (p_{i\sigma} + p_{\sigma i}) e_\sigma,$$
$$\dot{b}_{eik} = -\frac{1}{2} u_k \sum_{\sigma=1}^{N} (p_{i\sigma} + p_{\sigma i}) e_\sigma.$$

(4.7.8)

Note that since A_m, B_m are constant, we have $\dot{a}_{eij} = \dot{a}_{ij}$, $\dot{b}_{eik} = \dot{b}_{ik}$, where a_{ij}, b_{ik} are coefficients of $A(t), B(t)$.

Passing over now to nonlinear tracking, let us take (3.3.1), represented by the selecting equation (1.5.3″), but with the vector of adjustable parameters $\bar{\lambda}(t)$ *active* (see Section 1.3):

$$\dot{\bar{x}} = \bar{f}(\bar{x}, t, \bar{u}, \bar{w}, \bar{\lambda}).$$ (4.7.9)

The control vector $\bar{u}$ is to be selected by a program $\mathscr{P}(\bar{x}, t, \bar{\lambda})$ for some task different from tracking, which, however, must be tracking-compatible. To focus attention on tracking, we choose a simple task of stabilization in-the-large (see Section 3.4). Further, we let the adjustable vector $\bar{\lambda}(t)$ be a solution $\lambda(\bar{\lambda}^0, t_0, \cdot) \colon R_0^+ \to \Lambda$ of some *adaptive law*

$$\dot{\bar{\lambda}} = \bar{f}_a(\bar{x}, \bar{x}_m, t, \bar{\lambda})$$ (4.7.10)

to be designed such as to generate tracking in cooperation with the controller (4.7.8) (see Fig. 4.31).

The reference model is taken as the system

$$\dot{\bar{x}}_m = \bar{f}_m(\bar{x}_m, t, \bar{u}_m, \bar{\lambda}_m)$$ (4.7.11)

without uncertainty and with a given constant vector of parameters $\bar{\lambda}_m \in \Lambda \subset R^l$, within the same band Λ as for the adjustable $\bar{\lambda}(t)$. The control vector $\bar{u}_m(t), t \geq t_0$, is chosen by a program that makes (4.7.11) perform a specific task in state space for our convenience within the scope of Liapunov theory, that is,

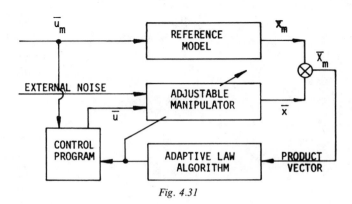

Fig. 4.31

4.7 MODEL REFERENCE ADAPTIVE CONTROL

conclude

$$\dot{V}_\eta(\bar{X}_m(t), \bar{y}(t)) \leq -c$$

which satisfies condition (v), thus closing our investigation in general terms. We now pass over to our specific shapes of (4.7.9) and consider (1.3.15″) or (1.3.24″). The same conclusions can be drawn for (1.5.16‴). Let us write the motion equations in terms of $\bar{\lambda}$ as

$$\sum_j a_{ij}(\bar{q}, \bar{r})\ddot{q}_j + K_i(\bar{q}, \dot{\bar{q}}, \bar{r}) - Q_i^{DD}(\bar{q}, \dot{\bar{q}}, \bar{\lambda}) - Q_i^P(\bar{q}, \bar{r})$$
$$= Q_i^F(\bar{q}, \dot{\bar{q}}, u_i) + Q_i^r(\bar{q}, \dot{\bar{q}}, t), \qquad (4.7.27)$$

or

$$\ddot{q}_i + \Gamma_i(\bar{q}, \dot{\bar{q}}, \bar{r}) + D_i^D(\bar{q}, \dot{\bar{q}}, \bar{r}, \bar{\lambda}) + \Pi_i(\bar{q}, \bar{r}, \bar{\lambda})$$
$$= F_i(\bar{q}, \dot{\bar{q}}, u_i, \bar{r}) + R_i^r(\bar{q}, \dot{\bar{q}}, \bar{r}, t). \qquad (4.7.28)$$

We then choose the model (4.7.11) to be in the simple, inertially decoupled form

$$\ddot{q}_{mi} - D_{mi}^D(\bar{q}_m, \dot{\bar{q}}_m, \bar{\lambda}_m) - \Pi_{mi}(\bar{q}_m, \bar{\lambda}_m) = \bar{u}_m. \qquad (4.7.29)$$

The total energy of the model is thus

$$H_m(\bar{q}_m, \dot{\bar{q}}_m) = \frac{1}{2}\sum_i \dot{q}_{mi}^2 + \sum_i \int_{q_{mi}^0}^{q_{mi}} Q_{mi}^P(\bar{q}_m)\, dq_{mi} \qquad (4.7.30)$$

and the power is

$$\dot{H}_m(\bar{q}_m, \dot{\bar{q}}_m) = \sum_i [u_{mi} + Q_{mi}^D(\bar{q}_m, \dot{\bar{q}}_m, \bar{\lambda}_m)]\dot{q}_{mi}. \qquad (4.7.31)$$

For the uniform asymptotic stability of $\bar{x}_m(t) \equiv 0$ required, we have $V_m(\bar{x}_m) \equiv H_m(\bar{q}_m, \dot{\bar{q}}_m)$, which is monotone increasing within the basic cup Δ_H. The control $u_m(t)$ must be chosen such as to provide (4.7.19), that is [see (3.4.13)], $\Sigma_i(u_{mi}\dot{q}_i - D_{mi}^D\dot{q}_{mi}) \leq -c$, $\dot{q}_i \neq 0$. By (2.3.3), $-D_{mi}^D\dot{q}_{mi} < 0$ and is decreasing, which means that the condition holds without control: $\bar{u}_m(t) \equiv 0$. This is obvious since without control the model is a monotone dissipative autonomous free system. The manipulator controller is then found from the control condition (4.7.25) with (4.7.28) substituted. As Γ_i, $i = 1, \ldots, n$, generate no power, we have

$$\min_{\bar{u}} \max_{\bar{r}} \sum_i [F_i(\bar{q}, \dot{\bar{q}}, u_i, \bar{r})\dot{q}_i + R_i^r(\bar{q}, \dot{\bar{q}}, \bar{r}, t)\dot{q}_i - D_i^D(\bar{q}, \dot{\bar{q}}, \bar{r}, \bar{\lambda})\dot{q}_i$$
$$+ \Pi_{mi}(\bar{q})\dot{q}_i - \Pi_i(\bar{q}, \bar{r})\dot{q}_i] \leq -\sum_i D_{mi}^D(\bar{q}_m, \dot{\bar{q}}_m, \lambda) \qquad (4.7.32)$$

for $\dot{q} \neq 0$, $i = 1, \ldots, n$, and on $\mathscr{C}^+\mathscr{M}_\eta$. The same formula but subject to $\max_{\bar{u}} \min_{\bar{w}}$ applies on $\mathscr{C}^-\mathscr{M}_\eta$. Hence in order to satisfy (4.7.32) it suffices to choose $\bar{u}$ satisfying on $\mathscr{C}^+\mathscr{M}_\eta$

$$\min_{\bar{u}} \max_{\bar{r}} \sum_i [F_i(\bar{q},\dot{\bar{q}},u_i,\bar{r})\dot{q}_i + R_i^r(\bar{q},\dot{\bar{q}},\bar{r},t)\dot{q}_i - D_i^D(\bar{q},\dot{\bar{q}},\bar{r},\bar{\lambda})q_i]$$
$$= -\sum_i D_{mi}^D(\bar{q}_{mi},\dot{\bar{q}}_m)\dot{q}_{mi} \tag{4.7.33}$$

subject to the *gravity–spring compensation* condition

$$\bar{\Pi}_m(\bar{q})\dot{\bar{q}} \leq \bar{\Pi}(\bar{q},\bar{r})\dot{\bar{q}} \quad \forall \bar{r}, \tag{4.7.34}$$

meaning that for sufficiently large t the power of potential forces of the manipulator must majorize such power of the model. Similarly on $\mathscr{C}^-\mathscr{M}_\eta$ we have (4.7.33) with $\min_{\bar{u}} \max_{\bar{w}}$ replaced by $\max_{\bar{u}} \min_{\bar{w}}$, and (4.7.34) with the inequality inverted. As $\Pi_m(\cdot)$ is given and bounded on bounded Δ_q, a suitable choice of masses and/or springs in designing the manipulator may secure (4.7.34), for all $\bar{r} \in \mathfrak{R}_1$. Here again (4.7.33) is as close to $\bar{u}$ as we can get without specifying the transmission characteristics F_i and at least to some extent the perturbations R_i^r.

On the other hand, (4.7.33) and its $\mathscr{C}^-\mathscr{M}_\eta$ counterpart are unfortunately not the only conditions imposed on $\bar{u}$. In order for the manipulator to be compatible with the model regarding equilibria, assuming that damping and perturbations vanish there [see (2.3.4), (2.3.9)], we must have

$$F_i(\bar{q}^e,\dot{\bar{q}}^e,u_i,\bar{r}) - \Pi_i(\bar{q}^e,\bar{r}) = u_{mi} - \Pi_{mi}(\bar{q}_m^e)$$

or

$$F_i(\bar{q}^e,\dot{\bar{q}}^e,u_i,\bar{r}) = u_{mi} + [\Pi_i(\bar{q}^e,\bar{r}) - \Pi_{mi}(\bar{q}_m^e) \quad \forall \bar{r}, \tag{4.7.35}$$

which complements the pair (4.7.33), (4.7.34) as the third control condition constraining $\bar{u}$. Observe, however, that on substituting $\bar{q}^e$, $\dot{\bar{q}}^e$ and (4.7.35) into (4.7.33) one obtains a single condition on $\bar{\Pi}_m$, $\bar{\Pi}$ which complements (4.7.34) and may be satisfied by a suitable spring design. It is granted that we may determine $\bar{u}$ from (4.7.33) alone.

The adaptive laws are immediately obtainable from (4.7.31) by substituting $\lambda_{n+i},\ldots,\lambda_N = \text{const}$, $u_{mi}(t) \equiv 0$, $i = 1,\ldots,n$, and $\dot{V}_m \bar{x}_m) \equiv \dot{H}(\bar{q}_m,\dot{\bar{q}}_m)$ into (4.7.22).

Example 4.7.1. In order to illustrate our discussion numerically, we consider again the RP manipulator of Section 1.1. Setting up the model as in (4.7.29),

$$D_{mi}^D = \lambda_{mi}\dot{q}_{mi}, \quad i = 1,2,$$
$$\Pi_{mi}(\bar{q}_m) = \Pi_i(\bar{q}_m), \quad i = 1,2 \quad \text{(same function)},$$
$$u_{mi} \equiv 0, \quad i = 1,2.$$

Recalling the calculations in Example 3.2.3 and the augmentation in Example 3.4.1, where m_2 was made uncertain, we see that the control program $\mathscr{P}(\bar{x},\bar{\lambda})$ is

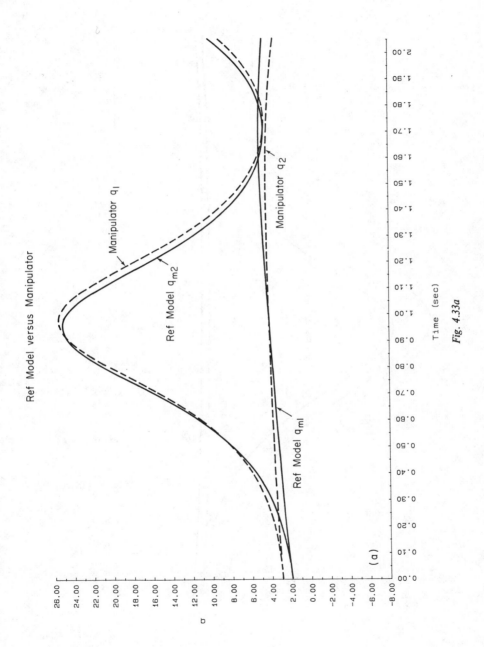

Fig. 4.33a

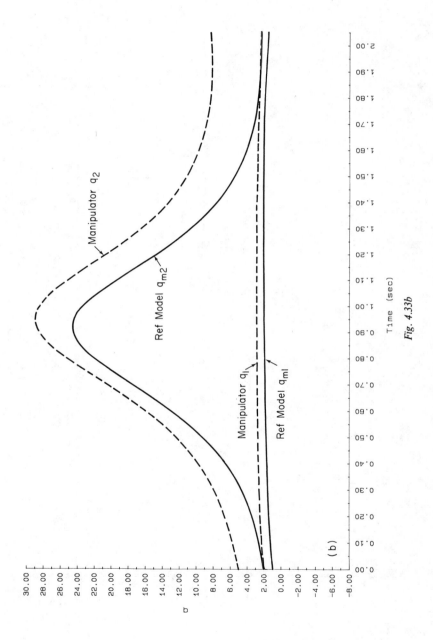

Fig. 4.33b

4.7 MODEL REFERENCE ADAPTIVE CONTROL

obtainable directly from (3.4.20). Letting $\varepsilon = 1$, we have $u_i = Q_i^r/B_i$, recognizing our tacit assumption that r_i specified by m_2 is the only uncertainty. The adaptive laws are

$$\dot{\gamma}_i = -\lambda_{mi} q_{mi}^2/(\text{sign } \gamma_i), \qquad i = 1, 2.$$

Quite reasonable convergence curves for $q_i \to q_{mi}$, $i = 1, 2$, are obtained by substituting $m_1 = 70$ kg, $m_2 = 36 - 40$ kg, $r_1 = 0.6$ m, $a = 20, b = 3, \lambda_{m1} = 5$, $\lambda_{m2} = 2, \gamma_1^0 = 60, \gamma_2^0 = 40$ (see Fig. 4.33a,b) for two different sets of initial states wherefrom the model and manipulator started. In Fig. 4.33b the convergence is visible after a longer time than it is in Fig. 4.33a. ∎

Chapter 5

Avoidance of Obstacles

5.1 CONDITIONS FOR AVOIDANCE

The research into the avoidance of obstacles by nonvisual means was concentrated for a considerable time on the geometry of both the obstacles and the gripper motion in configuration space. Lozano and Perez [1, 2] give a thorough mathematical treatment of configuration space methods and possibilities (see also Lozano, Perez, and Wesley [1] and Reibert and Horn [1]). The idea is to determine those parts of free space which a reference point of the gripper may occupy without collision. Rotation seems to be the greatest problem. Moravec [1] solved the problem partially for the two-dimensional work space. The so-called method of cones (of free approach) was introduced by Brooks [1, 2], and some algorithms are provided by Calm and Phillips [1] and Okhocimski and Platonov [1]. A relatively recent review of work in this group was edited by Brady, Hollerbach, and Johnson [1]. However, as the number of applications grows and thus the case studies multiply with the increasing variety of situations involved, this avenue of study, which by definition must refer to more or less particular cases, gets saturated with details and thus becomes less attractive. Methods that are less expensive than vision or overused sensing but that are generally applicable must be produced. We believe that this is the reason that research work is being brought more and more into the field of adaptive control (see Takegaki and Arimoto [1, 2], Timofeev [2], Tkachenko, Brovinskaya, and Kondratenko [1], and Udupa [1]) and toward the new trend which may be called the *push-off-potential method* (see Khatib and LeMaitre [1], Haass, Kuntze, and Schill [1], and Kuntze and Schill [1]), or a *safety zone method* (see Petrov and Sirota [1]). These two methods open the gate to using the wealth of control theory avoidance results based on *application of the Liapunov formalism* and the like (see Blaquière, Gerard, and Leitmann [1], Aggarwall and Leitmann [1], Leitmann and Skowronski [1, 2], Leitmann [4, 6], Leitmann and Liu [1], Kaskosz [1, 2], Chikriy [1, 2], Ostapenko [1], and Krogh [1, 2]) and/or *dynamic games*, (see Krassovski and Subbotin [1], Hajek [1], Hagedorn and

5.1 CONDITIONS FOR AVOIDANCE

Breakwell [1,2], Shinar and Gutman [1], Gutman [1,2], Foley and Schmitendorf [1], Merz [1,2], Olsder and Walter [1], Olsder and Breakwell [1], Paskov [1], Breakwell [1,2], Peng and Vincent [1], Vincent [1,2], Pozaritskii [1,2], and Pshenichnyi [1]) to name only a few results that are relevant to our following discussion.

We have discussed in Chapter 1 the models of objects and obstacles in Cartesian space and their transformation to Δ of state space (see Section 1.4). Here it will be assumed that the obstacles, either stationary or moving, are given as what we shall call *avoidance sets* in Δ. Note that we must avoid not only configurational obstacles in Δ_q but also some velocities. The conventional mechanical end-point switches at joints cannot reliably protect the robot unless the actual joint velocities are taken into account. The operation of the emergency brakes may occur either too early or too late.

Obviously it is best to be able to exclude the obstacles from Δ. This may, however, not be feasible for instance when Δ is not simply connected (see Section 1.4) or when the obstacles are moving (see Section 5.5). In all such cases we must use the controller to avoid them and this is our topic in this chapter.

Let $\mathscr{A}$ be a closed subset of Δ, not necessarily connected, termed the avoidance set containing (or equal to) the union of all grown obstacles in Δ, that is, the set into which for some $\mathscr{P}(\bar{x}, \bar{\lambda}, t)$ no motion of (3.3.1) must enter for any $\bar{w}$ (see Fig. 5.1). Then let Δ_ε be the closure of an open subset of Δ such that $\Delta_\varepsilon \supset \mathscr{A}$ and

$$\partial \Delta_\varepsilon \cap \partial \mathscr{A} \cap \mathring{\Delta} = \varnothing.$$

We term $\Delta_\mathscr{A} \triangleq \Delta_\varepsilon - \mathscr{A}$ the *safety zone*. This nomenclature is employed for the following reasons: If a motion avoids $\mathscr{A}$, then it cannot enter any of the grown obstacles (*antitargets*) and if a policy is implemented in $\Delta_\mathscr{A}$ that guarantees avoidance of $\mathscr{A}$, then a motion from outside of $\mathscr{A}$ cannot reach this set.

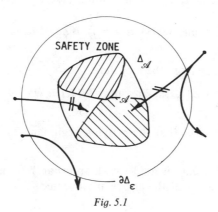

Fig. 5.1

The reader may now recall Definition 4.1.3 concerning avoidance and Definition 3.3.2 concerning strong controllability. Starting from anywhere outside Δ_ε, we want a control program $\mathscr{P}(\cdot)$ guaranteeing that, for any $\bar{w}$, that is, for any motion of $\mathscr{K}(\bar{x}^0, t_0)$, we have no contact with $\mathscr{A}$:

$$k(\bar{u}, \bar{x}^0, t_0, R_0^+) \cap (\mathscr{A} \times R) = \varnothing. \tag{5.1.1}$$

In terms of reaching (see Definition 4.1.1), this means that $T_\mathscr{R}$ fades away to infinity. The consequence of such a definition is that the corresponding region of strong controllability Δ_r is simply equal to $\Delta - \mathscr{A}$, so that the controllability is always *complete*.

Let us now introduce a function $V(\cdot): D_\mathscr{A} \to R$ with open $D_\mathscr{A} \supset \Delta_\mathscr{A}$ and

$$v_\mathrm{A}^+ \triangleq \sup V(\bar{x}) \,|\, \bar{x} \in \partial \mathscr{A}. \tag{5.1.2}$$

The following theorem is a slight modification of what was proved in Leitmann and Skowronski [1].

Theorem 5.1.1. *The system (3.3.1) is strongly controllable on $\Delta - \mathscr{A}$ for avoidance of $\mathscr{A}$ if there is $\Delta_\mathscr{A}$, $\mathscr{P}(\bar{x}, t)$ defined on $D_\mathscr{A} \times R$ and a C^1 function $V(\cdot): D_\mathscr{A} \to R$ such that*

(i) $V(\bar{x}) > v_\mathrm{A}^+, \forall \bar{x} \notin \mathscr{A}$
(ii) *for all $\bar{u} \in \mathscr{P}(\bar{x}, t)$,*

$$\nabla V(\bar{x}) \cdot \bar{f}(\bar{x}, t, \bar{u}, \bar{w}) \geq 0 \qquad \forall w(t) \in W. \tag{5.1.3}$$

Proof. Suppose some $k(\cdot) \in \mathscr{K}(\bar{x}^0, t_0)$, $(x^0, t_0) \notin \mathscr{A} \times R$, enters $\mathscr{A}$. Then there is a $t_1 > t_0$ such that $k(\bar{u}, \bar{x}^0, t_0, t_1) = x^1 \in \partial \mathscr{A}$ and, by (i), $V(\bar{x}^0) > V(\bar{x}^1)$ contradicting (ii).

Corollary 5.1.1. *Given $(\bar{x}^0, t_0) \in \Delta_\mathscr{A} \times R$, if there are $\tilde{\bar{u}}, \tilde{\bar{w}}$ such that*

$$\mathscr{L}(\bar{x}, t, \tilde{\bar{u}}, \tilde{\bar{w}}) = \max_{\bar{u}} \min_{\bar{w}} \mathscr{L}(\bar{x}, t, \bar{u}, \bar{w}) \geq 0, \tag{5.1.4}$$

then condition (ii) is met with $\bar{u} \in \mathscr{P}(\bar{x}, t)$, that is, we may deduce $\mathscr{P}(\cdot)$ from (5.1.4).

Proof follows directly from the fact that

$$\max_{\bar{u}} \min_{\bar{w}} \mathscr{L}(\bar{x}, t, \bar{u}, \bar{w}) \leq \max_{\bar{u}} \mathscr{L}(\bar{x}, t, \bar{u}, \bar{w}) \qquad \forall \bar{w} \in W.$$

Remark 5.1.1. Observe that the proving argument for Theorem 5.1.1 does not change if either

(a) the inequality in (i) is made weak and that in (ii) made strong or
(b) both inequalities of the above are inverted.

5.1 CONDITIONS FOR AVOIDANCE

This automatically applies to Corollary 5.1.1 as well, but with inverting the inequality (5.1.4) $\bar{u}$ becomes the minimizer again while $\bar{w}$ is back to its maximizing role as in reaching. We actually have done a similar step in Section 4.1 where we let $V(\bar{x}) = -\hat{H}(\bar{x})$ about an unstable equilibrium.

By now we hope the reader can see the energy interpretation of Theorem 5.1.1. First, note that without inverting the inequalities as in Remark 5.1.1, the noise $\bar{w}$ is now the minimizer, so to cancel its effect on $H(\bar{x}, w)$ we would have to use $V(\bar{x}) = +\hat{H}(\bar{x})$ for accumulation and $V(\bar{x}) = -\tilde{H}(\bar{x})$ for dissipation cases, the signs being opposite as they are in reaching. Suppose now that $\mathscr{A}$ is in a local cup of $\mathscr{H}$ and we descend toward it. To prevent collision we must brake, that is, insert a positive (accumulative) increment into the power balance $f_0(\bar{x}, t, \bar{u}, \bar{w}) \geq 0$ to make it "more positive" in order to offset the influence of $\bar{w}$. Since $\mathscr{L}(\cdot)$ is a field function on $\Delta_{\mathscr{A}}$, we must thus create an increase in the accumulative field in the safety zone that produces the braking. This is being made by an extra push-off potential (see Khatib and LeMaitre [1]). Such a potential is obviously a function of the position of the manipulator and thus requires sensing. The present push-off potential force algorithms are based on two methods of obstacle surveillance (see Kuntze and Schill [1]): *absolute* by fixed sensors mounted in the ceiling (see Paul [2]), or *relative* by sensor that are mobile with the arm (see Bejczy [2] and Stensloff [1]). Obviously the second method lends itself better to our purpose in that it requires a shorter sensing range and simpler algorithms based on direct measurement of the distance between the obstacle and the sensor. The power of the push-off force must increase inversely to the distance. Kuntze and Schill [1] give the following example of this potential field

$$\tilde{V}(\bar{x}) = \begin{cases} \dfrac{1}{d(\bar{x})} - \dfrac{1}{\varepsilon}, & \text{for} \quad d(\bar{x}) \leq \varepsilon, \\ 0 & \text{for} \quad d(\bar{x}) > \varepsilon \end{cases} \qquad (5.1.5)$$

on the safety zone, which results in the force $\nabla \tilde{V}(\bar{x})$. Here $d(\bar{x})$ is the distance and $\varepsilon > 0$ is a small constant.

Naturally, if $\mathscr{A}$ is located not in the cup but, say, over a threshold of energy, we consider $\mathscr{H}$, and to reach $\mathscr{A}$ the system must accumulate energy. So to prevent collision, our push-off becomes a push-down rather than the push-up of the previous case. We take $V(\bar{x}) = -\tilde{H}(\bar{x})$ and need an increment in outflux of the energy supplied by the push-off potential force. Nothing changes in the algorithms except the sign of the force.

While a desirable $\mathscr{A}$ is stipulated, its neighborhood may not be well determined enough to allow $\mathscr{P}(\cdot)$ and $V(\cdot)$ to be defined on it. Either $\partial \mathscr{A}$ may be unpredictable or $\Delta_{\mathscr{A}}$ may be disjoint. For instance, envision the collection of obstacles stretching indefinitely with Δ bounded. Then there will be a "leak" in $\Delta_{\mathscr{A}}$ at the intersection of some grown obstacle and $\partial \Delta$ if Δ is open.

In other cases there may be no room at all for introducing the safety zone other than the boundary contours of the obstacles. Alternatively, we may not want any safety zone, having inserted the push-off potential field in the boundaries of obstacles. This is the case, for instance, when the push-off potential is implemented as a repelling electromagnetic or magnetic field formed on the surfaces concerned to repel the joints which are magnetized with the opposite charge.

In either of the above cases, it may be convenient to narrow the sufficient conditions to the avoidance set itself, that is, to make $\partial'\mathscr{A} = \partial\mathscr{A} \cap \mathring{\Delta}$ the safety zone and avoid $\mathring{\mathscr{A}}$. This way we arrive at the concept of *rejection* defined already in Definition 4.1.4. Introducing $V(\cdot): \bar{\mathscr{A}} \to R$ with

$$v_A^- \triangleq \inf V(\bar{x}) \,|\, \bar{x} \in \partial\mathscr{A}, \qquad (5.1.6)$$

we may adjust Theorem 5.1.1 to the following:

Theorem 5.1.2. *The system (3.3.1) is strongly controllable for rejection from $\mathscr{A}$ if there is a $\mathscr{P}(\bar{x}, t)$ on $\bar{\mathscr{A}}$ and a C^1 function $V(\cdot): \bar{\mathscr{A}} \to R$ such that*

(i) $V(\bar{x}) < v_A^-$,
(ii) *for all* $\bar{u} \in \mathscr{P}(\bar{x}, t)$,

$$\nabla V(\bar{x}) \cdot \bar{f}(\bar{x}, t, \bar{u}, \bar{w}) \geq 0 \qquad (5.1.7)$$

for all $\bar{w}(t) \in W$.

Proof. Motions may enter $\mathscr{A}$ only through $\partial'\mathscr{A}$, thus we let $(\bar{x}^0, t_0) \in \partial'\mathscr{A} \times R$ and suppose there is a $t_1 > t_0$ such that $k(x^0, t_0, t_1) = x^1 \in \mathring{\mathscr{A}}$. Then by (i), we have $V(x^1) < V(x^0)$, contradicting (ii).

Remark 5.1.2. Note that Remark 5.1.1 applies here as well.

Corollary 5.1.2. *If $\mathscr{A}$ is defined by a V level*

$$\mathscr{A}: V(\bar{x}) \leq \text{const} = v_A^-, \qquad (5.1.8)$$

then condition (5.1.7) is both necessary and sufficient for strong controllability for rejection.

Sufficiency follows from Theorem 5.1.2. To prove necessity, let us contradict (5.1.7). Then there are $\tilde{\bar{u}}, \tilde{\bar{w}}$ generating $\dot{V}(t) < 0$ for $\partial'\mathscr{A} \times R$, which by continuity of $V(\cdot)$ means that the motions enter $\mathring{\mathscr{A}}$ (there exists an $\varepsilon > 0$ such that $V(x, t) < v_A^- \Rightarrow k(t + \varepsilon) \in \mathring{\mathscr{A}}$).

Corollary 5.1.2 implies that within the accuracy of the V choice, (5.1.7) determines the safe $\mathscr{A}$, that is, the safe envelope for obstacles: Given $\mathscr{P}(\cdot), V(\cdot)$,

$$\bar{\mathscr{A}} \triangleq \{\bar{x} \in \Delta \,|\, \nabla V(\bar{x}) \cdot \bar{f}(\bar{x}, t, \bar{u}, \bar{w}) \geq 0 \,\forall \bar{w} \in W\}. \qquad (5.1.9)$$

5.1 CONDITIONS FOR AVOIDANCE

The discussion of Section 4.4 that followed (4.4.18) applies here as well. We obviously may have a "better" $\mathscr{A}$ by a change of the Liapunov function. In this case, the smaller $\mathscr{A}$ is, the better. Obviously it is best if $\mathscr{A}$ "fits" into the union of obstacles exactly, but try to find a Liapunov function that smoothes all the corners involved in the obstacle geometry!

From Definition (5.1.9) it follows that for the same Liapunov function, an extra push-off potential added to the field enlarges $\bar{\mathscr{A}}$, while any decline in the power of the push-off potential force reduces the safety margin about the obstacles produced by $\mathring{\mathscr{A}}$. This obviously may be compensated for by the controller, which may apply the brake. It is in fact for the designer to decide whether he wishes the role of the controller or of the push-off field to be larger. The advantage of leaving the larger role with the controller is that the extra power may be used elsewhere as well.

Obviously, the repelling may refer equally well to $\Delta - \mathscr{A}$, and hence both $\mathring{\mathscr{A}}$ and $\overline{\Delta - \mathscr{A}}$ can be avoided at the same time. Generalizing, we let $\mathscr{A}_1, \ldots, \mathscr{A}_l$, $l < \infty$, be a sequence of subsets in Δ such that $\bigcap_\sigma \mathring{\mathscr{A}}_\sigma = \emptyset$, $\bigcup_\sigma \mathscr{A}_\sigma = \Delta$, $\sigma = 1, \ldots, l$, representing an avoidance archipelago (see Fig. 5.2). Then if we can find $V_\sigma(\cdot)$, $\sigma = 1, \ldots, l$, such that

$$\mathscr{A}_\sigma = \{\bar{x} \in \Delta \,|\, V_\sigma(\bar{x}) < v_{A\sigma}^-\} \tag{5.1.9'}$$

for some $v_{A1}^-, \ldots, v_{Al}^-$ and that (5.1.7) is satisfied on $\mathscr{B} = \bigcap_\sigma \bar{\mathscr{A}}_\sigma$, we may consider $\mathscr{B}$ the *repelling mesh* and each $\mathscr{A}_\sigma$ strongly invariant under $\mathscr{P}(\bar{x}, t)$. Consequently $\mathscr{B}$ is positively strongly invariant under $\mathscr{P}(\cdot)$ and we can use its shape to plan a work path of the system by applying suitable push-off potential forces. We leave it to the reader to interface such a repelling archipelago with $\mathscr{H}$.

Remark 5.1.3. Corollary 5.1.1 applies to repelling without change.

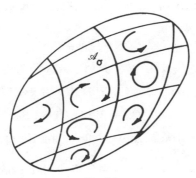

Fig. 5.2

For either avoiding or repelling, (5.1.4) gives the controller. To obtain it we may use the same technique as in the cases of stabilization, reaching, or capture, but with the inequality inverted and the roles of $\bar{u}$ and $\bar{w}$ interchanged. The general method leading to the control condition (4.4.17) has been illustrated before by examples both in Section 3.4 and in Chapter 4.

Let us now show how to use the L–G controller, obviously with the inverted sign. The following linear example encompasses a fair class of manipulator models still in use in both design and research. We borrow it from Leitmann and Skowronski [1].

Example 5.1.1. Consider the linear system

$$\dot{\bar{x}} = A\bar{x} + B\bar{u} + C\bar{w},$$

with A, B, C constant matrices with appropriate dimensions. Suppose that $-A$ is stable (that is, with negative real parts of its eigenvalues). Choose $V(\bar{x}) \triangleq \bar{x}^T P \bar{x}$, where P is positive-definite, symmetric, and comes from the Liapunov matrix equation (see Example 3.2.2). If $\mathscr{A} \triangleq \{\bar{x} \mid \bar{x}^T P \bar{x} \leq \text{const}$, condition (i) of Theorem 5.1.1 holds. Furthermore, if there is a matrix D such that $C = BD$ and if

$$U \triangleq \{\bar{u} \in R^n \mid \|\bar{u}\| < \rho_1 = \text{const} > 0\}$$

and

$$W \triangleq \{\bar{w} \in R^s \mid \|\bar{w}\| < \rho_2 = \text{const} > 0\},$$

with $\rho_1 \geq \|D\|\rho_2$, then condition (ii) of Theorem 5.1.1 is met and the corresponding avoidance control program is

$$\mathscr{P}(x) \triangleq (B^T P \bar{x} / \|B^T P \bar{x}\|)\rho_1 \qquad (5.1.10)$$

for all $\bar{x}$ such that $B^T P \bar{x} \neq 0$. For $B^T P \bar{x} = 0$, the control $\bar{u}$ may take any admissible value: $\mathscr{P}(\bar{x}) = U$, see (3.4.7).

The satisfaction of (i) follows at once, whereas that of (ii) is readily seen by considering

$$\mathscr{L}(\bar{x}, t, \bar{u}, \bar{w}) = 2\bar{x}^T P [A\bar{x} + B\bar{u} + C\bar{w}]$$
$$= \bar{x}^T (PA + A^T P)\bar{x} + 2\bar{x}^T P B \bar{u} + 2\bar{x}^T P B D \bar{w}$$
$$= -\bar{x}^T Q \bar{x} + 2[\bar{x}^T P B \bar{u} + \bar{x}^T P B D \bar{w}]$$
$$\geq -\bar{x}^T Q \bar{x} + 2[\bar{x}^T P B \bar{u} - \|B^T P \bar{x}\| \cdot \|D\| \rho_2],$$

where Q is the symmetric negative-definite matrix of the Liapunov matrix equation (see Example 3.2.2).

Suppose $-A$ is not stable, but $(-A, -B)$ is stabilizable, that is, there exists an E such that $(-A - BE)$ is stable, then the controller (5.1.10) can be replaced

5.1 CONDITIONS FOR AVOIDANCE

by

$$\mathscr{P}(\bar{x}) \triangleq E\bar{x} + (B^T P \bar{x}/\|B^T P \bar{x}\|)\|D\|\rho_2, \qquad (5.1.10')$$

and for this program to be admissible U must be such that $\mathscr{P}(\bar{x}) \subset U$ for all $(\bar{x}, t) \in \Delta_{\mathscr{A}} \times R$. In any event, $\mathscr{P}(\cdot)$ must be at least piecewise continuous.

In order to illustrate the case in more detail, let us specify further $A = \begin{pmatrix} 0 & 1 \\ 0 & 0 \end{pmatrix}$, $B = C = \begin{pmatrix} 0 \\ 1 \end{pmatrix}$ and $W: |\bar{w}| \leq 1$. Since $-A$ is not stable, but $(-A, -B)$ is stabilizable, we determine first the linear part of $\mathscr{P}(\cdot)$. This is readily deduced to be $E\bar{x}$ with, for instance, $E = (-1, 1)$. Then, if $Q = -\begin{pmatrix} 1 & 0 \\ 0 & 1 \end{pmatrix}$, the solution of the Liapunov matrix equation with A replaced by $A + BE$ is

$$P = \begin{pmatrix} \frac{3}{2} & -\frac{1}{2} \\ -\frac{1}{2} & 1 \end{pmatrix}$$

so that the nonlinear part of $\mathscr{P}(\cdot)$ is $\text{sign}(-\frac{1}{2}x_1 + x_2)$. Thus, provided $\mathscr{A} \triangleq \{\bar{x} \mid \bar{x}^T P \bar{x} \leq a = \text{const} > 0\}$, the avoidance control is given by

$$\mathscr{P}(\bar{x}) = -x_1 + x_2 + \text{sign}(-\tfrac{1}{2}x_1 + x_2). \qquad (5.1.10'')$$

∎

Example 5.1.2. Now take the R manipulator of Example 4.1.2 and form $\mathscr{A} = \{\bar{x} \mid H(\bar{x}) \leq 1\}$, and $\Delta_\varepsilon = \{\bar{x} \mid a - bx_1^2 \geq 0\} = \Delta_H$; whence $\Delta_{\mathscr{A}} = \Delta_H - \mathscr{A}$. Obviously, condition (i) of both Theorem 5.1.1 and 5.1.2 holds. Condition (ii) is satisfied by $u = [lx_2|x_2| \mp N/x_2]$, $l > 0$, as shown before, and thus we have either avoidance or rejection depending upon where the field $\dot{V} = lx_2^2|x_2| \geq 0$ (see Example 4.4.1) is defined. Observe also that the above accumulative field may be used to define the set $\mathscr{A}$, easily to cover the cup $\mathscr{A} = \Delta_H$ or far beyond that, depending on how powerful a controller we use or on how much push-off potential power will be inserted into the applied forces. ∎

The original version of sufficient conditions proposed in Leitmann and Skowronski [1] featured a nonstationary Liapunov function. Adopting it, we have

Theorem 5.1.1'. *The system (3.3.1) is strongly controllable on $\Delta - \mathscr{A}$ for avoidance of $\mathscr{A}$ if there is $\Delta_{\mathscr{A}}$, $\mathscr{P}(\bar{x}, t)$ defined on $\Delta_{\mathscr{A}} \times R$ and a C^1 function $V(\cdot): D_{\mathscr{A}} \times R \to R$ such that for all $(\bar{x}, t) \in \Delta_{\mathscr{A}} \times R$,*

  (i) $V(\bar{x}, t) > V(\bar{x}_a, \tau)$, $\bar{x}_a \in \partial \mathscr{A}$, $\tau > t$,
  (ii) *for all $\bar{u} \in \mathscr{P}(\bar{x}, t)$,*

$$\partial V/\partial t + \nabla V(\bar{x}, t) \cdot \bar{f}(\bar{x}, t, \bar{u}, \bar{w}) \geq 0 \qquad (5.1.11)$$

for all $\bar{w}(t) \in W$.

Proof. Suppose some $k(\cdot) \in \mathscr{K}(\bar{x}^0, t_0)$, $\bar{x}^0 \notin \mathscr{A}$, $t_0 \in R$, enters $\mathscr{A}$. Then there is $t_1 > t_0$ such that $k(\bar{u}, \bar{x}^0, t_0, t_1) = \bar{x}^1 \in \partial \mathscr{A}$ and, by (i), $V(x^0, t_0) > V(\bar{x}^1, t_1)$, contradicting (ii).

By the same argument as for Corollary 5.1.1, we immediately have

Corollary 5.1.1'. *Given* $(\bar{x}, t) \in \Delta_{\mathscr{A}} \times R$, *if there are* $\tilde{\bar{u}}, \tilde{w}$ *such that*

$$\mathscr{L}'(\bar{x}, t, \tilde{\bar{u}}, \tilde{w}) = \max_{\bar{u}} \min_{w} \mathscr{L}'(\bar{x}, t, \bar{u}, w) \geq 0, \tag{5.1.12}$$

then condition (ii) of Theorem 5.1.1' is met with $\bar{u} \in \mathscr{P}(\bar{x}, t)$, *that is, we may deduce* $\mathscr{P}(\cdot)$ *from (5.1.12)*.

The max–min and the L–G controllers deduced before now become particular cases. There is no difference in applying the technique in case $V(\cdot)$ is nonstationary, however, the stationary $V(\cdot)$ obviously better fits our energy interpretation at present. Later, however, we will need the nonstationary version for adaptive control.

From the last example, we see the advantage of defining $\Delta_{\mathscr{A}}$ by V levels. There is, however, an obvious difficulty in fitting the necessarily smooth V level into the boundary contour of an obstacle with many possible corners. We already mentioned this when discussing the size of $\mathscr{A}$ determined by (5.1.9). A way of solving the problem is to admit a V function which is only piecewise smooth. Following the work by Leitmann [4], we shall briefly discuss such a possibility.

Definition 5.1.1. A *denumerable decomposition* $\mathfrak{D}$ of a set $\Omega \subset R^N \times R$ is a denumerable collection of disjoint subsets whose union is Ω:

$$\mathfrak{D} \triangleq \{\Omega_\sigma \mid \sigma \in \Sigma\}, \tag{5.1.13}$$

where Σ is a denumerable index set of disjoint subsets.

Definition 5.1.2. A function $V(\cdot): \Omega \to R$ is said to be of class C^1 with respect to $\mathfrak{D}$ if and only if for each σ there is a pair $\mathscr{W}_\sigma, V_\sigma(\cdot)$ such that $\mathscr{W}_\sigma$ is an open set containing Ω_σ and $V(\cdot): \mathscr{W}_\sigma \to R$ is C^1 satisfying $V_\sigma(\bar{x}, t) = V(x, t)$ for all $(\bar{x}, t) \in \Omega_\sigma$.

The following lemma was proved by Stalford [1].

Lemma 5.1.3. *Let* $V(\bar{x}, t): \Omega \to R$ *be* C^1 *with respect to* $\mathfrak{D}$ *and* $\{\mathscr{W}_\sigma, V(\cdot) \mid \sigma \in \Sigma\}$ *be a collection of pairs associated with* $V(\cdot)$. *Consider* $k(\cdot): [t_0, t_1] \to \Omega$ *to be absolutely continuous and* $T_\sigma = \{t \in [t_0, t_1] \mid k(t) \in \Omega_\sigma\}$, $\sigma \in \Sigma$, *and suppose that for each* σ

$$\dot{V}(k(t), t) \geq 0 \tag{5.1.14}$$

almost everywhere on T_σ. *Then a function* $\tilde{V}(\cdot): [t_0, t_1] \to R$ *defined by* $\tilde{V}(t) = V(k(t), t)$ *is absolutely continuous and monotone nondecreasing.* Now we can adjust Theorem 5.1.1'.

Theorem 5.1.3. *The system (3.3.1) is strongly controllable on $\Delta - \mathscr{A}$ for avoidance of $\mathscr{A}$ if there are $\Delta_{\mathscr{A}} \supset \mathscr{A}, \mathscr{P}(\bar{x}, t)$ defined on $\Delta_{\mathscr{A}} \times R$, and a function $V(\cdot): \Omega_{\mathscr{A}} \to R, \Omega_{\mathscr{A}} = \Delta_{\mathscr{A}} \times R$, continuous and of class C^1 with respect to a denumerable decomposition $\mathfrak{D}$ of $\Omega_{\mathscr{A}}$ such that for all $(\bar{x}, t) \in \Omega_{\mathscr{A}}$,*

(i) $V(\bar{x}, t) > V(\bar{x}_a, \tau)$ for all $\bar{x}_a \in \partial \mathscr{A}, \tau > t$,
(ii) for all $(\bar{x}, t) \in \Omega_\sigma, \sigma \in \Sigma$, and all $\bar{u} \in \mathscr{P}(\bar{x}, t)$,

$$\partial V_\sigma / \partial t + \nabla V_\sigma(\bar{x}, t) \cdot \bar{f}(\bar{x}, t, \bar{u}, \bar{w}) \geq 0 \tag{5.1.15}$$

for all $\bar{w}(t) \in W$.

The proof is analogous to that of Theorem 5.1.1', differing only in the fact that Lemma 5.1.3 is used. Suppose some motion of $\mathscr{K}(\bar{x}^0, t_0), (\bar{x}^0, t_0) \notin \mathscr{A} \times R$ enters $\mathscr{A}$. Then there is $t_1 > t_0$ such that $k(\bar{u}, \bar{x}^0, t_0, t_1) = \bar{x}^1 \in \partial \mathscr{A}$ and, by (i), $V(\bar{x}^0, t_0) > V(\bar{x}^1, t_1)$. This is contradicted by the fact that by (ii) and Lemma 5.1.3 the function $V(k(t), t)$ is nondecreasing on $[t_0, t_1]$.

Corollary 5.1.3. *Given $(\bar{x}, t) \in \Omega_\sigma, \sigma \in \Sigma$, if there are $\tilde{\bar{u}}, \tilde{\bar{w}}$ such that*

$$\mathscr{L}_\sigma(\bar{x}, t, \tilde{\bar{u}}, \tilde{\bar{w}}) = \max_{\bar{u}} \min_{\bar{w}} \mathscr{L}_\sigma(\bar{x}, t, \bar{u}, \bar{w}) \geq 0, \tag{5.1.16}$$

then condition (ii) of Theorem 5.1.3 holds with $\bar{u} \in \mathscr{P}(\bar{x}, t)$ which thus could be determined by (5.1.16).

The proof follows by the same argument as that used for Corollary 5.1.1. Leitmann [4] worked out an example that not only illustrates the use of Theorem 5.1.3 and Corollary 5.1.3 but at the same time also indicates how the theory of pursuit–evasion games can be used to define a controller for the avoidance of obstacles by a planar manipulator.

Example 5.1.3. Consider the representing point G of a gripper moving with constant speed v_G on a horizontal planar workbench confronted with the obstacle $\mathfrak{T}_A$ as shown in Fig. 5.3, also moving with the constant speed p. Fixing the Cartesian planar coordinate system $0\xi_1\xi_2$ to the obstacle (see Fig. 5.3), we specify the obstacle as

$$\mathfrak{T}_A : \xi_1^2 + \xi_2^2 \leq l^2,$$

$l = \text{const}$, and the velocity vector $\bar{p}$ along ξ_2. The obstacle may not change its speed, but its orientation angle w is unpredictable within a bound $W = [-N, N]$, with $N > 0$. The gripper must also keep its speed constant but may maneuver with its orientation angle u, which is left unconstrained. From Fig. 5.3, we can write the components of the relative velocity as

$$\dot{\xi}_1 = v_G \sin u + w\xi_2, \quad \dot{\xi}_2 = v_G \cos u - w\xi_1 - p.$$

The equations are known as the kinematic equations of the so-called

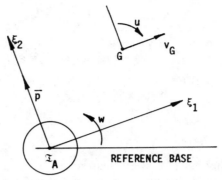

Fig. 5.3

homicidal chauffeur game (see Isaacs [1]). We introduce the avoidance set as

$$\mathscr{A} \triangleq \begin{cases} (\xi_1, \xi_2) \,|\, \xi_1^2 + \xi_2^2 \leq l, \, \xi_2 \leq |\xi_1| \tan \alpha, \\ (\xi_1, \xi_2) \,|\, |\xi_1| \tan \alpha \leq \xi_2 \leq -|\xi_1| \cot \alpha + l \csc \alpha \end{cases};$$

for α see Fig. 5.4. Let Δ_ε be defined as is $\mathscr{A}$ but with l replaced by $l + \varepsilon$, $l = \text{const} > 0$. Let X_σ be the projection of Ω_σ onto $R^{N=2}$, that is,

$$\Omega_\sigma = X_\sigma \times R.$$

Consider the decomposition of $\Delta_\mathscr{A}$ defined by

$$X_1 \triangleq \{(\xi_1, \xi_2) \,|\, l^2 < \xi_1^2 + \xi_2^2 < (l + \varepsilon)^2, \, \xi_2 \leq |\xi_1| \tan \alpha\}$$
$$X_2 \triangleq \{(\xi_1, \xi_2) \,|\, \xi_1 \leq 0, \, \xi_2 > -\xi_1 \tan \alpha,$$
$$\xi_1 \cos \alpha + l \csc \alpha < \xi_2 \leq \xi_1 \cos \alpha + (l + \varepsilon) \csc \alpha\}$$
$$X_3 \triangleq \{(\xi_1, \xi_2) \,|\, 0 < \xi_1, \, \xi_2 > \xi_1 \tan \alpha,$$
$$-\xi_1 \cot \alpha + l \csc \alpha < \xi_2 \leq -x_1 \cot \alpha + (l + \varepsilon) \csc \alpha\}.$$

To apply Theorem 5.1.3, let

$$V_1(\bar{x}, t) \triangleq \xi_1^2 + \xi_2^2 \qquad \text{for } (\xi_1, \xi_2) \in X_1,$$
$$V_2(\bar{x}, t) \triangleq (\xi_2 - \xi_1 \cot \alpha) l \sin \alpha \qquad \text{for } (\xi_1, \xi_2) \in X_2,$$
$$V_3(\bar{x}, t) \triangleq (\xi_2 - \xi_1 \cot \alpha) l \sin \alpha \qquad \text{for } (\xi_1, \xi_2) \in X_3.$$

Condition (i) is seen to be satisfied. On the use of Corollary 5.1.3, it is readily shown that (ii) holds if

$$\sin \alpha \leq v_G/p, \qquad N \leq (v_G - p \sin \alpha)/(l + \varepsilon) \cot \alpha,$$

5.2 AVOIDANCE WITH STIPULATED HANDLING TIME

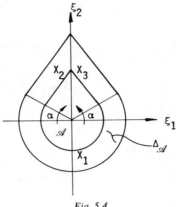

Fig. 5.4

with the controller given by

$$\sin u = \xi_1/\sqrt{\xi_1^2 + \xi_2^2}, \qquad \cos u = \xi_2/\sqrt{\xi_1^2 + \xi_2^2} \quad \text{on} \quad X_1,$$
$$\sin u = -\cos \alpha, \qquad \cos u = \sin \alpha \qquad \text{on} \quad X_2,$$
$$\sin u = \cos \alpha, \qquad \cos u = \sin \alpha \qquad \text{on} \quad X_3.$$

Exercises and Comments

5.1.1 What is the difference, if any, between the avoidance of Definition 3.3.2 specified by (5.1.1) and reaching with $T_\mathcal{R} \to \infty$?

5.1.2 Compare Theorems 5.1.1 and 5.1.1'. Which of the two has more restrictive conditions and in what sense? Can Theorem 5.1.1 be proved for nonstationary $V(\cdot)$?

5.1.3 Give a practical example of a manipulator arm traveling along a repelling mesh B (see Fig. 5.2) placed somewhere on $\mathcal{H}$. Find the controller.

5.1.4 Can we prove Theorem 5.1.3 for a stationary $V(\cdot)$ in Theorem 5.1.1? Note that the nonstationary $V(\cdot)$ is not very convenient for building the potential push-off force field in $\Delta_\mathcal{A}$ of any obstacles.

5.2 AVOIDANCE WITH STIPULATED HANDLING TIME

Requiring complete avoidance of obstacles all the time, or for an unspecified time anyway, may be unduly demanding and a costly proposition in manipulator work. We have seen in Chapter 4 that the time factor in such work is essential and a free path is needed for stipulated rather than indefinite time intervals. It is the stipulated *real-time avoidance* of obstacles which we

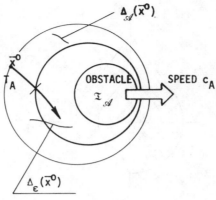

Fig. 5.5

propose to consider in this section, following the work of Leitmann and Skowronski [2].

Let us make Δ_ε depend upon $\bar{x}^0$. As before, $\Delta_\varepsilon(\bar{x}^0) \supset \mathscr{A}$, $\partial\Delta_\varepsilon(\bar{x}^0) \cap \partial\mathscr{A} \cap \mathring{\Delta} = \varnothing$, but also $\bar{x}^0 \in \partial\Delta_\varepsilon(\bar{x}^0)$, that is, we consider motions that emanate from $\partial\Delta_\varepsilon$ (see Fig. 5.5). Then the safety zone is $\Delta_\mathscr{A}(\bar{x}^0) \triangleq \Delta_\varepsilon(\bar{x}^0) - \mathscr{A}$, with $\mathscr{A}$ still fixed.

Definition 5.2.1. The system (3.3.1) is strongly controllable at $(\bar{x}^0, t_0)$, $\bar{x}^0 \notin \mathscr{A}$, for real-time avoidance of $\mathscr{A}$ during stipulated $T_A^+ > 0$, if and only if there is a control program $\mathscr{P}(\bar{x}, t)$ such that for every motion of $\mathscr{K}(\bar{x}^0, t_0)$, we have

$$k(\bar{u}, \bar{x}^0, t_0, [t_0, t_0 + T_A^+]) \cap (\mathscr{A} \times R) = \varnothing. \tag{5.2.1}$$

See Fig. 5.6, which also compares other avoidances, that is, avoidance for all times (see Definition 4.1.3), real-time avoidance with unspecified $T_\mathscr{A}$ (see Definition 5.4.1), and ultimate avoidance (see Definition 5.2.2).

Let

$$\rho(\bar{x}^0, \mathscr{A}) = \min_{\bar{x} \in \mathscr{A}} \|\bar{x}^0 - \bar{x}\|$$

be the distance between $\bar{x}^0$ and $\mathscr{A}$ and let $T_A^+ \cdot c_A \triangleq \rho(\bar{x}^0, \mathscr{A})$, c_A being the safe rate of approach to $\mathscr{A}$ during T_A^+. Given c_A the corresponding region of strong controllability, say, Δ_{AT} depends upon T_A^+: $\Delta_{AT}(T_A^+) = \Delta - \Delta_\varepsilon(\bar{x}^0)$ and clearly $\Delta_{AT}(T_{A1}^+) \subset \Delta_{AT}(T_{A2}^+)$ for any two T_{A1}^+, T_{A2}^+ such that $T_{A1}^+ \leq T_{A2}^+$. It agrees with the fact that the (infinite time) avoidance $T_A^+ \to \infty$ implies any real-time avoidance. On the other hand, in this section as much as previously, it is the safety zone $\Delta_\mathscr{A}(\bar{x}^0)$ which matters.

Let $C\mathscr{A} \triangleq \Delta - \mathscr{A}$ and introduce a C^1 function $V(\cdot): D_A \to R$, open $D_A \supset C\mathscr{A}$, such that for some $\Delta_\varepsilon(\bar{x}^0)$, we may recall

$$v_A^+ \triangleq \sup V(\bar{x}) \mid \bar{x} \in \partial\mathscr{A}, \tag{5.1.2}$$

5.2 AVOIDANCE WITH STIPULATED HANDLING TIME

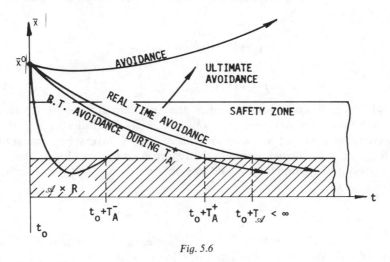

Fig. 5.6

and define
$$v_\varepsilon^- \triangleq \inf V(\bar{x}) \,|\, \bar{x} \in \partial \Delta_\varepsilon. \tag{5.2.2}$$

Theorem 5.2.1. *The system (3.3.1) is strongly controllable on some $\Delta_0 \times R$ for real-time avoidance of $\mathcal{A}$ during T_A^+ if there is $\mathcal{P}(\cdot)$ defined on $D_A \times R$, C^1 function $V(\cdot): D_A \to R$ and, given $\bar{x}^0 \notin \mathcal{A}$, there is $\Delta_\varepsilon(\bar{x}^0)$ such that*

(i) $0 < v_A^+ < v_\varepsilon^- < \infty$,
(ii) *for all $\bar{u} \in \mathcal{P}(\bar{x}, t)$,*

$$\nabla V(\bar{x}) \cdot \bar{f}(\bar{x}, t, \bar{u}, \bar{w}) \geq (v_\varepsilon^- - v_A^+)/T_A^+, \tag{5.2.3}$$

for all $\bar{w}(t) \in W$.

Proof. Consider a motion from $\mathcal{K}(\bar{x}^0, t_0)$, for $\bar{x}^0 \in \partial \Delta_\varepsilon(\bar{x}^0)$ at any t_0. If it does not cross $\partial \mathcal{A}$, then it does not collide with $\mathcal{A}$ at all. Thus suppose it does intersect $\partial \mathcal{A}$. Let t_1 be the time of the first intersection: $k(\bar{u}, \bar{x}^0, t_0, t_1) = \bar{x}^1 \in \partial \mathcal{A}$. Integrating (5.2.3) along this motion, we obtain

$$t_1 - t_0 \geq \frac{V(\bar{x}^0) - V(\bar{x}^1)}{v_\varepsilon^- - v_A^+} T_A^+. \tag{5.2.4}$$

But in view of (i),

$$\frac{V(\bar{x}^0) - V(\bar{x}^1)}{v_\varepsilon^- - v_A^+} \geq 1,$$

thus yielding $t_1 - t_0 \geq T_A^+$, which concludes the proof.

Observe that in view of Definition 5.2.1 and (5.2.4), the avoidance of Section 5.1 is recovered by letting $T_A^+ \to \infty$. Now recall (5.1.6): $v_A^- \triangleq \inf V(\bar{x}) \,|\, \bar{x} \in \partial \mathcal{A}$,

and define
$$v_\varepsilon^+ \triangleq \sup V(\bar{x}) \mid \bar{x} \in \partial\Delta_\varepsilon(\bar{x}^0) \tag{5.2.5}$$
Then
$$V(\bar{x}^1) - V(\bar{x}^0) \geq v_A^- - v_\varepsilon^+ \triangleq -c_A T_A^+ \tag{5.2.6}$$

so that $-c_A \leq 0$ is the lower bound on the time derivative of $V(\bar{x}(t))$ along a motion from $\partial\Delta_\varepsilon$ intersecting $\partial\mathscr{A}$ at $x^1 = k(\bar{u}, \bar{x}^0, t_0, t)$. Thus $-c_A$ is a *safe rate of decrease of* $\rho(\bar{x}^0, \mathscr{A})$ *measured in terms of* $V(\bar{x})$. Conversely, as mentioned, given c_A and T_A^+, one can estimate the region $\Delta_{AT}(T_A^+)$ (see Fig. 5.5). On the other hand, given c_A and $(\bar{x}^0, t_0)$, one may obtain $T_A = (v_\varepsilon' - v_A')/c_A$. Finally, on utilizing

$$c_A = \frac{v_\varepsilon^+ - v_A^-}{T_A^+} \geq \frac{v_\varepsilon^- - v_A^+}{T_A^+} \geq 0 \tag{5.2.7}$$

in (5.2.3), we see that $c_A = 0$ implies indefinite time avoidance, that is, $T_A^+ \to \infty$. Obviously, if $\partial\Delta_\varepsilon(\bar{x}^0)$ and $\partial\mathscr{A}$ are V-level curves of $V(\cdot)$, then $v_A^- = v_A^+$ and $v_\varepsilon^- = v_\varepsilon^+$.

The controller for avoidance in T_A^+ can be found from the following corollary to Theorem 5.2.1.

Corollary 5.2.1. *Given* $\bar{x} \in \mathscr{A}$, $t \in R$, *if there is a pair* $\tilde{\tilde{u}}, \tilde{w}$ *such that*

$$\mathscr{L}(\bar{x}, t, \tilde{\tilde{u}}, \tilde{w}) = \max_{\bar{u}} \min_{\bar{w}} \mathscr{L}(\bar{x}, t, \bar{u}, \bar{w}) + \frac{v_\varepsilon^- - v_A^+}{T_A^+} \geq 0, \tag{5.2.8}$$

then condition (ii) of Theorem 5.2.1 is met with $\bar{u} \in \mathscr{P}(\bar{x}, t)$. *Thus the program is definable from (5.2.8)*.

The reader might have observed already that replacing real-time avoidance of $\mathscr{A}$ during T_A^+ by the same of the interior $\mathring{\mathscr{A}}$ will not produce any basic changes either in definitions or in sufficient conditions concerned. As much as for avoiding $\mathscr{A}$ Theorem 5.2.1 incorporates Theorem 5.1.1, for $\mathring{\mathscr{A}}$ it would have to do the same with Theorem 5.1.2 dealing with rejection. Then the Liapunov function is defined on $\mathscr{A}$ instead of $\Delta_{\mathscr{A}}$, and the control program specified by the previous Corollary 5.1.1 as the safety zone collapses to $\partial\mathscr{A}$. This means, however, that the stipulation of the time interval T_A^+ becomes irrelevant.

The above adaptations are immediate. We leave them to the reader, illustrating the technique in the following example.

Example 5.2.1. To complete the sequential reaching of Example 4.3.1, we may prefer to make the gripper avoid something on its way, with the avoidance being essential only until it reaches the target specified for each step of the itinerary (see Fig. 4.14).

5.2 AVOIDANCE WITH STIPULATED HANDLING TIME

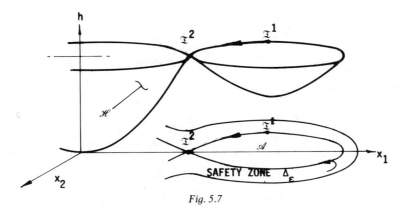

Fig. 5.7

Suppose the gripper has already reached $\mathfrak{T}^1$ on level h_H at time $t_{\mathscr{R}1} = t_0 + T^+_{\mathscr{R}1} = 10$ and attempts $\mathfrak{T}^2$ before $t_{\mathscr{R}2} = t_0 + 2T^+_{\mathscr{R}1} = 20$. We do not want it to go below h_H since then it will later have to waste energy to recover the target level. Hence for the time interval $t_{\mathscr{R}2} - t_{\mathscr{R}1} \leq T^+_{\mathscr{R}}$ it should avoid $\mathring{\Delta}^{k=2}_H$. We thus let $T^+_A \triangleq T^+_{\mathscr{R}} = 10$.

While on this assignment, the "conservative" control $u = \pm N/x_2, x_2 \lessgtr 0$, is used and is calculated from (4.1.29), with zero flux of energy—which produces some difficulties in realization as mentioned before. Apart from the remedies proposed in Example 4.3.1, we may, perhaps more successfully, use real-time avoidance of $\Delta^{k=2}_H$.

We have $\mathscr{A} = \Delta^{k=2}_H$ and want to avoid $\mathring{\mathscr{A}}$, thus in fact to attain *rejection from such* $\mathscr{A}$. Condition (i) of Theorem 5.2.1 as adjusted by Theorem 5.1.2 and Corollary 5.1.2 is obviously satisfied over the second energy cup $\Delta^{k=2}_H$ after we let $V(\bar{x}) \equiv H(\bar{x})$. All we need to check is whether the accumulative power field $\mathscr{H}^+$ in the interior $\mathring{\mathscr{A}} = \mathring{\Delta}^{k=2}_H$ is strong enough to satisfy (5.2.3) over the safety zone, which in practice reduces to $\partial \mathscr{A}$ (see Fig. 5.7). This means that Corollary 5.1.1 will be used instead of Corollary 5.2.1, thus generating no change in the controller $u = \pm N/x_2$, which was imposed before in Example 4.3.1. Observe that the *gripper* in fact *moves along the repelling mesh* (see Fig. 5.2), since the same controller prevents it as well from climbing above the conservative route on h_H. ∎

Example 5.2.2. Consider an RP manipulator whose motion is described by the variables shown in Fig. 5.8, with the gripper reference point G and an obstacle point $\mathfrak{T}_A$ considered an *inertialess* object in planar motion on a horizontal bench, with velocities of constant magnitudes v_G and $v_{\mathscr{A}}$, respectively. The kinematic equations of motion of the gripper are

$$\dot{\mathscr{R}} = v_G \cos(\theta + u), \qquad \dot{\theta}_G = (v_G/\mathscr{R})\sin(\theta + u), \qquad (5.2.9)$$

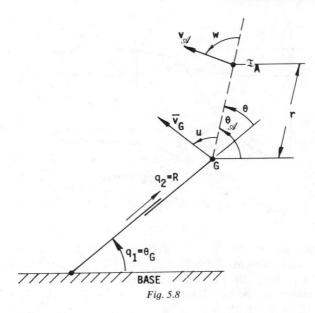

Fig. 5.8

while those of the obstacle relative to the gripper are

$$\dot{r} = v_{\mathscr{A}} \cos w - v_G \cos u, \qquad \dot{\theta}_{\mathscr{A}} = (1/r)(v_{\mathscr{A}} \sin w - v_G \sin u). \quad (5.2.10)$$

Assuming $v_{\mathscr{A}} > v_G$ we wish that, in spite of w, the gripper would avoid a circular zone of radius $r_{\mathscr{A}}$ about $\mathfrak{T}_{\mathscr{A}}$ during the interval of duration T_A^+. Hence $\mathscr{A}: r(t) \leq r_{\mathscr{A}}$. We choose the Liapunov function in terms of push-off potential, that is, the relative distance from the obstacle [see Eq. (5.1.5)]. In our present simple case, we do not have to do the transformation of variables but simply let $V = r(t)$. Then also $v_A^+ = v_A^- = r_{\mathscr{A}}$. By Corollary 5.2.1,

$$-v_{\mathscr{A}} + v_G \geq -(r^0 - r_{\mathscr{A}})/T_A^+, \quad (5.2.11)$$

where $r^0 = r(t_0)$. Whence $u = -\pi$ defines the program. ∎

Now let us suppose that it is desired to assure, against the noise $\bar{w}$, that no motion from anywhere in $\Delta \times R$ (including $\mathscr{A} \times R$) is allowed to stay in some $\mathscr{A}$ for any time *after* a certain real-time interval which is either to be determined a posteriori or stipulated. We begin with the first alternative.

Definition 5.2.2. The system (3.3.1) is strongly controllable at $(\bar{x}^0, t_0) \in \Delta \times R$ for *ultimate avoidance of* $\mathscr{A}$ if and only if there is a $\mathscr{P}(\bar{x}, t)$ and a constant $T_{\mathscr{A}} > 0$ such that for each motion of $\mathscr{K}(\bar{x}^0, t_0)$ we have

$$k(\bar{u}, \bar{x}^0, t_0, [t_0 + T_{\mathscr{A}}, \infty)) \cap (\mathscr{A} \times R) = \varnothing. \quad (5.2.12)$$

The motions must either stay out of $\mathscr{A}$ or leave this set before the time

5.2 AVOIDANCE WITH STIPULATED HANDLING TIME

$t_0 + T_\mathscr{A}$, that is, the complement $C\mathscr{A}$ must be strongly positively invariant under $\mathscr{P}(\cdot)$.

Clearly, there is no need for a safety zone in the above definition. Neither may we narrow the controllability concerned to any set smaller than Δ. The designed $\mathscr{A}$ should be checked against some sufficient conditions. We quote the latter from Leitmann and Skowronski [2]. Recall v_A^- and v_A^+ defined by (5.1.6) and (5.1.2), respectively.

Theorem 5.2.2. *The system (3.3.1) is strongly controllable on $\Delta \times R$ for ultimate avoidance of $\mathscr{A}$ if there is a $\mathscr{P}(x,t)$ defined on $\mathscr{A} \times R$ and a C^1 function $V(\cdot): D_A \to R$, $D_A(\text{open}) \supset \mathscr{A}$, such that*

  (i) $0 < V(\bar{x}) < v_A^+, \bar{x} \in \mathscr{A}$,
  (ii) $V(\bar{x}) \geq v_A^-, \bar{x} \notin \mathscr{A}$,
  (iii) *for all $\bar{u} \in \mathscr{P}(\bar{x}, t)$, there is $c_A > 0$ such that*

$$\nabla V(\bar{x}) \cdot \bar{f}(\bar{x}, t, \bar{u}, \bar{w}) \geq c_A \qquad (5.2.13)$$

for all $\bar{w}(t) \in W$.

Proof. First we show that no motion from $C\mathscr{A} \times R$ can enter $\mathscr{A}$. Indeed, suppose some motion of $\mathscr{K}(\bar{x}^0, t_0)$, $\bar{x}^0 \notin \mathscr{A}$, $t_0 \in R$, does cross $\partial \mathscr{A}$. Then there is a $t_1 \geq t_0$ such that $k(\bar{u}, \bar{x}^0, t_0, t_1) = \bar{x}^1 \in \partial \mathscr{A}$. Then by (ii) we have $V(\bar{x}^1, t_1) \leq V(\bar{x}^0, t_0)$, thus contradicting (iii).

Then we must show that motions from $\mathscr{A} \times R$ leave $\mathscr{A}$ after some $t_0 + T_\mathscr{A}$. Indeed, integrating (5.2.13) along such a motion, we obtain

$$t - t_0 \leq c_A^{-1}(V(\bar{x}) - V(\bar{x}^0)). \qquad (5.2.14)$$

By (i), $V(\bar{x}) - V(\bar{x}^0) \leq v_A^+$, and thus, letting

$$T_\mathscr{A} = v_A^+/c_A, \qquad (5.2.15)$$

we obtain $t - t_0 \leq T_\mathscr{A}$, which completes the proof.

The discussion of (5.2.15) is similar to that of (5.2.6) and (5.2.7), we leave it to the reader.

Corollary 5.2.2. *Corollary 5.2.1 holds here with (5.2.8) replaced by the condition*

$$\mathscr{L}(\bar{x}, t, \tilde{\bar{u}}, \tilde{\bar{w}}) = \max_{\bar{u}} \min_{\bar{w}} \mathscr{L}(\bar{x}, t, \bar{u}, \bar{w}) \geq c_A. \qquad (5.2.16)$$

Now consider the second alternative, namely, when $T_\mathscr{A}$ is stipulated, say as $T_\mathscr{A} = T_A^-$.

Definition 5.2.3. The system (3.3.1) is strongly controllable at $(\bar{x}^0, t_0) \in$

$\Delta \times R$ for *ultimate avoidance of* $\mathscr{A}$ *after* T_A^- if and only if Definition 5.2.2 holds for stipulated T_A^-.

From Theorem 5.2.2, we have

Theorem 5.2.3. *Definition 5.2.3 is satisfied if the conditions of Theorem 5.2.2 hold with* $c_A = v_A^+/T_A^- > 0$.

Remark 5.2.1. Note that Corollary 5.2.2 holds for $c_A = v_A^+/T_A^- > 0$.

Note that Theorem 5.2.2 is more demanding than Theorem 5.1.1, but then we obtain a stronger result: We not only avoid $\mathscr{A}$, but if need be, we can escape from it.

Example 5.2.3. We continue Example 5.2.2, but with the objective changed: We want the gripper G to reach the target $\mathfrak{T}: \mathscr{R}(t) = \mathscr{R}_G$ in minimum time and to avoid $\mathfrak{T}_{\mathscr{A}}$ thereafter, that is, after the time $t_0 + T_A^-$, where

$$T_A^- = (\mathscr{R}_G - \mathscr{R}^0)/v_G,$$

with $\mathscr{R}^0 = \mathscr{R}(t_0)$. We choose $\mathscr{A}$, $V(\cdot)$, and v_A^+ as in Example 5.2.2. Then condition (iii) of Theorem 5.2.2 with c_A from Theorem 5.2.3 becomes

$$v_{\mathscr{A}} \cos w - v_G \cos u \geq r_{\mathscr{A}} v_G/(\mathscr{R}_G - \mathscr{R}^0) \qquad \forall w,$$

that is,

$$-v_{\mathscr{A}} - v_G \cos u \geq r_{\mathscr{A}} v_G/(\mathscr{R}_G - \mathscr{R}^0)$$

or

$$\cos(u + \pi) \geq \frac{r_{\mathscr{A}}}{\mathscr{R}_G - \mathscr{R}^0} + \frac{v_{\mathscr{A}}}{v_G}.$$

This condition yields a control program for u, provided that the right-hand side ≤ 1, that is,

$$r_{\mathscr{A}} < \mathscr{R}_G - \mathscr{R}^0, \qquad v_{\mathscr{A}} < v_G. \quad \blacksquare$$

Exercises

5.2.1 In Example 5.2.1, find conditions which assure that the h_H-level motion from $\mathfrak{T}^1$ to $\mathfrak{T}^2$ proceeds along the repelling mesh $\mathscr{B}$.

5.2.2 Form sufficient conditions for real-time avoidance of $\mathscr{A}$ during T_A^+, and specify means of obtaining a corresponding control program.

5.2.3 Adjust calculations of Example 5.2.2 to the case which is not inertialess, that is, in terms of our RP manipulator in Section 1.1.

5.2.4 Discuss the meaning of (5.2.15) following our discussion on (5.2.7).

5.3 REACHING WITH CAPTURE WHILE AVOIDING OBSTACLES

While the manipulator arm reaches, captures, or both, it meets nonremovable obstacles in Δ and the controller must be adjusted to avoidance as well. Here again the timing is usually essential, because we want to avoid before we reach or capture. We investigate in this section two cases: reaching $\mathfrak{T}$ with avoidance of $\mathscr{A}$ and capture in $\mathfrak{T}_C$ while avoiding $\mathscr{A}$. After reaching, avoidance may or may not be required—it is of no interest to reaching. After capture, avoidance normally will not be required since we may reasonably assume that the target is disjoint from obstacles. However, this may not be the case when obstacles move and may hit the arm even if it is in the configuration included in the target (remember that the target refers to all joints). This case would require some other combination of the elementary objectives. We leave it to the reader. Many other such combinations may be needed in case studies.

Let us begin with the objective of reaching a stipulated $\mathfrak{T}$ with avoidance of an envelope of grown obstacles $\mathscr{A}$ assumed given and such that $\mathfrak{T} \cap \mathscr{A} = \varnothing$. Such reaching could also be called "reaching first" $\mathfrak{T}$ before $\mathscr{A}$ since there is so far no requirement as to what the arm should do after the reaching.

Definition 5.3.1 System (3.3.1) is strongly controllable at $(\bar{x}^0, t_0)$, $\bar{x}^0 \notin \mathscr{A}$, for reaching $\mathfrak{T}$ with avoidance of $\mathscr{A}$ or briefly *reaches first* if and only if there is a $\mathscr{P}(\cdot)$ and a constant $T_F > 0$ such that for each motion of $\mathscr{K}(\bar{x}^0, t_0)$ we have

$$k(\bar{u}, \bar{x}^0, t_0[t_0, t_0 + T_F)) \cap (\mathscr{A} \times R) = \varnothing \qquad (5.3.1)$$

and

$$k(\bar{u}, \bar{x}^0, t_0, t_0 + T_F) \in \mathfrak{T}. \qquad (5.3.2)$$

This means that T_F is conditioned by reaching and the corresponding regions are enclosed in those of reaching during T_F with $\mathscr{A}$ subtracted, $\Delta_F \subset \Delta_{\mathscr{R}T} - \mathscr{A}$, and the avoidance time adjusted. It also means that for the avoidance part, we should take the stipulated real-time avoidance (Definition 5.2.1), replacing T_A by T_F. The fact that the avoidance time must match (be equal to or larger than) the interval before reaching gives an additional condition from which Δ_F may be determined. It will be seen below to follow from the calculation of the intervals concerned.

For the sufficient conditions, we combine Theorems 4.3.1 and 5.2.1. The situation may be seen in Fig. 5.9. Given the sets $\mathfrak{T}, \mathscr{A}$, the candidate Δ_0 for Δ_F such that $\Delta_0 \subset \Delta_{\mathscr{R}} - \mathscr{A}$ and $\Delta_\varepsilon(\bar{x}^0) \supset \mathscr{A}$, we adjust the complements $C_F \mathfrak{T} = \Delta_0 - \mathfrak{T}$, $C\mathscr{A} = \Delta_0$, and then let open $D \supset \overline{C_F \mathfrak{T}}$, $D_A \supset \Delta_0$. Further we consider two functions $V_{\mathscr{R}}(\cdot): D \to R$ and $V_A(\cdot): D_A \to R$ such that $v_{\mathfrak{T}}, v_0$ are defined by (4.1.3) in terms of $V_{\mathscr{R}}(\cdot)$, and v_A, v_ε, v'_A, and v'_ε are defined by (5.2.2), (5.2.5) in terms of $V_A(\cdot)$ (see Fig. 5.9).

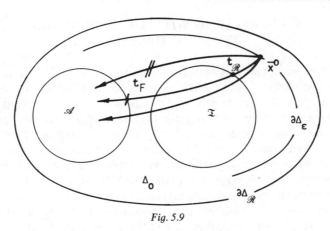

Fig. 5.9

Granted the above, from Theorems 4.3.1 and 5.2.1, we obtain the following sufficient conditions needed to check Δ_0.

Theorem 5.3.1. *Definition 5.3.1 holds if there is a $\mathscr{P}(\cdot)$ defined on $(D \cup D_A) \times R$, two C^1 functions $V_{\mathscr{R}}(\cdot): D \to R$ and $V_A(\cdot): D_A \to R$, and, given $\bar{x}^0 \notin \mathscr{A}$, $t_0 \in R$, there is a $\Delta_\varepsilon(\bar{x}^0)$ such that for all $(\bar{x}, t) \in \Delta_0 \times R$,*

(i) $V_{\mathscr{R}}(\bar{x}) > v_{\mathfrak{T}}, \bar{x} \notin \mathfrak{T}$;
(ii) $V_{\mathscr{R}}(\bar{x}) \leq v_0, \bar{x} \in C_F \mathfrak{T}$;
(iii) $0 < v_A^+ < v_\varepsilon^- < \infty$;
(iv) *for all $(\bar{x}, t) \in D \times R$, $\bar{u} \in \mathscr{P}(\bar{x}, t)$, there is a constant $T_F(\bar{x}, t, k(\cdot)) > 0$ such that*

$$\int_{t_0}^{t_0 + T_F} [\nabla V_{\mathscr{R}}(\bar{x}) \cdot \bar{f}(\bar{x}, t, \bar{u}, \bar{w})] \, dt \leq -(v_0 - v_{\mathfrak{T}}) \tag{5.3.3}$$

for all $\bar{w}(t) \in W$;
(v) *for all $(\bar{x}, t) \in D_A \times R$, $\bar{u} \in \mathscr{P}(\bar{x}, t)$,*

$$\nabla V_A(\bar{x}) \cdot \bar{f}(\bar{x}, t, \bar{u}, \bar{w}) \geq (v_\varepsilon^- - v_A^+)/T_F \tag{5.3.4}$$

for all $\bar{w}(t) \in W$.

Clearly, both Corollaries 4.1.4 and 5.2.1 must be used together to determine the control program.

As an illustrating example, we can use here our Example 5.2.1, provided we can allow the fact that avoidance is replaced by rejection, with corresponding adjustments of Definition 5.3.1, and Theorem 5.3.1.

We turn now to the objective of capture in a stipulated $\mathfrak{T}_C \subset \mathfrak{T}$ during an arbitrary time while avoiding $\mathscr{A}$, $\mathfrak{T}_C \cap \mathscr{A} = \varnothing$. We adapt here the results of

5.3 REACHING WITH CAPTURE WHILE AVOIDING OBSTACLES

Getz and Leitmann [1], Skowronski [19], Stonier [2], and Skowronski and Stonier [1], which were geared for a slightly different purpose.

Definition 5.3.2. The system (3.3.1) is strongly controllable on $\Delta_0 \times R$ for capture in $\mathfrak{T}_C$ with avoidance of $\mathscr{A}$, briefly *capture with avoidance*, if and only if there is a $\mathscr{P}(\cdot)$ and $T_{AC} > 0$ such that for each motion of $\mathscr{K}(\bar{x}^0, t_0)$, $(x^0, t_0) \in \Delta_0 \times R$, we have

$$k(\bar{u}, \bar{x}^0, t_0, R_0^+) \cap (\mathscr{A} \times R) = \varnothing \qquad (5.3.5)$$

and

$$k(\bar{u}, \bar{x}^0, t_0, [t_0 + T_{AC}, \infty)) \subset \mathfrak{T}_C \times R. \qquad (5.3.6)$$

Let $\Delta_0 \subset \Delta_C - \mathscr{A}$ be the candidate set for the region Δ_{AC} of strong controllability for capture with avoidance, and recall that $C\mathfrak{T}_C \triangleq \Delta_0 - \mathfrak{T}_C$, $\Delta_\mathscr{A} \triangleq \Delta_0$, with $\Delta_\varepsilon \supset \mathscr{A}$ (see Fig. 5.10). Then let open $D \supset \overline{C\mathfrak{T}_C}$, $D_\mathscr{A} \supset \overline{\Delta}_\mathscr{A}$, and introduce two functions $V_C(\cdot): D \to R$ and $V_\mathscr{A}(\cdot): D_\mathscr{A} \to R$, with $v'_\mathfrak{T}, v'_0$ defined by (4.4.7) in terms of $V_C(\cdot)$, and v_A^- and v_A^+ defined by (5.1.6) and (5.1.2), respectively, in terms of $V_\mathscr{A}(\cdot)$.

Theorem 5.3.2. *Definition 5.3.2. holds if there is a $\mathscr{P}(\cdot)$ defined on $(D \cup D_\mathscr{A}) \times R$ and two C^1 functions $V_C(\cdot)$, $V_\mathscr{A}(\cdot)$ as above such that*

(i) $V_C(\bar{x}) \geq v'_\mathfrak{T}$ *for all* $\bar{x} \notin \mathfrak{T}_C$;
(ii) $V_C(\bar{x}) \leq v'_0$ *for all* $\bar{x} \in C\mathfrak{T}_C$;
(iii) $V_\mathscr{A}(\bar{x}) > v_A^+$, *for all* $\bar{x} \notin \mathscr{A}$, $\qquad (5.3.7)$
(iv) *for all* $\bar{u} \in \mathscr{P}(\bar{x}, t)$, *there is a* $c > 0$ *such that*

$$\nabla V_C(\bar{x}) \cdot \bar{f}(\bar{x}, t, \bar{u}, \bar{w}) \leq -c \qquad (5.3.8)$$

for all $\bar{w}(t) \in W$;

(v) *for all* $\bar{u} \in \mathscr{P}(\bar{x}, t)$, *there is a* $c_\mathscr{A} \geq 0$ *such that*

$$\nabla V_\mathscr{A}(\bar{x}) \cdot \bar{f}(\bar{x}, t, \bar{u}, \bar{w}) \geq c_\mathscr{A} \qquad (5.3.9)$$

for all $\bar{w}(t) \in W$.

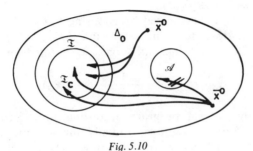

Fig. 5.10

The proof of avoidance follows by the same argument as that for Theorem 5.1.1, capture follows from Theorem 4.4.1. Independently, conditions (i)–(v) satisfy those imposed by Getz and Leitmann [1], which were proved to be sufficient for the same hypothesis. The proof gives also the definition of the region Δ_{AC}, which is similar to that derived in what follows.

From (4.4.10) (see Remark 4.4.2), we have

$$T_{AC} \leq c^{-1}(v_0' - v_{\mathfrak{T}}'), \tag{5.3.10}$$

and by the same argument as for (5.2.6),

$$T_{AC} = c_{\mathscr{A}}^{-1}(V(\bar{x}^0) - v_A^-). \tag{5.3.11}$$

Comparing (5.3.10) and (5.3.11),

$$(v_0' - v_{\mathfrak{T}}')/c \geq (V(\bar{x}^0) - v_A^-)/c_{\mathscr{A}}, \tag{5.3.12}$$

which, given the rates c and $c_{\mathscr{A}}$, determines $\Delta_0 \subset \Delta_{AC}$.

Then the control program $\mathscr{P}(\cdot)$ may be found by satisfying simultaneously both Corollaries 4.4.1 and 5.1.1 with added $c_{\mathscr{A}}$.

Example 5.3.1. We consider the manipulator of Examples 5.2.2 and 5.2.3 with slight adjustment of the objective. We wish the gripper to reach the target $\mathfrak{T}_C : \mathscr{R}(t) = \mathscr{R}_G$ with capture, while avoiding $\mathscr{A} : r(t) \leq r_{\mathscr{A}}$, during the time of reaching T_{AC}. To apply Theorem 5.3.2 we choose $V_C = \mathscr{R}$ and $V_{\mathscr{A}} = r$, the latter as before. Conditions (5.3.8) and (5.3.9) become

$$v_G \cos(\theta + u) \leq -c, \tag{5.3.13}$$

$$-v_{\mathscr{A}} - v_G \cos u \geq -c_{\mathscr{A}}, \tag{5.3.14}$$

and the strongly winning Δ_0 is determined by

$$r - r_{\mathscr{A}} > (c_A/c)(\mathscr{R} - \mathscr{R}_G). \tag{5.3.15}$$

Condition (5.3.15) implies

$$-1 \leq \cos(\theta + u) \leq -c/v_G < 0 \tag{5.3.16}$$

so that $c = v_G/(1 + \delta)$, $\delta \in [0, \infty)$.

Given c, the set Δ_0 is maximized by choosing the smallest $c_{\mathscr{A}}$ such that (5.3.14) is met for all possible θ. This results in $c_{\mathscr{A}} = v_{\mathscr{A}} + v_G/(1 + \delta)$, whence $c_{\mathscr{A}}/c = 1 + ((1 + \delta)v_A/v_G)$. Hence, the largest Δ_0 results from $\delta = 0$, namely,

$$\Delta_{AC} : r - r_{\mathscr{A}} > \frac{v_G + V_{\mathscr{A}}}{v_G}(\mathscr{R} - \mathscr{R}_G). \tag{5.3.17}$$

The corresponding control program according to (5.3.16) is given by $\cos(\theta + u) = -1$, that is, G moves radially inward and is assumed to have a collision-free approach, provided it moves in Δ_{AC}. ■

5.4 ADAPTIVE AVOIDANCE

Exercises and Comments

5.3.1 In a manner similar to that used in Exercise 5.2.1, the reader is asked to form, in the general case, sufficient conditions for real-time first reaching during T_F but with avoidance of $\mathring{\mathscr{A}}$ instead of $\mathscr{A}$.

5.3.2 Condition (5.3.12) gives the strong region for the property concerned but this region is dependent upon $\bar{x}^0$, which is obvious owing to the fact that T_{AC} depends upon $\bar{x}^0$. In Example 5.3.1 we have made a search for the maximal Δ_0 which specifies Δ_{AC}. Can the argument of this search be generalized?

5.3.3 Compare the situation in Example 5.2.1 and the problem of first reaching.

5.4 ADAPTIVE AVOIDANCE

Following the work by Corless, Leitmann, and Skowronski [1], we introduce in this section adaptive controllers for the system (3.3.1) but with the selecting equation (1.5.3″) augmented to the case which generalizes (3.4.21). While doing this, we may consider as well a case of avoidance slightly different from those investigated so far, namely, real-time avoidance, but without stipulating $T_\mathscr{A}$.

First, let us consider (3.3.1) with the selecting equation

$$\dot{\bar{x}} = \bar{f}(\bar{x}, t, \bar{u}, \bar{u}^A, \bar{w}) \tag{5.4.1}$$

with $\bar{u}$ as before, while $\bar{u}^A(t) \in U^A \subset R^n$ is an added control vector for later use in adaptation [see Eq. (3.4.21)] with the closed and bounded set U^A. Moreover, $\bar{w}(t) \in W(\bar{x}(t), t)$ where $W(\cdot): \Delta \times R \to$ (all subsets of W) is an unknown function in a known class $\mathscr{W}$ and $\bar{f}(\cdot)$ is *uncertain* as well within a known class $\mathscr{F}$. The sets $\mathscr{A}$ and $\Delta_\mathscr{A}$ are left the same as defined in Section 5.1.

Definition 5.4.1. The system (3.3.1) with selector (5.4.1) is strongly controllable on $\Delta - \mathscr{A}$ for *real-time avoidance* of $\mathscr{A}$ if and only if there is a $\mathscr{P}(\cdot)$ and a constant $T_\mathscr{A} > 0$ such that for each motion of $\mathscr{K}(\bar{x}^0, t_0), \bar{x}^0 \notin \mathscr{A}, t_0 \in R$, we have

$$k(\bar{u}, \bar{u}^A, \bar{x}^0, t_0, [t_0, t_0 + T_\mathscr{A})) \cap (\mathscr{A} \times R_0^f) = \emptyset. \tag{5.4.2}$$

Here $R_0^f = [t_0, t_f], t_f < \infty$.

It is seen from the definition of $\Delta_\mathscr{A}$ that if $k(\cdot)$ is continuous with some $k(t_2) \in \mathscr{A}$, then there exist $t_3, t_4 \in [t_0, t_2], t_3 < t_4$, such that $k(t) \in \Delta_\mathscr{A}$ for all $t \in [t_3, t_4)$ and $k(t_4) \in \partial\mathscr{A}$, that is, every motion entering $\mathscr{A}$ must pass the zone.

We may now rearrange Theorem 5.1.1′. Recalling that open $D_\mathscr{A} \supset \Delta_\mathscr{A}$, we have

Theorem 5.4.1 *The system (3.3.1) with selecting equation (5.4.1) is strongly controllable on $\Delta - \mathcal{A}$ for avoidance of $\mathcal{A}$ if there is $\Delta_\mathcal{A}$, a program $\mathcal{P}(\cdot)$ defined on $D_\mathcal{A} \times R$, and a C^1 function $V(\cdot): D_\mathcal{A} \times R \to R$ such that*

(i) *if $(x^\mathcal{A}, t_\mathcal{A}) \in \partial \mathcal{A} \times R$, then*

$$V(\bar{x}, t) \to \infty \quad \text{as} \quad (\bar{x}, t) \to (x^\mathcal{A}, t_\mathcal{A}); \quad (5.4.3)$$

(ii) *for each $W(\cdot) \in \mathcal{W}$, $\bar{f}(\cdot) \in \mathcal{F}$ and for all $(\bar{x}, t) \in \Delta_\mathcal{A} \times R$ we have*

$$\partial V/\partial t + \nabla V(\bar{x}, t) \cdot \bar{f}(\bar{x}, t, \bar{u}, \bar{u}^A, \bar{w}) \leq 0 \quad (5.4.4)$$

for all $\bar{u} \in \mathcal{P}(\bar{x}, t)$, $u^A \in U^A$, and $w \in W(\bar{x}, t)$.

Proof. For any $W(\cdot)$, $\bar{f}(\cdot)$ consider a motion from the corresponding $\mathcal{K}(\bar{x}^0, t_0), (\bar{x}^0, t_0) \notin \mathcal{A} \times R$, and suppose there is a $t_2 \in R_0^f$ with $k(t_2) \in \mathcal{A}$. Since $\Delta_\mathcal{A}$ is a safety zone for $\mathcal{A}$ and $k(\cdot)$ is continuous, there are $t_3, t_4 \in [t_0, t_2]$, $t_3 < t_4$, such that $k(t) \in \Delta_\mathcal{A}$ for $t \in [t_3, t_4)$ and $k(t_4) \in \partial \mathcal{A}$. By (i),

$$\lim_{t \to t_4} V(\bar{x}(t), t) = \infty.$$

This contradicts (ii) since (ii) yields $\dot{V}(\bar{x}(t), t) \leq 0$; hence

$$V(\bar{x}(t), t) \leq V(\bar{x}(t_3) t_3) \quad \forall t \in [t_3, t_4). \quad (5.4.5)$$

Corollary 5.4.1. *Given $(\bar{x}, t) \in \Delta_\mathcal{A} \times R$, if there are $\tilde{\bar{u}}, \tilde{\bar{u}}^A, \tilde{\bar{w}}, \tilde{\bar{f}}(\cdot)$ such that*

$$\mathcal{L}'(\bar{x}, t, \tilde{\bar{u}}, \tilde{\bar{u}}^A, \tilde{\bar{w}}, \tilde{\bar{f}}(\cdot)) = \min_{\bar{u}} \sup_{\bar{u}^A, \bar{w}, \bar{f}} \mathcal{L}'(\bar{x}, t, \bar{u}, \bar{u}^A, \bar{w}, \bar{f}(\cdot)) \leq 0, \quad (5.4.6)$$

then $\bar{u} \in \mathcal{P}(\bar{x}, t)$ satisfies (5.4.4), and (5.4.6) may be used to define $\mathcal{P}(\cdot)$.

Now let us augment (5.4.1) to the case

$$\dot{\bar{x}} = \bar{f}(\bar{x}, t, \bar{u}, \bar{u}^A, \bar{w}) + B^A(\bar{x}, t) \bar{g}^A(\bar{x}, t, \bar{u}, \bar{u}^A, \bar{w}^A), \quad (5.4.7)$$

where $\bar{w}^A(t)$ is defined identically as $\bar{w}$, $B^A(\bar{x}, t)$ is some known $N \times n$ matrix, and $\bar{g}^A$ is an unknown n-vector function within a known class $\mathcal{G}$. Then let us introduce a safety zone $\Delta^A_\mathcal{A}$ such that for each $\bar{w}(\cdot) \in \mathcal{W}$, $\bar{g}(\cdot) \in \mathcal{G}$ there are scalars $\beta_0 > 0$, $\beta > 0$ such that for all $(\bar{x}, t) \in \Delta^A_\mathcal{A}$,

$$(\bar{u}^A)^T \bar{g}(\bar{x}, t, \bar{u}, \bar{u}^A, \bar{w}^A) \geq \beta_0 \|\bar{u}^A\| [\|\bar{u}^A\| - \beta] \quad (5.4.8)$$

for all $\bar{u} \in U$, $\bar{u}^A \in U^A$, $\bar{w}^A \in W(\bar{x}, t)$.

Denoting, as before in Section 3.4, $\alpha^A(\bar{x}, t) \triangleq [B^A(\bar{x}, t)]^T \nabla^T V(\bar{x}, t)$, we introduce an adaptive parameter $\lambda(t) \in R$ defined by the adaptive law

$$\dot{\lambda} = l \|\alpha(\bar{x}, t)\|, \quad (5.4.9)$$

5.4 ADAPTIVE AVOIDANCE

for some constant $l > 0$, and the adaptive controller

$$\mathscr{P}^A(\bar{x}, \lambda, t) \triangleq \left\{-\frac{\alpha(\bar{x}, t)}{\|\alpha(\bar{x}, t)\|}\lambda\right\}, \quad \alpha(\bar{x}, t) \neq 0, \quad (5.4.10)$$

for all $(\bar{x}, \lambda, t) \in \Delta_{\mathscr{A}}^S \times R^+ \times R$, where $\Delta_{\mathscr{A}}^S \triangleq \Delta_{\mathscr{A}} \cap \Delta_{\mathscr{A}}^A$.
Now we can state the following:

Theorem 5.4.2. *Suppose that $\mathscr{P}(\bar{x}, t)$ and $V(\bar{x}, t)$ satisfy Theorem 5.4.1 and there are $\Delta_{\mathscr{A}}^A$ satisfying (5.4.8), $\mathscr{P}^A(\bar{x}, t)$ of (5.4.10), and the adaptive law (5.4.9), then Definition 5.4.1 holds for (3.3.1) with the selector (5.4.7).*

Proof. The proof is the same as that for Theorem 5.4.1 except that we consider now a motion $k(\bar{u}, \bar{u}^A, \lambda, \bar{x}^0, t_0, \cdot): R_0^f \to \Delta \times R$ of $\mathscr{K}(\bar{x}^0, t_0)$ enlarged by the classes $\mathscr{F}, \mathscr{G}, \mathscr{W}$; the set $\Delta_{\mathscr{A}}$ is replaced by $\Delta_{\mathscr{A}}^A$, and the boundedness of $V(\bar{x}(t), t) = V(t)$ is demonstrated as follows (recall (5.4.5)). Define V^A: $[t_3, t_4) \to R$ by

$$V^A(t) = V(t) + (\beta_0/2l)(\lambda(t) - \beta)^2, \quad (5.4.11)$$

where $\beta_0 > 0$ and $\beta \geq 0$ are chosen to satisfy (5.4.8). Then

$$\dot{V}^A(t) = \dot{V}(t) + (\beta_0/l)(\lambda - \beta)\dot{\lambda}$$
$$= \partial V/\partial t + \nabla V(\bar{x}, t)\dot{\bar{x}} + (\beta_0/l)(\lambda - \beta)\dot{\lambda},$$

which upon substituting (5.4.7), (5.4.9), (5.4.10), and (ii) of Theorem 5.4.1 gives

$$\dot{V}^A(t) \leq 0 + \nabla V(\bar{x}, t)B^T(\bar{x}, t)\bar{g}(\bar{x}, t, \bar{u}, \bar{u}^A, \bar{w}^A) + (\beta_0/l)(\lambda - \beta)\dot{\lambda}$$
$$= \alpha(\bar{x}, t)^T\bar{g}(\bar{x}, t, \bar{u}, \bar{u}^A, \bar{w}^A) + \beta_0\|\alpha(\bar{x}, t)\|(\lambda - \beta). \quad (5.4.12)$$

If $\alpha(\bar{x}, t) = 0$, then clearly $\dot{V}^A(t) \leq 0$. If $\alpha(\bar{x}, t) \neq 0$, then utilizing (5.4.9) and (5.4.10),

$$\alpha(\bar{x}, t) = -\left(\frac{\|\alpha(\bar{x}, t)\|}{\lambda}\right)\bar{u}^A,$$

and it follows from (5.4.12) and (5.4.8) that

$$\dot{V}^A(t) \leq -\left(\frac{\|\alpha(\bar{x}, t)\|}{\lambda}\right)(\bar{u}^A)^T\bar{g}(\bar{x}, t, \bar{u}, \bar{u}^A, \bar{w}) + \beta_0\|\alpha(\bar{x}, t)\|(\lambda - \beta)$$
$$\leq -\left(\frac{\|\alpha(\bar{x}, t)\|}{\lambda}\right)\beta_0\|\bar{u}^A\|(\|\bar{u}^A\| - \beta) + \beta_0\|\alpha(\bar{x}, t)\|(\lambda - \beta)$$
$$= -\beta_0\|\alpha(\bar{x}, t)\|(\lambda - \beta) + \beta_0\|\alpha(\bar{x}, t)\|(\lambda - \beta) = 0.$$

Thus, on $[t_3, t_4)$, $\dot{V}^A(t) \leq 0$. Hence $V^A(t) \leq V^A(t_3)$ for $t \in [t_3, t_4)$. Using (5.4.11) and the fact that $(\beta_0/2l)(\lambda - \beta)^2$ is nonnegative, we obtain $V(t) \le V^A(t_3)$ for all $t \in [t_3, t_4)$, which was to be shown.

It may be demonstrated that the conditions of Theorem 5.4.1 follow from those of Theorem 5.1.1'. To do so, denote $V(\cdot)$ of Theorem 5.4.1 by $\hat{V}(\cdot)$ and show that there is a C^1 function $\phi: R \to R$ such that for all $t \in R$, $\hat{V}(\bar{x}^1, t) = \phi(t)$, $\forall x^1 \in \partial \mathcal{A}$, $\dot{\phi}(t) \le 0$. Then letting $D_{\mathcal{A}} \times R = \{(\bar{x}, t) \in \hat{D}_A \times R \mid \hat{V}(\bar{x}, t) - \phi(t) \ne 0\}$, we define our $V(\cdot): D_{\mathcal{A}} \times R \to R$ by

$$V(\bar{x}, t) \triangleq [\hat{V}(\bar{x}, t) - \phi(t)]^{-1}. \tag{5.4.13}$$

Now suppose that $\hat{V}(\cdot)$ and $\mathcal{P}(\bar{x}, t)$ satisfy Theorem 5.1.1' for (3.3.1) specified by (5.4.1) and that $V(\cdot)$ is given by (5.4.13). Then

$$\alpha(\bar{x}, t) = -[\hat{V}(\bar{x}, t) - \phi(t)]^{-2} \hat{\alpha}(\bar{x}, t) = -V(\bar{x}, t)^2 \hat{\alpha}(\bar{x}, t),$$

where

$$\hat{\alpha}(\bar{x}, t) \triangleq [B^A(\bar{x}, t)]^T \nabla^T \hat{V}(\bar{x}, t). \tag{5.4.14}$$

Also,

$$\mathcal{P}^A(\bar{x}, \lambda, t) = \{(\hat{\alpha}(\bar{x}, t)/\|\hat{\alpha}(\bar{x}, t)\|)\lambda\}, \quad \hat{\alpha}(\bar{x}, t) \ne 0, \tag{5.4.15}$$

and

$$\dot{\lambda} = l[\hat{V}(\bar{x}, t) - \phi(t)]^{-2} \|\hat{\alpha}(\bar{x}, t)\|. \tag{5.4.16}$$

Example 5.4.1. To illustrate the introduced method, consider the uncertain linear system

$$\dot{\bar{x}} = A\bar{x} + B\bar{u} + B^A \bar{u}^A + C\bar{w} + C^A \bar{w}^A \tag{5.4.17}$$

with $\bar{u}(t) \in U$, $\bar{u}^A(t) \in U^A$, $\bar{w}(t) \in W$ known, $\bar{w}^A(\cdot): R \to R^1$ an unknown bounded function with unknown bound, and A, B, B^A, C, C^A known matrices of appropriate dimensions. Thus there is a known $\rho_W \in R^+$ and unknown $\rho_{WA} \in R^+$ such that

$$\|\bar{w}(t)\| \le \rho_W, \|\bar{w}^A(t)\| \le \rho_{WA}. \tag{5.4.18}$$

We make the following assumptions:

(i) $-A$ is stable,
(ii) there are matrices D and D^A such that $C = BD$ and $C^A = B^A D^A$. Thus we rewrite (5.4.17) as

$$\dot{\bar{x}} = A\bar{x} + B(\bar{u} + D\bar{w}) + B^A(\bar{u}^A + D^A \bar{w}^A), \tag{5.4.19}$$

which is of the form (5.4.7), with the obvious meaning of terms. Now choose any negative-definite $N \times N$ matrix Q and solve the Liapunov matrix equation to get $\hat{V} = \bar{x}^T P \bar{x}$. Then define $\mathcal{A} \triangleq \{\bar{x} \mid \bar{x}^T P \bar{x} \le a\}$, and taking any

$\varepsilon > 0$ choose $\Delta_{\mathscr{A}} \triangleq \{\bar{x} \mid a < \bar{x}^T P \bar{x} < a + \varepsilon\}$, which satisfies condition (i) of Theorem 5.1.1. Now using

$$\mathscr{P}(\bar{x}, t) = (B^T P \bar{x} / \|B^T P \bar{x}\|)\rho, \qquad B^T P \bar{x} \neq 0, \qquad (5.4.20)$$

where $\rho \geq \|D\|\rho_W$, we satisfy (ii) as well. To show that (5.4.8) holds, we let $\Delta_{\mathscr{A}}^A = R^N$ and recall $\|\bar{w}^A\| \leq \rho_{WA}$. Then

$$(\bar{u}^A)^T \bar{g}(\bar{x}, t, \bar{u}, \bar{u}^A, \bar{w}^A) = (\bar{u}^A)^T [\bar{u}^A + D^A \bar{w}^A]$$
$$\geq \|\bar{u}^A\| [\|\bar{u}^A\| - \|D^A\|\rho_{WA}],$$

which satisfies (5.4.8), with $\beta_0 = 1$ and $\beta = \|D^A\|\rho_{WA}$. Using (5.4.14)–(5.4.16), we have

$$\mathscr{P}^A(\bar{x}, \lambda, t) = \left\{\frac{(B^A)^T P \bar{x}}{\|(B^A)^T P \bar{x}\|} \lambda\right\}, \qquad (B^A)^T P \bar{x} \neq 0, \qquad (5.4.21)$$

$$\dot{\lambda} = \frac{l \|(B^A)^T P \bar{x}\|}{(\bar{x}^T P \bar{x} - a)^2} \qquad (5.4.22)$$

for all $(\bar{x}, \lambda, t) \in \Delta_{\mathscr{A}} \times R^+ \times R$.

5.5 AVOIDANCE OF MOVING OBSTACLES

When the arm has to avoid an obstacle resting on a conveyor, we face, perhaps, the simplest case of the objective in question. This obviously only occurs as long as the motion of the conveyor is well defined, which by no means always happens. The case of two arms avoiding each other is probably at the other end of the range of difficulty for the problems concerned. We will pose these cases here in terms of the *avoidance of some obstacle path*, as a study symmetric to that of target-path tracking in Section 4.6.

Let us suppose that the obstacle path to be avoided is given in terms of the vector $\bar{x}_A(t) = (\bar{q}_A(t), \dot{\bar{q}}_A(t))$ moving in Δ with $t \in R_0^+$ and thus producing a curve $\bar{x}_A(R_0^+)$. In a manner similar to $\bar{x}_m(R_0^+)$, the curve $\bar{x}_A(R_0^+)$ may be augmented into $\Delta \times R$, but if this curve was given in $\Delta \times R$ its projection into Δ may not be unique.

We employ the method of *relative coordinates* investigated in Section 4.6. Again, consider first the system without uncertainty (1.5.3′), specified either by (1.3.24′) or by (1.5.16″). Substituting $\bar{x}_A(t)$ into the above equations, we determine the open-loop controller $\bar{u}_A(t)$ of the (3.1.1) type. Then we substitute this into the state equations concerned and, given the same initial conditions as those for the path starting point $\bar{x}_A^0 = (\bar{q}_A^0, \dot{\bar{q}}_A^0)^T$, we obtain the motion $\bar{x}(t) = k(\bar{u}_A, \bar{x}_A^0, t_0, t)$, $t \geq t_0$, in $\Delta \times R$ called the *obstacle motion*, denoted briefly $\bar{k}_A(t)$. Here again we must have the proviso that the $\bar{x}_A(t)$ is compatible with a solution to the equation concerned. Examples were given in Section 4.6.

Then we define the same relative vector $\bar{z}(t) = \bar{x}(t) - \bar{k}_A(t)$, $t \geq t_0$, as before, which yields [see Eq. (4.6.1)]

$$\dot{\bar{z}} = \bar{f}(\bar{z} + \bar{k}_A(t), t, \bar{u}) - \bar{f}(\bar{k}_A(t), t, \bar{u}_A). \tag{5.5.1}$$

By the same argument as that presented in Section 4.6 we now want the zero motion $\bar{z}(t) \equiv 0$ to be completely unstable, or the controllability for avoidance of some $\mathscr{A}$ enclosing the antitarget $\mathfrak{T}_{\mathscr{A}} = \{\bar{z}(t) \equiv 0\}$. In a manner similar to that used in Section 4.6, we observe that substituting a given $\bar{w}$ makes the controllability a particular case of strong controllability. Thus all the objectives and methods discussed in this chapter so far, that is, Sections 5.1–5.4, apply directly and may be successfully used in our case.

Now referring to (1.3.24′) we have from (5.5.1),

$$\ddot{\bar{q}}_z + \bar{\Gamma}(\bar{q}_z + \bar{q}_A, \dot{\bar{q}}_z + \dot{\bar{q}}_A) + \bar{D}(\bar{q}_z + \bar{q}_A, \dot{\bar{q}}_z + \dot{\bar{q}}_A) + \bar{\Pi}(\bar{q}_z, \bar{q}_A)$$
$$= \bar{F}(\bar{q}_z + \bar{q}_A, \dot{\bar{q}}_z + \dot{\bar{q}}_A, \bar{u}) + \bar{R}(\bar{q}_z + \bar{q}_A, \dot{\bar{q}}_z + \dot{\bar{q}}_A, t, \bar{u}_A). \tag{5.5.2}$$

Then, letting $V(\bar{z}) = H(\bar{z})$, we obtain the main condition of all the theorems regarding the accumulative power, as

$$f_0(\bar{z}, \bar{x}_A, t, \bar{u}, \bar{u}_A) = \bar{F}(\bar{q}_z + \bar{q}_A, \dot{\bar{q}}_z + \dot{\bar{q}}_A, \bar{u})\dot{\bar{q}}_z$$
$$- \bar{R}(\bar{q}_z + \bar{q}_A, \dot{\bar{q}}_z + \dot{\bar{q}}_A, t, \bar{u}_A)\dot{\bar{q}}_z$$
$$\geq 0. \tag{5.5.3}$$

Adjusting this condition to the various versions of the corollaries concerned, we obtain the maximum or the L–G controllers for the said avoidance.

Example 5.5.1. Inverting the roles from tracking to avoidance in Example 4.6.1, we again consider

$$\ddot{q} + u + aq - bq^3 = 0 \tag{5.5.4}$$

or

$$\dot{x}_1 = x_2,$$
$$\dot{x}_2 = -ax_1 + bx_1^3 - u, \tag{5.5.5}$$

with the objective being to avoid the obstacle trajectory

$$\frac{1}{2}x_{A2}^2 + \frac{a}{2}x_{A1}^2 - \frac{b}{4}x_{A1}^4 = h_c \leq \frac{a^2}{4b} \tag{5.5.6}$$

of an obstacle moving in Δ_H (see Eq. (4.6.7) and Fig. 4.25). Substituting (5.5.6) into (5.5.5),

$$\dot{x}_{A1} = x_{A2},$$
$$\dot{x}_{A2} = -ax_{A1} + bx_{A1}^3 - u_A, \tag{5.5.7}$$

5.5 AVOIDANCE OF MOVING OBSTACLES

with $u_A \equiv 0$ [see Eq. (4.6.8)], which produces the following form of Eqs. (5.5.1):

$$\dot{z}_1 = z_2,$$
$$\dot{z}_2 = -az_1 + bz_1^3 + 3b(z_1 x_{A1}^2 + z_1^2 x_{A1}) - u. \tag{5.5.8}$$

We want the avoidance of $\mathscr{A} : z_1^2 + z_2^2 < \eta$, with η suitable to include $\mathfrak{T}_A = \{\bar{z}(t) \equiv 0\}$ in $\mathscr{A}$, either for all $t \in R_0^+$, or for real time, or for stipulated real time, depending upon the need. Let us focus attention on the first. Again, letting $V(\bar{z}) = \frac{1}{2} z_2^2 + \frac{1}{2} a z_1^2 - \frac{1}{4} b z_1^4$, we need only verify (5.5.3), the remaining conditions of Theorem 5.5.1 obviously being satisfied. This verification gives the control condition

$$3b(z_1 x_A^2 + z_1^2 x_A) z_2 - u z_2 \geq 0, \tag{5.5.9}$$

suggesting the choice of the program defined by

$$u = 3b(z_1 x_A^2 + z_1^2 x_A). \tag{5.5.10}$$

It must be remembered, however, that for our present purpose the initial conditions $\bar{z}^0 = \bar{z}(t_0)$ and $\dot{\bar{z}}^0 = \dot{\bar{z}}(t_0)$ must be such as to place $\bar{z}^0$ and $\dot{\bar{z}}^0$ outside $\mathscr{A}$. The discussion following (4.6.12) is entirely applicable for avoidance as well. We leave this discussion to the reader. ∎

We also have here the problem of $\ddot{\bar{q}}_A$ appearing in the controller and the analogous can be used to get rid of it. The same refers to the linearization problem [see Eq. (4.6.15)].

Passing over to uncertain systems of the (3.3.1)-type, we use the product space method with inverted role of the diagonal $\mathscr{M}$ and its neighborhood. These sets must now be avoided, by motions of the product system

$$\dot{\bar{x}}_A = f(\bar{x}_A, t, \bar{u}_A(\bar{w}, t), \bar{w}),$$
$$\dot{\bar{x}} = f(\bar{x}, t, \bar{u}, \bar{w}) \tag{5.5.11}$$

in $\Delta^2 = \Delta \times \Delta \subset R^{2N}$. We define $X_A \triangleq (\bar{x}, \bar{x}_A)^T$ and

$$\mathscr{A}^2 \triangleq \{X_A \in \Delta^2 \,|\, \|x - \bar{x}_A\| < \eta\} \tag{5.5.12}$$

(see Fig. 4.26), replacing $\mathfrak{T}_C$ by $\mathscr{A}$. The technique does not change. We use the avoidance theorems of Sections 5.1–5.4 in terms of (5.5.11), its product motions $k(\bar{u}, \bar{u}_A, \bar{X}_A^0, t_0, \cdot) \in \mathscr{K}(X_A^0, t_0)$ in $\Delta^2 \times R \subset R^{2N+1}$, the avoidance set (5.5.12), and some Liapunov function similar to that used in Sections 4.6 and 4.7, defined outside $\mathscr{A}^2$ somewhere in Δ^2.

The reader may now recall Examples 5.2.2 and 5.2.3. The obstacle represented by $\mathfrak{T}_A$ is moving with the specified motion (5.2.10) represented by $\bar{x}_A = (r, \theta_\mathscr{A}, \dot{r}, \dot{\theta}_\mathscr{A})^T$, while the gripper moves according to (5.2.9), the motion represented by $\bar{x} = (\mathscr{R}, \theta, \dot{\mathscr{R}}, \dot{\theta})^T$. Observe also that it was convenient in Examples 5.2.2 and 5.2.3 to take r, θ as relative variables. The product-space vector is then given by $(r, \theta_{\mathcal{A}}, \mathcal{R}, \theta, \dot{r}, \dot{\theta}_{\mathcal{A}}, \dot{\mathcal{R}}, \dot{\theta})^T \triangleq \bar{X}_A$, and we have taken the Liapunov function simplified to $V(\bar{X}_A) = r$, which suited our purpose.

Path avoidance in product space is a subcase of a wider problem when the motion of an obstacle is given not in terms of a specified path (which becomes a solution of some equation), but in terms of a dynamical system—some state equations independent of our manipulator, such as another working manipulator or another moving object. Then we have a *model reference* (usually) *adaptive avoidance*, symmetric to Model Reference Adaptive Control (MRAC). Since the corresponding technique covers that of path avoidance, we describe them together below.

The obstacle model is given by a specified well-defined vector $\bar{f}_A(\bar{x}_A, t, \bar{u}_A, \bar{\lambda}_A)$, generally different from $\bar{f}(\cdot)$ thus leading to the generalization of (5.5.11) in terms of the following system of equations

$$\dot{\bar{x}}_A = \bar{f}_A(\bar{x}_A, t, \bar{u}_A, \bar{\lambda}_A),$$
$$\dot{\bar{x}} = \bar{f}(\bar{x}, t, \bar{u}, \bar{w}, \bar{\lambda}), \quad (5.5.13)$$

with given $\bar{\lambda}_A = $ const and adjustable $\lambda(t) \in \Lambda \subset R^l$. Note that we still have here a simplified case, for in the instance of two cooperating arms both are essentially on an equal basis, that is, $\bar{\lambda}_A$ should be made adjustable as well. There is no literature on the case, even for linear systems.

Following the same technique as for MRAC in Section 4.7 (see Figs. 4.31 and 4.32) we choose adaptive laws

$$\dot{\bar{\lambda}} = \bar{f}_a(\bar{X}_A, t, \bar{\lambda}), \quad (5.5.14)$$

and a signal adaptive control program

$$\bar{u} \in \mathcal{P}(\bar{X}_A, t, \bar{\lambda}, \bar{u}_A), \quad (3.5.15)$$

to generate strong controllability for avoidance (real-time avoidance, stipulated time avoidance, ultimate avoidance, etc.) of the set

$$\mathcal{A}^3 \triangleq \{(\bar{X}_A, \bar{\gamma}_A) \in \Delta^2 \times \Lambda \,|\, \|\bar{x} - \bar{x}_A\| < \eta, \|\bar{\gamma}_A\| < \eta\}, \quad (5.5.16)$$

where $\bar{\gamma}_A(t) \triangleq \bar{\lambda}(t) - \lambda_A$, in the set $\Delta^2 \times \Lambda$ of the product space R^{2N+l}, by the product motions of the product system consisting of (5.5.13) and (5.5.14).

We may use Theorem 5.1.1 to give us the following sufficient conditions. Define $\Delta^2_{\mathcal{A}} \times \Lambda \triangleq \Delta_{\mathcal{A}} \times \Delta_{\mathcal{A}} \times \Lambda$, $D^3_{\mathcal{A}} \triangleq D_{\mathcal{A}} \times D_{\mathcal{A}} \times \Lambda$, and recall (5.1.2). We have

Theorem 5.5.1. *The (3.3.1)-like system with selecting equations (5.5.13) and (5.5.14) is strongly controllable on $(\Delta^2 \times \Lambda) - \mathcal{A}^3$ for avoidance of suitable $\mathcal{A}^3$ if there are $\Delta^2_{\mathcal{A}} \times \Lambda$, $\mathcal{P}(\bar{X}_A, t, \bar{u}_A, \bar{\lambda})$, and a C^1 function $V(\cdot): D^3_{\mathcal{A}} \to R$ such that*

(i) $V(\bar{X}_A, \bar{\gamma}_A) > v_A^+ \quad \forall (\bar{X}_A, \bar{\gamma}_A) \notin \mathcal{A}^3$;

5.5 AVOIDANCE OF MOVING OBSTACLES

(ii) for all $\bar{u} \in \mathscr{P}(\bar{X}_A, t, \bar{u}_A, \bar{\lambda})$,

$$\nabla V(\bar{X}_A, \bar{\gamma}_A) \cdot (\bar{f}(\bar{x}, t, \bar{u}, \bar{w}, \bar{\lambda}), \bar{f}_A(\bar{x}_A, t, \bar{u}_A, \bar{\lambda}_A), f_a(\bar{X}_A, t, \bar{\lambda})) \geq 0 \quad (5.5.17)$$

for all $\bar{w}(t) \in W$.

The proof follows by the same argument as for Theorem 5.1.1, namely, contradiction between (i) and (ii) upon crossing $\partial \mathscr{A}^3$. The controller is found below in conjunction with the adaptive law.

To focus attention on a specific case, let us assume that the antitarget model is energy accumulative:

$$\dot{H}_A(\bar{x}_A) \geq 0, \quad (5.5.18)$$

where $H_A(\bar{x}_A)$ is the total energy of such a model. Then we let $l \leq N$ and choose

$$V(\bar{X}_A, \bar{\gamma}_A) \triangleq |H_A(\bar{x}) - H_A(\bar{x}_A)| + \bar{a} \cdot \bar{y}_A + k\bar{a}_\eta(\bar{\gamma}_A - \eta),$$

where $\bar{a}_\eta = (\operatorname{sign} \gamma_{A1} - \eta_1, \ldots, \operatorname{sign} \gamma_{AN} - \eta_N)$, which satisfies (i) if for suitable $k > 0$

$$\hat{a}(\bar{\gamma} - \eta) \geq \eta \quad (5.5.20)$$

where $\bar{\bar{a}} \triangleq (a + k\bar{a}_\eta)$. The adaptive condition

$$\bar{\bar{a}} \dot{\bar{\gamma}}_A = \dot{H}_A(\bar{x}_A) \quad (5.5.21)$$

and the control condition

$$0 \leq \nabla H_A(\bar{x}) \bar{f}(\bar{x}, t, \bar{u}, \bar{w}, \bar{\lambda}) \leq 2\dot{H}_A(\bar{x}_A), \quad \forall \bar{w} \in W \quad (5.5.22)$$

produce the desired $\dot{V}(t) \geq 0$ outside $\mathscr{A}^3$ satisfying (ii). The adaptive condition (5.5.21) is implied by the adaptive laws

$$\dot{\gamma}_{Ai} = \frac{1}{\hat{a}_i} \frac{\partial H_A(\bar{x}_A)}{\partial x_{Ai}} f_{Ai}(\bar{x}_A, t, \bar{u}_A, \bar{\lambda}_A), \quad i = 1, \ldots, l = N. \quad (5.5.23)$$

The control condition (5.5.22) is satisfied by the control program $\mathscr{P}(X_A, t, \bar{\lambda})$ defined by

$$\max_{\bar{u}} \min_{\bar{w}} \nabla H_A(\bar{x}) \cdot \bar{f}(\bar{x}, t, \bar{u}, \bar{w}, \bar{\lambda}) = \dot{H}_A(\bar{x}_A) \quad (5.5.24)$$

The case of $l < N$ is analogous to (4.7.23). We cannot get closer to $\bar{u}$ than the above unless $H_A(\cdot)$ and $\bar{f}(\cdot)$ are specified. Observe that by (5.5.18) and (5.5.23), $\dot{\gamma}_{Ai}(t) \geq 0$, whence the values of $\bar{\lambda}(t)$ and $\bar{\lambda}_A$ diverge.

Exercises and Comments

5.5.1 Expand Example 5.5.1 to the case of stipulated real-time avoidance of the moving obstacle. Combine a case of path following with another of path avoidance for the same example. Discuss the case in which the obstacle path is defined parametrically in t.

5.5.2 Granted that (5.5.11) is a subcase of (5.5.13), using the (5.5.13) technique can we obtain more specific results for path avoidance than for the model-reference adaptive avoidance?

5.5.3 Apply Theorem 5.5.1 when the reference model is dissipative: $\dot{H}_A(\bar{x}_A) = f_0(\bar{x}_A, t, u_A, \lambda_A) \leq 0$.

5.5.4 Specify (5.5.24) in terms of our RP manipulator of Section 1.1. Obtain a similar controller for the case in Exercise 5.5.3.

Chapter 6

Adaptive Identification

6.1 IDENTIFICATION OF STATES AND PARAMETERS

Another alternative to the penalizing worst-case design as a means to combat uncertainties is to identify them, possibly on-line, adaptively, while the manipulator performs its task. This thus works as an adjoint procedure to any other type of control, be it point-to-point, path tracking, or model-reference control, for both types of objectives, either reaching or avoidance.

The uncertainty is not confined to parameters. The states of the system, needed for both feedback control and adaptive laws, may not be observable and may thus be uncertain. Moreover, as we have seen from Section 3.3 onward, uncertain parameters in the system produce nonunique motions which alone make the states uncertain within the attainability sets. Consequently, we face the problem of *simultaneous identification of both parameters and states*, while at the same time securing the control objectives—whatever they may be.

Traditionally, stochastic methods are applied to such a task (see Eykhoff [1]), but they may prove to be unfeasible not only because of Murphy's law mentioned at the opening of our discussion in Section 3.3: Whatever bad may happen, will, but also because there are many situations in which we can make no probabilistic assumptions about the random variables involved and/or in which the process is observed and controlled during a single run, etc. Then it seems useful to employ the theory of *adaptive identifiers*, also called predictors or estimators, which has worked well for a long time in many other applications (see Narendra and Tripathi [1]). The theory treats the uncertain state and parameters of the investigated plant as if given, and constructs an auxiliary mathematical system termed an identifier whose states and parameters converge ultimately to those of the manipulator. At present, the theory is mainly linear and identifies constant parameters (see Narendra [1]). It uses the technique of subtracting the equations of the manipulator and the identifier, thus producing an "error" equation, similar to the linear model tracking in Section 4.7. Then the asymptotic stability of the zero-error produces the desired convergence. For genuinely nonlinear systems, such subtraction of equations is impractical, if not impossible, even if the identifier is made linear, which is not advisable for the sake of the multiple equilibria in manipulator equations. Moreover, the parameters are often variable.

Thus we propose to achieve the required convergence in the same way that it was done in Section 4.7—by stabilizing the product system about the "diagonal" in the product space of the two state spaces, that is, those of the manipulator and the identifier. Note that since many of the tasks of the manipulator are often within a stipulated time and with desired accuracy, the time of identification and the "closeness" to the diagonal must also be stipulated.

Although there is a vast literature on adaptive identification (see Landau [1, 2], Narendra [1], and Rajbman [1] for review) and the problem is more than 20 years old, there has been very little done for the cases in which linearization is impossible—such as, unfortunately, in our manipulators of recent design. There are obviously attempts to kill the nonlinearity by a powerful controller while performing the discussed identification but such killing makes the identification method not much better than the worst-case design, which penalizes us in power usage. Envisage the situation in which the controller must push the manipulator over several energy thresholds to the selected stable equilibrium just in order to justify the linear mathematical model of the problem (see Belusov and Furasov [1], Hwang and Seinfeld [1], Kon, Elliot, and Tarn [1], Saridis and Lobbia [1], Bradshaw and Porter [1], and Griffith and Kumar [1]).

In the general terms of (1.5.3), state identification (observability) was discussed by Kostiukovski [1] as early as in 1968, followed by Roitenberg [1] and Drenick [1, 2] (see also Schmitendorf [1] and Wilson and Leitmann [1] for discussions with respect to optimal control). A marked attack on nonlinear observers was made by Galperin [1] and Nikonov [1]. The product-space method was introduced by Skowronski [20, 22, 23, 25] and applied to manipulator control by Skowronski [27] and Skowronski and Pszczel [1]. We shall follow this line here.

In a manner similar to that in Section 4.7, it seems instructive to refer briefly to the linear *model reference adaptive identification* technique which already has a long tradition. Consider the system (see Kroc [1])

$$\dot{\bar{x}} = A\bar{x} + B\bar{u} + \mathcal{R}(t) \tag{6.1.1}$$

with the output relation

$$\bar{y} = C\bar{x}, \tag{6.1.2}$$

where A, B are constant matrices of appropriate dimension with unknown but bounded coefficients, C is a known $n \times N$ matrix, $\mathcal{R}(t)$ is an $N \times 1$

6.1 IDENTIFICATION OF STATES AND PARAMETERS

perturbation matrix, and $\bar{y}(t)$ is a measured output or read-out n vector. To avoid, at this stage, any nonidentificational problems, we assume $\bar{u}$ given for a specified objective. We search for $\bar{x}(t)$ as well as the constant parameters in A, B, and the search is made within the known bounded sets $\Delta, \mathfrak{A}, \mathfrak{B}$, respectively. To this end, we *design* the identifier

$$\dot{\bar{x}}_p = (A_p(t) + KC)\bar{x}_p + K\bar{y} + B_p(t)\bar{u} + \mathcal{R}(t) \qquad (6.1.3)$$

with $A_p(t)$, $B_p(t)$ of the same dimensions as A, B and K some constant $N \times n$ matrix. Now choose $\bar{x}_p^0 = \bar{x}_p(t_0)$ from Δ and parameters of $A_p^0 = A_p(t_0)$ and $B_p^0 = B_p(t_0)$ from $\mathfrak{A}$ and $\mathfrak{B}$, respectively. Since the $\Delta, \mathfrak{A}, \mathfrak{B}$ are bounded, $\bar{x}_p^0$ and the parameters of A_p^0, B_p^0 may not be further from $\bar{x}^0$ and those of A, B than within the accuracy of the diameters of $\Delta, \mathfrak{A}, \mathfrak{B}$. Granted this, we should show that the mismatch error between states and parameters of (6.1.1) and (6.1.3) are below a stipulated $\eta > 0$ for all $t \geq t_0 + T_\eta$, $T_\eta < \infty$, perhaps also stipulated. The linear techniques used so far replace the above requirement by the demand that these errors tend to zero as $t \to \infty$. Also, either state or parameter identification is attempted, never both. Subtracting (6.1.1) from (6.1.3), with (6.1.2) substituted, we obtain the error equation [cf. (4.7.3)]

$$\dot{\bar{e}} = (A_p(t) - KC)\bar{e} + A_e\bar{x} + B_e\bar{u} \qquad (6.1.4)$$

where $\bar{e}(t) \triangleq \bar{x}_p(t) - \bar{x}(t)$, $A_e(t) = A_p(t) - A$, $B_e(t) = B_p(t) - B$. Then we require that *either*, given the parameters $A_e \equiv 0$ and $B_e \equiv 0$, some $\bar{e}(t_0) = \bar{e}^0 \in \Delta$ implies $\bar{e}(t) \to 0$ as $t \to \infty$ (state identification) *or*, given $e(t) \equiv 0$, parameters from some $A_e^0 = A_e(t_0)$ and $B_e^0 = B_e(t_0)$ imply $A_e(t), B_e(t) \to 0$ as $t \to \infty$ (parameter identification).

Let us look at state identification (observability). With $A_p(t) \equiv A$ and $B_p(t) \equiv B$, the identifier (6.1.3) becomes the well-known Luenberger observer (see Luenberger [1, 2])

$$\dot{\bar{x}}_p = (A - KC)\bar{x}_p + K\bar{y} + B\bar{u}, \qquad (6.1.5)$$

with the error equation

$$\dot{\bar{e}} = (A - KC)\bar{e}. \qquad (6.1.6)$$

It then suffices to select K such as to ensure that $(A - KC)$ will be a stable matrix thus producing asymptotic stability of $\bar{e}(t) \equiv 0$.

For parameter identification, we want asymptotic stability of the origin in the space $\{\bar{e}, A_e, B_e\}$. Following Narendra [1], it is possible to do it with the Liapunov function (4.7.5) and the adaptive laws (4.7.8). Note, however, that by using this method only stability is attained (the Liapunov derivative is not negative-definite). It requires, according to Narendra [1], an additional "intensification" of $\bar{u}$ to attain asymptotic stability. We left this problem open in Section 4.7 as we do now, since it is rather pointless for our illustrative purpose. The procedure is illustrated in Fig. 6.1, which is symmetric to Fig. 4.24.

Observe that the same result as above may be obtained using our *product-space method*, namely, by direct analysis of the asymptotic stability of the "diagonal" $\{\bar{x}, \bar{x}_p, A, B, A_p, B_p | \bar{x} = \bar{x}_p, A = A_p, B = B_p\}$ in the product space $\{\bar{x}, \bar{x}_p, A_e, B_e\}$. We shall show this on the subcase of state identification $A_e = 0$, $B_e = 0$. Indeed, let the Liapunov function be $V(\bar{x}, \bar{x}_p) \triangleq (\bar{x} - \bar{x}_p)^T(\bar{x} - \bar{x}_p)$. The Liapunov derivative is then

$$\dot{V}(\bar{x}(t), \bar{x}_p(t)) = (\dot{\bar{x}} - \dot{\bar{x}}_p)^T(\bar{x} - \bar{x}_p) + (\bar{x} - \bar{x}_p)^T(\dot{\bar{x}} - \dot{\bar{x}}_p).$$

Substituting Eqs. (6.1.1), (6.1.2), and (6.1.5) into the derivative we have

$$\dot{V}(\bar{x}(t), \bar{x}_p(t)) = -(\bar{x} - \bar{x}_p)^T Q (\bar{x} - \bar{x}_p),$$

with $-Q = (A_p - KC)^T + (A_p - KC)$. With Luenberger's K the matrix Q becomes positive definite, yielding the same result as before: $\bar{x} - \bar{x}_p \to 0$, $t \to \infty$, which means that the product-space method works on linear systems as well, and the Luenberger method becomes a special case of our product-space approach.

We turn now to the nonlinear system (1.5.3'')

$$\dot{\bar{x}} = \bar{f}(\bar{x}, t, \bar{u}, \bar{w}). \tag{6.1.7}$$

Because of $\bar{u} \in \mathcal{P}(\bar{x}, t)$, (6.1.7) is still a selecting equation of the contingent form (3.3.1), but now we take $\bar{w}(t)$ as if it were known, so the uncertainty "adaptively" disappears. The control vector $\bar{u}$ is to be selected by $\mathcal{P}(\bar{x}, t)$ as usual, that is, to produce some control objective of the system. The state observation available is given by the read-out (output) vector $\bar{y}(t) \in Y \subset R^n$, which is in general defined by

$$\bar{y}(t) \in Y(\bar{x}(t), t), \qquad t \geq t_0, \tag{6.1.8}$$

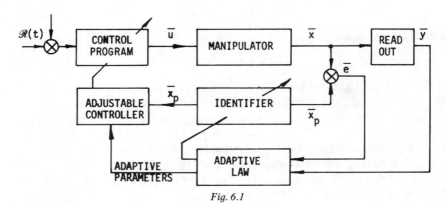

Fig. 6.1

6.1 IDENTIFICATION OF STATES AND PARAMETERS

where $Y(\cdot): \Delta \times R \to$ (all subsets of Y) estimates the *measurement noise* by the *known* band Y. In particular, $\bar{y}$ may be represented by some n-tuple of observable components of the state vector $\bar{x}$. It will also be assumed that we may have a noisy estimate of the change of $\bar{w}(t)$. Let the measurement be given by

$$\dot{\bar{w}} = \bar{f}_w(\bar{s}, t) \qquad (6.1.9)$$

with the noise $\bar{s}(t) \in \mathfrak{S}$, where $\mathfrak{S}$ is a known bounded set in R^l. We introduce the auxiliary system

$$\dot{\bar{x}}_p = \bar{f}_p(\bar{x}_p, t, \bar{u}, \bar{\mu}, \bar{y}), \qquad (6.1.10)$$

with the adaptive law

$$\dot{\bar{\mu}} = \bar{f}_\mu(\bar{x}_p, \bar{x}, \bar{y}, t), \qquad (6.1.11)$$

where $\bar{f}_p(\cdot)$, $\bar{f}_\mu(\cdot)$ are known (designed) vector functions with appropriate dimensions, $\bar{x}_p(t) \in \Delta \subset R^N$ is a known state vector of the identifier, and $\bar{\mu}(t) \in W$ is a vector of adjustable parameters of the same dimension as $\bar{w}$, that is $s + n$. We let $s + n = l$ without loss of generality. Note that since $\bar{y}$ is noisy [see Eq. (6.1.8)], Eqs. (6.1.10) and (6.1.11) are selecting equations of some contingent forms corresponding to Y of (6.1.8).

Let us now form a vector $\bar{Z}(t) \triangleq (\bar{x}(t), \bar{w}(t), \bar{x}_p(t), \bar{\mu}(t))^T$ in $\Delta \times W \times \Delta \times W \triangleq \Delta^2 \times W^2 \triangleq \mathscr{Z} \subset R^M$, $M = 2(N + l)$. Collecting the right-hand sides of (6.1.7), (6.1.9), (6.1.10), and (6.1.11) into one vector $\bar{F}(\bar{Z}, t, \bar{u}, \bar{s}, \bar{y})$, we may write all these equations as a product system

$$\dot{\bar{Z}} = \bar{F}(\bar{Z}, t, \bar{u}, \bar{s}, \bar{y}) \qquad (6.1.12)$$

subject to the set-valued control program $\mathscr{P}(\bar{x}, t)$ and the noise in (6.1.8) and (6.1.9). This makes (6.1.12) the selecting equation of the system

$$\dot{\bar{Z}} \in \{\bar{F}(\bar{Z}, t, \bar{u}, \bar{s}, \bar{y}) \,|\, \bar{u} \in \mathscr{P}(\bar{x}, t), \bar{s} \in \mathfrak{S}, \bar{y} \in Y(\bar{x}, t)\}. \qquad (6.1.13)$$

As before for (3.3.1), given $\bar{Z}^0 = \bar{Z}(t_0)$, for suitable $\bar{F}(\cdot)$, $\mathscr{P}(\cdot)$, and $Y(\cdot)$, we have absolutely continuous solutions $k(\bar{u}, \bar{Z}^0, t_0, \cdot): R_0^+ \to \mathscr{Z}$, within the class $\mathscr{K}(\bar{Z}^0, t_0)$. We again call them *product motions*. Note that the noise in (6.1.8) and (6.1.9) could, in some cases, be included in $\bar{w}$ by simply increasing the dimension of this vector. In general, however, the noise in (6.1.10) and (6.1.11), if present at all, must be fought against by the previous methods, unless we want an identifier to the identifier.

Let us now split up $\bar{Z}(t)$ into two $(N + l)$ vectors: $\bar{Z}(t) = (\bar{z}(t), \bar{z}_p(t))^T$, where $\bar{z}(t) \triangleq (\bar{x}(t), \bar{w}(t))^T \in \Delta \times W$, and $\bar{z}_p(t) \triangleq (\bar{x}_p(t), \bar{\mu}(t))^T \in \Delta \times W$. Then we define the "diagonal" set

$$\mathscr{M}' = \{\bar{Z} \in \mathscr{Z} \,|\, \|\bar{z} - \bar{z}_p\| = 0\}, \qquad (6.1.14)$$

where $\|\cdot\|$ is a norm in R^M. Moreover, given $\eta > 0$, let

$$\mathcal{M}'_\eta = \{\bar{Z} \in \mathcal{L} \mid \|\bar{z}(t) - \bar{z}_p(t)\| \leq \eta \ \forall t\}, \tag{6.1.15}$$

and let Δ_0 be some stipulated subset of Δ. We then have the following

Definition 6.1.1. The auxiliary models (6.1.10) and (6.1.11) represent an identifier of (6.1.7) with accuracy to within $\eta > 0$, briefly called an η *identifier* on $\Delta_0 \times R$, if and only if, given $\bar{f}, Y, \mathfrak{S}$, there are admissible $\bar{f}_p(\cdot), \bar{f}_\mu(\cdot), \mathscr{P}(\cdot)$ and a constant $T_\eta > 0$ such that for each product motion $k(\cdot) \in \mathscr{K}(Z^0, t_0)$, $(Z^0, t_0) \in \mathscr{L}_0$, $\mathscr{L}_0 \triangleq \Delta_0^2 \times W^2$, of (6.1.13), we have

$$k(\bar{u}, \bar{Z}^0, t_0, [t_0 + T_\eta, \infty)) \subset (\mathcal{M}'_\eta \cap \mathscr{L}_0) \times R. \tag{6.1.16}$$

Figure 4.27 may serve as a guide in envisaging the above in space, but the reader should be aware of the differences since our present $\mathcal{M}'$, $\mathcal{M}'_\eta$ are in different spaces. As a very simplified version, we propose Fig. 6.2.

Let $\mathcal{N}'(\partial \mathscr{L}_0)$ be a neighborhood of the boundary of $\mathscr{L}_0$ and let $\mathcal{N}'_\varepsilon \triangleq \mathcal{N}'(\partial \mathscr{L}_0) \cap \mathscr{L}_0$, $C\mathcal{M}'_\eta \triangleq \mathscr{L}_0 - \mathcal{M}_\eta$, while open $D_\eta^4 \supset \overline{C\mathcal{M}'_\eta}$ such that $D_\eta^4 \cap \mathcal{M}' = \emptyset$. We introduce two functions $V'_s(\cdot): \mathcal{N}'_\varepsilon \to R$ and $V'_\eta(\cdot): D_\eta^4 \to R$ with positive constants

$$v'_s \triangleq \inf V'_s(\bar{Z}) \mid \bar{Z} \in \partial \mathscr{L}_0,$$
$$v'^{-}_\eta \triangleq \inf V'_\eta(\bar{Z}) \mid \bar{Z} \in \partial \mathcal{M}_\eta \cap \overline{C\mathcal{M}_\eta}, \tag{6.1.17}$$
$$v'^{+}_\eta \triangleq \sup V'_\eta(\bar{Z}) \mid \bar{Z} \in \partial \mathscr{L}_0 \cap \overline{C\mathcal{M}_\eta}.$$

Theorem 6.1.1. *Given suitable Δ_0, η, the auxiliary system (6.1.10) and (6.1.11) is an η identifier on $\Delta_0 \times R$ of (6.1.7) if there are $\bar{f}_p(\cdot), \bar{f}_\mu(\cdot)$, two at least piece wise C^1 functions $V'_s(\cdot)$ and $V'_\eta(\cdot)$, and a winning control program*

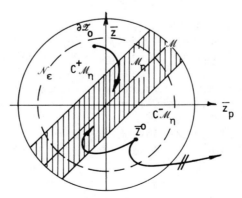

Fig. 6.2

6.1 IDENTIFICATION OF STATES AND PARAMETERS

$\mathscr{P}(\cdot)$, selected for some objective, can be such that for all $\bar{Z}$ in corresponding domains of definition

(i) $V'_S(\bar{Z}) < v'_S$ for $\bar{Z} \in \mathcal{N}'_\varepsilon$;
(ii) for each $\bar{u} \in \mathscr{P}(\bar{x}, t)$,

$$\nabla V'_S(\bar{Z}) \cdot \bar{F}(\bar{Z}, t, \bar{u}, \bar{s}, \bar{y}) \leq 0 \tag{6.1.18}$$

for all $\bar{s} \in \mathfrak{S}, \bar{y} \in Y, t \geq t_0$;

(iii) $0 \leq V'_\eta(\bar{Z}) \leq v'^+_\eta$, for $\bar{Z} \in \overline{C\mathcal{M}'_\eta}, v'^+_\eta > 0$;
(iv) $V'_\eta(\bar{Z}) \leq v'^-_\eta, \bar{Z} \in \mathcal{M}'_\eta \cap D^4_\eta$;
(v) for each $\bar{u} \in \mathscr{P}(\bar{x}, t)$, there is a constant $c > 0$ such that

$$\nabla V'_\eta(\bar{Z}) \cdot \bar{F}(\bar{Z}, t, \bar{u}, \bar{s}, \bar{y}) \leq -c \tag{6.1.19}$$

for all $\bar{s} \in \mathfrak{S}, \bar{y} \in Y, t \geq t_0$.

Proof. No product motion leaves $\mathscr{Z}_0$ unless through $\mathcal{N}'_\varepsilon$ so we consider $k(\cdot) \in \mathscr{K}(\bar{Z}^0, t_0), (\bar{Z}^0, t_0) \in \mathcal{N}'_\varepsilon \times R$, and suppose it crosses $\partial \mathscr{Z}_0$. Then there is a $t_1 \geq t_0$ such that $k(t_1) \in \partial \mathscr{Z}_0$ and, by (i), $V'_S(k(t_1)) \geq v'_S > V'_S(\bar{Z}^0)$, which contradicts (ii). Hence no product motion leaves $\mathscr{Z}_0$.

Next we show that product motions from $C\mathcal{M}'_\eta$ may not stay indefinitely in this set. Indeed, integrating (6.1.19) along such an arbitrary motion over $[t_0, t]$, we get the estimate

$$t \leq t_0 + c^{-1}[V'_\eta(\bar{Z}^0) - V'_\eta(\bar{Z})]. \tag{6.1.20}$$

Then by (iii), $V'_\eta(\bar{Z}^0) \leq v'^+_\eta, V'_\eta(\bar{Z}) \geq 0$, whence $V'_\eta(\bar{Z}^0) - V'_\eta(\bar{Z}) \leq v'^+_\eta$, which allows us to rewrite (6.1.20) as $t \leq t_0 + v'^+_\eta/c$. Thus there is a constant

$$T_\eta = v'^+/c \tag{6.1.21}$$

depending only upon the diameter of $C\mathcal{M}'_\eta$ and quite independent of particular motions such that for $t > t_0 + T_\eta$ the product motions leave $C\mathcal{M}'_\eta$. Since the product motions must not leave $\mathscr{Z}_0$, they must enter $\mathcal{M}'_\eta$ and be there at some $t_2 = t_0 + T_\eta + \tau, \tau > 0$. There is no return to $C\mathcal{M}'_\eta$. Indeed, if there were $t_3 > t_2$ such that $k(t_3) \in \overline{C\mathcal{M}'_\eta}$, then, by (iv), $V'_\eta(k(t_3)) \geq v'^-_\eta \geq V'_\eta(k(t_2))$, which contradicts (v).

Corollary 6.1.1. *Given Δ_0, η, T_η, the auxiliary system (6.1.10) and (6.1.11) is an η identifier in stipulated time T_η on $\Delta_0 \times R$ for the system (6.1.7) if Theorem 6.1.1 holds with $c = v'^+_\eta/T_\eta$.*

Clearly, we use the conditions for identification together with some sufficient conditions for the control objective concerned. To be more specific, let us write

Definition 6.1.2. The system (3.3.1) with selector (6.1.7) and output (6.1.8) is *strongly controllable* on $\Delta_0 \times R$ *for an objective property Q with adaptive identification of uncertainties*, if and only if there is a $\mathscr{P}(\bar{Z}, t)$ such that for each product motion of $\mathscr{K}(Z^0, t_0), (Z^0, t_0) \in \mathscr{L}_0 \times R$ we have Q and Definition 6.1.1 holds.

Proposition 6.1.1. Definition 6.1.2 holds if system (6.1.13) satisfies on $\mathscr{L}_0 \times R$ some sufficient conditions for Q and Theorem 6.1.1.

We shall apply the above general results to our manipulator in Section 6.4. In the meantime, it seems instructive to give a simple example of how the method is used.

Example 6.1.1. Consider (1.5.16''') in the simple single-DOF form

$$\dot{x}_1 = \frac{\partial H(\bar{x})}{\partial x_2}, \tag{6.1.22}$$

$$\dot{x}_2 = -\frac{\partial H(\bar{x})}{\partial x_1} + Q^D(\bar{x}) + B(\bar{x})u + w,$$

and assume that we can read out the momentum, that is, $y(t) \equiv x_2(t)$, without noise. We want to stabilize it below some h_B level in the basic energy cup and identify adaptively the perturbation $w(t) \in [0, 1]$, $\dot{w} = -b$, $b > 0$. Thus Δ_0 is the desired Δ_B corresponding to $h_B > 0$ and both objectives may be referred to the same set.

We design the identifier as

$$\dot{x}_{p1} = \frac{\partial H(\bar{x}_{p1}, y)}{\partial y}, \tag{6.1.23}$$

$$\dot{x}_{p2} = -\frac{\partial H(x_{p1}, y)}{\partial x_{p1}} - y,$$

and the adaptive law as

$$\dot{\mu} = -\dot{H}(x_{p1}, y), \qquad \mu \in [0, 1]. \tag{6.1.24}$$

Observe that both our objectives follow from Theorem 6.1.1 satisfied on $\mathscr{L}_0 = \Delta_B \times [0, 1] \times \Delta_B \times [0, 1]$, $\Delta_B \subset \Delta_H$. Take

$$V'_S \triangleq H(x_1, x_2) + H(x_{p1}, y) + w + \mu \tag{6.1.25}$$

and

$$V'_\eta \triangleq \begin{cases} |H(x_1, x_2) - H(x_{p1}, y)| + |w - \mu|, & \bar{Z} \in C\mathscr{M}'_\eta, \\ |w - \mu|, & \bar{Z} \in \mathscr{M}'_\eta, \end{cases} \tag{6.1.26}$$

[See Eqs. (4.7.20) and (4.7.21)]. Since we are within a local energy cup, conditions (i), (iii), and (iv) hold automatically and all we must show is (ii) and (v), which is similar to our task in Section 4.7. Calculate

$$\dot{V}'_s = \dot{H}(x_1, x_2) + \dot{H}(x_{p1}, y) - b + \dot{\mu}.$$

By (2.4.12), for $x_2 \neq 0$, we have

$$\dot{H}(x_1, x_2) = Q^D(x_1, x_2)x_2 + B(x_1, x_2)x_2 u + wx_2, \quad (6.1.27)$$

$$\dot{H}(x_{p1}, y) = -y^2. \quad (6.1.28)$$

For positive damping $Q^{DD} \cdot x_2 < 0$ and the controller

$$u = \begin{cases} -\dfrac{1}{B}\left(\dfrac{d}{x_2} + 1\right), & x_2 > 0, \\ -\dfrac{d}{Bx_2}, & x_2 < 0, \end{cases} \quad (6.1.29)$$

where $d > b$, makes $\dot{H}(x_1, x_2) < -d$ and, with (6.1.24), $\dot{V}'_s < 0$ as required by (ii). Differentiating (6.1.26), for all $\bar{Z} \in C\mathcal{M}'_\eta$, we obtain

$$\dot{V}'_\eta = \begin{cases} \dot{H}(x_1, x_2) - \dot{H}(x_{p1}, y) - b - \dot{\mu} & \text{for } \bar{Z} \subset C^+\mathcal{M}'_\eta, \\ \dot{H}(x_{p1}, y) - \dot{H}(x_1, x_2) + \dot{\mu} + b & \text{for } \bar{Z} \in C^-\mathcal{M}'_\eta, \end{cases} \quad (6.1.30)$$

where

$$\begin{aligned} C^+\mathcal{M}'_\eta &: H(x_1, x_2) \geq H(x_{p1}, y), & w > \mu, \\ C^-\mathcal{M}'_\eta &: H(x_1, x_2) < H(x_{p1}, y), & w < \mu. \end{aligned} \quad (6.1.31)$$

Substituting (6.1.24) and $\dot{H}(x_1, x_2) < -d$, $d > 0$ into (6.1.30) and choosing $\bar{Z}^0 \in C^+\mathcal{M}'_\eta$, we obtain $V'_\eta \leq -c$ and $c \triangleq d + b$, which satisfies (v). ∎

6.2 IDENTIFICATION OF VARIABLE PAYLOADS

Let us investigate the manipulator equations in the inertially decoupled form (1.3.24″) of Section 3.3 such that

$$\ddot{q}_i + \Phi_i(\bar{q}, \dot{\bar{q}}) + \Pi_i(\bar{q}, \bar{r}) = B_i(\bar{q}, \dot{\bar{q}})u_i + R^r_i(\bar{q}, \dot{\bar{q}}, \bar{r}, t), \quad (6.2.1)$$

where $\Phi_i(\cdot)$ are the nonpotential and $\Pi_i(\cdot)$ the potential characteristics as defined in Section 1.3, with $\Phi_i = \Gamma_i + D_i$, $\Pi_i = G_i + \Psi_i$, the brackets being ignored. Moreover, $F_i(\bar{q}, \dot{\bar{q}}, t) = B_i(\bar{q}, \dot{\bar{q}})u_i$ are the inputs from actuator gear and $R^r_i(\bar{q}, \dot{\bar{q}}, \bar{r}, t)$ are external perturbations. Both Π_i and R^r_i are subject to uncertainty represented by the vector $\bar{r}(t) \in \Re_1 \subset R^s$, where $\Re_1$ is a known bounded noise band.

The case of gravity, external load, or both, being uncertain, seems to be the most frequent case in the industrial work of manipulators.

Let us assume that we read out the velocities, that is, $y_i = \dot{q}_i(t)$, $t \geq t_0$, $i = 1, \ldots, n$, and that we know the change

$$\dot{\bar{r}} = g(t), \qquad (6.2.2)$$

[See Eq. (6.1.9)], where $g(t)$, $t \geq t_0$, is a measured quantity. As our basic objective, we want strong controllability on some $\Delta_0 \subset \Delta_H$ for reaching with capture in a stipulated subtarget $\mathfrak{T}_C$ about the basic equilibrium during a stipulated time T_C. Note that Π_i includes uncertainty, hence so does $H(\bar{x}, \bar{r})$. As a side objective, we also want the η-identification of $\bar{r}(t)$ during stipulated $T_\eta < T_C$, possibly well before capture (see Skowronski [22]).

We may thus combine Theorem 6.1.1 with Theorem 4.4.3 to produce the required joint objective (see Fig. 6.3). To this end we define

$$\mathscr{Z}_0 \triangleq \Delta_0^2 \times \mathfrak{R}_1^2, \qquad \mathfrak{T}_C^4 \triangleq \mathfrak{T}_C \times \mathfrak{T}_C \times \mathfrak{R}_1^2,$$

$$\mathfrak{R}_1^2 \triangleq \mathfrak{R}_1 \times \mathfrak{R}_1, \qquad C\mathfrak{T}_C^4 \triangleq \mathscr{Z}_0 - \mathfrak{T}_C^4,$$

with the remaining notation being the same as in Section 6.1. Then we propose the identifier in the form

$$\ddot{q}_{pi} + \Phi_{pi}(\bar{q}_p, y) + \Pi_{pi}(\bar{q}_p, \bar{\mu}) = 0, \qquad (6.2.3)$$

with $\Pi_{pi}(\cdot)$ *the same functions* as $\Pi_i(\cdot)$, thus producing the same total energy function $H(\cdot)$, except that $H(\bar{x}, \bar{r})$ is uncertain while $H(\bar{x}_p, \bar{\mu})$ is known. Note that $\bar{\mu}$ is now s-dimensional: $l = s$. For the sake of identification, we take $H(\bar{z})$, $H(\bar{z}_p)$.

We shall need three Liapunov functions: $V_C(\cdot): D_C^4 \to R$ for Theorem 4.4.3 and $V_S(\cdot): \mathscr{N}_\varepsilon' \to R$, $V_\eta(\cdot): D_\eta^4 \to R$ for Theorem 6.6.1, with open $D_C^4 \supset \bar{C}\mathfrak{T}_C^4$,

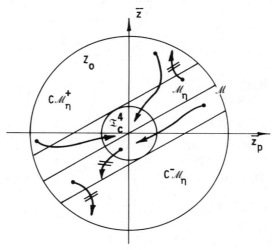

Fig. 6.3

6.2 IDENTIFICATION OF VARIABLE PAYLOADS

$D_\eta^4 \supset \overline{C\mathcal{M}_\eta'}$, $D_\eta^4 \cap \mathcal{M}' = \emptyset$ and the bounds v_0', v_x', v_s', $v_\eta'^-$, $v_\eta'^+$ defined by (4.4.7) and (6.1.17), respectively (see Fig. 6.3). Then we choose

$$V_C(\bar{Z}) \triangleq V_S(\bar{Z}) \triangleq H(\bar{z}) + H(\bar{z}_p), \tag{6.2.4}$$

$$V_\eta(\bar{Z}) \triangleq \begin{cases} |H(\bar{z}) - H(\bar{z}_p)|, & \bar{Z} \in C\mathcal{M}_\eta', \\ 0, & \bar{Z} \notin C\mathcal{M}_\eta'. \end{cases} \tag{6.2.5}$$

With such a choice, conditions (i) and (ii) of Theorem 4.4.3 and conditions (i), (iii), and (iv) of Theorem 6.1.1 hold immediately if we note that $\Delta_0 \subset \Delta_H$. Moreover, (4.4.13) of condition (iii) in Theorem 4.4.3 implies (6.1.18) of condition (ii) in Theorem 6.1.1; thus only (iii) of Theorem 4.4.3 and (v) of Theorem 6.1.1 have to be verified. From (6.2.4), we have

$$\dot{V}_C(t) = \dot{H}(\bar{z}) + \dot{H}(\bar{z}_p) \tag{6.2.6}$$

while (6.2.5) gives

$$\dot{V}_\eta(t) = \begin{cases} \dot{H}(\bar{z}) - \dot{H}(\bar{z}_p) & \text{for } \bar{Z} \in C_z^+ \mathcal{M}_\eta'', \\ \dot{H}(\bar{z}_p) - \dot{H}(\bar{z}) & \text{for } \bar{Z} \in C_z^- \mathcal{M}_\eta'', \end{cases} \tag{6.2.7}$$

where

$$C_z^+ \mathcal{M}_\eta' : H(\bar{z}) - H(\bar{z}_p) \geq 0,$$

$$C_z^- \mathcal{M}_\eta' : H(\bar{z}) - H(\bar{z}_p) < 0.$$

Since $H(\bar{x}, \bar{r})$, $H(\bar{x}_p, \bar{\mu})$ depend upon variable $\bar{r}$, $\bar{\mu}$, we have

$$\dot{H}(\bar{x}, \bar{r}) = f_0(\bar{x}, t, \bar{u}, \bar{r}) + \sum_i \frac{\partial \mathcal{V}}{\partial r_i} g_i(t), \tag{6.2.8}$$

$$\dot{H}(\bar{x}_p, \bar{\mu}) = f_0(\bar{x}_p, t, \bar{\mu}) + \sum_i \frac{\partial \mathcal{V}}{\partial \mu_i} \dot{\mu}_i, \tag{6.2.9}$$

with the partial derivatives obviously depending upon the shape of the potential energy $\mathcal{V}(\bar{q}, \bar{r})$ or $\mathcal{V}(\bar{q}_p, \mu)$ concerned. Introducing the adaptive condition

$$\sum_i \frac{\partial \mathcal{V}}{\partial \mu_i} \dot{\mu}_i = -f_0(\bar{x}_p, t, \bar{\mu}), \quad i = 1, \ldots, s \tag{6.2.10}$$

we reduce $\dot{H}(\bar{x}_p, \bar{\mu})$ to zero. Then, in view of (6.2.6), (6.2.8), (6.2.9), and (6.2.10), condition (iii) of Theorem 4.4.3 becomes

$$\min_{\bar{u}} \max_{\bar{r}} [f_0(\bar{x}, t, \bar{u}, \bar{r}) + \sum_i \frac{\partial \mathcal{V}}{\partial r_i} g_i(t)] \leq \frac{\tilde{h}_0 - \tilde{h}_x}{T_C}, \tag{6.2.11}$$

which serves as the control condition for defining $\mathcal{P}(\cdot)$.

Note that the initial condition $\bar{z}_p^0 = \bar{z}_p(t_0)$ is of our choice and that we may choose it so that the initial product state $\bar{Z}^0 = (\bar{z}^0, \bar{z}_p^0) \in C_z^+ \mathcal{M}_\eta'$. Then in view of (6.2.7)–(6.2.11), we have $\dot{V}_\eta(t) \leq -(\tilde{h}_0 - \tilde{h}_x)/T_C$. Hence, condition (v) of Theorem 6.1.1 holds if

$$(\tilde{h}_0 - \tilde{h}_x)/T_C \geq v_\eta^+/T_\eta, \qquad (6.2.12)$$

see Corollary 6.1.1. This closes our investigation in the general case. We conclude that the min–max controller calculated from (6.2.11) subject to (6.2.12) and the said choice of $\bar{z}_p^0$ together with adaptive laws implying (6.2.10), secure the joint objective of capture with identification.

Since $f_0(\bar{x}_p, t, \bar{\mu}) \equiv \sum_{i=1}^{n} -\Phi_{pi}(\bar{q}_p, \bar{y})\dot{\bar{q}}_i$, to obtain (6.2.10), we need

$$\dot{\mu}_i = \Phi_{pi}(\bar{q}_p, \bar{y})\dot{\bar{q}}_i/(\partial \mathcal{V}/\partial \mu_i), \qquad i = 1, \ldots, n, \qquad (6.2.13)$$

which represents the adaptive laws concerned for $s = n$. For $s \neq n$, we apply the same discussion as used for (4.7.22). In view of our assumption that $T_\eta < T_C$, we shall need $v_\eta^+ \leq \tilde{h}_0 - \tilde{h}_x$ in order to satisfy (6.2.12). By (6.2.5) this means that

$$[\tilde{H}(\bar{x}, \bar{\gamma}) - H(\bar{x}_p, \bar{\mu})]_{\bar{Z} \in \partial \mathcal{M}_\eta'} \leq \tilde{h}_0 - \tilde{h}_x. \qquad (6.2.14)$$

Substituting (6.2.1) into (6.2.11), we obtain

$$\min_{\bar{u}} \sum_i B_i(\bar{q}, \dot{\bar{q}})u_i + \max_{\bar{r}} \sum_i \left[R_i^r(\bar{q}, \dot{\bar{q}}, \bar{r}, t)\dot{q}_i + \frac{\partial \mathcal{V}}{\partial r_i} g_i(t) \right]$$

$$\leq -\frac{\tilde{h}_0 - \tilde{h}_x}{T_C} + \sum_i \Phi_i(\bar{q}, \dot{\bar{q}})\dot{q}_i, \qquad i = 1, \ldots, n, \qquad (6.2.15)$$

as the control conditions for the program $\mathcal{P}(\cdot)$.

As an exercise, the reader is asked to calculate the program and solve the adaptive laws (6.2.13) for our RP manipulator of Section 1.1 and the R manipulator of Example 4.1.2.

6.3 CONSTANT PARAMETER IDENTIFICATION AND MODEL TRACKING

In the case of constant uncertainty parameters: $\bar{w}(t) \equiv \text{const} \in W$, our study simplifies considerably. First, we have $\dot{\bar{w}}(t) \equiv 0$ in (6.1.9), the noise being avoided. Second, we may reduce the dimension of the product space. Indeed, we designate $\bar{\gamma}_p(t) = \bar{\mu}(t) - \bar{w}$, with $\dot{\bar{\gamma}}_p = \dot{\bar{\mu}} = f_\mu(\bar{x}, \bar{x}_p, y, t)$.

This allows us to take $\bar{Z}_p(t) \triangleq (\bar{x}(t), \bar{x}_p(t), \bar{\gamma}_p(t))$ in R^{2N+l} instead of the previous R^{2N+2l}. We may also redefine

$$\mathcal{M}_{\eta p} \triangleq \{\bar{Z}_p \in \mathcal{Z} \mid \|\bar{x} - \bar{x}_p\| < \eta, \|\bar{\gamma}_p\| < \eta\}, \qquad (6.3.1)$$

6.3 CONSTANT PARAMETER IDENTIFICATION AND MODEL TRACKING

with all further formal consequences. Neither the definitions of identification nor Theorem 6.1.1 and its corollaries change, but their implementation becomes simpler. We shall apply the case to the combined objective of Model Reference Adaptive Control (MRAC) and adaptive identification for the general type of our manipulator equations. Let us write them as follows:

$$\ddot{q}_i + \Phi_i(\bar{q}, \dot{\bar{q}}, \bar{\lambda}, \bar{w}) + \Pi_i(\bar{q}, \bar{\lambda}, \bar{w}) = B_i(\bar{q}, \dot{\bar{q}})u_i + R_i^r(\bar{q}, \dot{\bar{q}}, \bar{w}, t),$$
$$i = 1, \ldots, n, \quad (6.3.2)$$

with the known meaning of particular terms, and subject to the assumptions of Sections 2.3 and 3.3. The reference model to be tracked is then set up as

$$\ddot{q}_{mi} + \Phi_{mi}(\bar{q}_m, \dot{\bar{q}}_m, \bar{\lambda}_m) + \Pi_{mi}(\bar{q}_m, \bar{\lambda}_m) = u_{mi}, \quad i = 1, \ldots, n, \quad (6.3.3)$$

with Φ_{mi} representing positive damping characteristics that vanish at the equilibria of the model.

We let the manipulator output (read-out) vector $y(t)$ [see Eq. (6.1.8)] be given in terms of the measured velocities $\dot{q}_1(t), \ldots, \dot{q}_n(t)$ and we propose the identifier (6.1.10) in the general form

$$\ddot{q}_{pi} + \Phi_{pi}(\bar{q}_p, \bar{y}, \bar{\mu}) + \Pi_{pi}(\bar{q}_p, \bar{\mu}) = R_i(\bar{q}_p, \bar{y}, \bar{\mu}, t), \quad (6.3.4)$$

with the functions $\Phi_{pi}(\cdot)$, $\Pi_{pi}(\cdot)$, and $R_i(\cdot)$ to be designed suitable for our identification, but compatible with (6.3.2). The block scheme in Fig. 6.4 displays the procedure underlying the control for our combined objective of

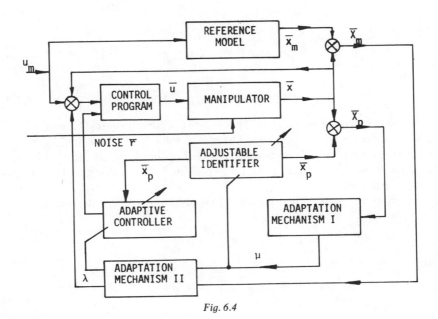

Fig. 6.4

model tracking and identification. It stands as follows: Given the working region Δ and the parameter bands W, Λ as well as the control band U, given the structure of the manipulator with $\Phi_i(\cdot)$, $\Pi_i(\cdot)$ $B_i(\cdot)$ adjustable through $\bar{\lambda}(t)$, and given the reference model (6.3.3), we want to design the predictor (6.3.4), adaptive laws for $\bar{\lambda}(t)$ and $\bar{\mu}(t)$, as well as a feedback and signal adaptive control program $\mathscr{P}(\bar{x}, \bar{x}_p, \bar{\lambda}, \bar{\mu})$ so that both Definition 4.7.1 (adaptive stabilization and tracking) and Definition 6.1.1 (identification) hold simultaneously for the two product systems (6.3.2), (6.3.3) and (6.3.2), (6.3.4).

We introduce the vectors

$$\bar{X}_m(t) \triangleq (\bar{x}(t), \bar{x}_m(t))^T \in \Delta^2,$$

$$\bar{\gamma}_m(t) \triangleq \bar{\lambda}(t) - \bar{\lambda}_m \in \Lambda \subset R^l,$$

$$\bar{Z}_m(t) \triangleq (\bar{X}_m(t), \bar{\gamma}_m(t))^T \in \mathscr{Z}_m = \Delta^2 \times \Lambda \subset R^{2N+l}$$

for tracking, and

$$\bar{X}_p(t) \triangleq (\bar{x}(t), \bar{x}_p(t))^T \in \Delta^2,$$

$$\bar{\gamma}_p(t) \triangleq \bar{\mu}(t) - \bar{w} \in W, \qquad \bar{w} = \text{const},$$

$$\bar{Z}_p(t) \triangleq (\bar{X}_p(t), \bar{\gamma}_p(t))^T \in \mathscr{Z}_p = \Delta^2 \times W \subset R^{2N+s} \qquad \text{for identification.}$$

It does not narrow generality to assume $l, s = N$ as long as $l, s \le N$, which is most frequently the case, cf. (4.7.22).

Moreover, let us recall (6.3.1) about the following diagonal

$$\mathscr{M}_p \triangleq \{\bar{Z}_p \in \mathscr{Z} \mid \|\bar{x} - \bar{x}_p\| = 0, \|\bar{\gamma}_p\| = 0\}, \tag{6.3.5}$$

and redefine

$$\mathscr{M}_{\eta m} \triangleq \{\bar{Z}_m \in \mathscr{Z} \mid \|\bar{x} - \bar{x}_m\| < \eta, \|\bar{\gamma}_m\| < \eta\}, \tag{6.3.6}$$

about

$$\mathscr{M}_m \triangleq \{\bar{Z}_m \in \mathscr{Z} \mid \|\bar{x} - \bar{x}_m\| = 0, \|\bar{\gamma}_m\| = 0\}. \tag{6.3.7}$$

We shall use simultaneously Theorem 4.7.1 for verifying Definition 4.7.1 and Theorem 6.1.1 for verifying Definition 6.1.1, with the same $\Delta_0^2 \times \Lambda$ in both cases, and with $T = \max(T_{\eta m}, T_{\eta p})$ being the duration after which the objective is achieved. Clearly, $T_{\eta m}$, $T_{\eta p}$ specifies T_η for tracking and identification, respectively.

In order to satisfy Theorem 4.7.1, we need the Liapunov functions (see Section 4.7)

$$V_s(\bar{X}_m, \bar{\gamma}_m) = H_m(\bar{x}) + H_m(\bar{x}_m) + \bar{a} \cdot \bar{\gamma}_m \tag{6.3.8}$$

and

$$V_\eta(\bar{X}_m, \bar{\gamma}_m) = \begin{cases} |H_m(\bar{x}) - H_m(\bar{x}_m)| + \bar{a} \cdot \bar{\gamma}_m, & \bar{Z}_m \notin \mathscr{M}_{\eta m}, \\ \bar{a} \cdot \bar{\gamma}_m, & \bar{Z}_m \in \mathscr{M}_{\eta m}, \end{cases} \tag{6.3.9}$$

6.3 CONSTANT PARAMETER IDENTIFICATION AND MODEL TRACKING

where $\bar{a} = (\operatorname{sign} \gamma_{m1}, \ldots, \operatorname{sign} \gamma_{mN})^T$ and $H_m(\cdot)$ is the total energy function of the model (6.3.3). We calculate

$$H_m(\bar{x}_m) = \frac{1}{2}\sum_i \dot{q}_{mi} + \frac{1}{2}\sum_i \int_{q_{mi}^0}^{q_{mi}} \Pi_i(\bar{q}_m, \bar{\lambda}_m)dq_{mi}, \quad (6.3.10)$$

with the derivative

$$\dot{H}_m(\bar{x}_m) = \sum_i u_{mi}\dot{q}_{mi} - \sum_i \Phi_{mi}(\bar{q}_m, \dot{\bar{q}}_m, \bar{\lambda}_m)\dot{q}_{mi}.$$

We also need the adaptive laws (4.7.23):

$$\dot{\gamma}_{mi} = \frac{\dot{q}_{mi}}{\operatorname{sign} \gamma_{mi}}[u_{mi} - \Phi_{mi}(\bar{q}_m, \dot{\bar{q}}_m, \bar{\lambda}_m)], \quad i = 1, \ldots, n, \quad (6.3.11)$$

with the model controller for $\bar{u}_m$ selected such as to make the model monotone dissipative:

$$u_{mi} = \begin{cases} \Phi_{mi}(\bar{q}_m, \dot{\bar{q}}_m, \bar{\lambda}_m) - c_i/\dot{q}_{mi}, & \dot{q}_{mi} \neq 0, \\ 0, & \dot{q}_{mi} = 0, \end{cases} \quad (6.3.12)$$

where $\Phi_{mi}(\bar{q}_m, \dot{\bar{q}}_m, \bar{\lambda}_m)\dot{q}_{mi} > 0$, $\dot{q}_{mi} \neq 0$, and $c_i > 0$, $i = 1, \ldots, n$.

From (4.7.33), in view of (6.3.10), we obtain the tracking control condition

$$\min_{\bar{u}} \max_{\bar{w}} \sum_i [B_i(\bar{q}, \dot{\bar{q}})\dot{q}_i u_i + R_i^r(\bar{q}, \dot{\bar{q}}, \bar{w}, t)\dot{q}_i - \Phi_i(\bar{q}, \dot{\bar{q}}, \bar{\lambda}, \bar{w})\dot{q}_i] = -c \quad (6.3.13)$$

on $\mathscr{C}^+\mathscr{M}_\eta$, subject to (4.7.34) and (4.7.35). Similarly on $\mathscr{C}^-\mathscr{M}$ we have (6.3.13) with min–max replaced by max–min and (4.7.34) with the inequality inverted.

According to our discussion in Section 4.7, the above assumptions yield the model tracking specified by Definition 4.7.1.

In turn, to satisfy Theorem 6.6.1, we need the Liapunov functions

$$V_s'(\bar{X}_p, \bar{\gamma}_p) = H_m(\bar{x}) + H_m(\bar{x}_p) + \bar{a} \cdot \bar{\gamma}_p,$$

and

$$V_\eta'(\bar{X}_p, \bar{\gamma}_p) = \begin{cases} |H_m(\bar{x}) - H_m(\bar{x}_p)| + \bar{a} \cdot \bar{\gamma}_p, & \bar{Z}_p \notin \mathscr{M}_{\eta p}, \\ \bar{a} \cdot \bar{\gamma}_p, & \bar{Z}_p \in \mathscr{M}_{\eta p}, \end{cases}$$

with the identifier characteristics $\Phi_{pi}(\cdot)$, $\Pi_{pi}(\cdot)$ designed the same as $\Phi_{mi}(\cdot)$, $\Pi_{mi}(\cdot)$, and with

$$R_i(\bar{q}_p, \bar{y}, \bar{\mu}, t) \triangleq u_{mi}(t).$$

Then the adaptive identification is completely symmetric to tracking, and the adaptive laws

$$\dot{\gamma}_{pi} = (y_i/\operatorname{sign} \gamma_{pi})[u_{mi} - \Phi_{pi}(\bar{q}_p, \bar{y}, \bar{\mu})], \quad i = 1, \ldots, n,$$

together with the controller specified by (6.3.13) satisfy Theorem 6.1.1, yielding Definition 6.1.1.

Using the same numerical simulation data as in Example 4.7.1, we may obtain for our unit RP manipulator the twofold convergence of $\bar{x}(t)$ with $\bar{x}_m(t)$ and with $\bar{x}_p(t)$ (see Fig. 6.5 with $q_1(t)$, $q_{m1}(t)$, $q_{p1}(t)$ recorded for illustration).

246   6 ADAPTIVE IDENTIFICATION

Quite obviously, for some reference models and some manipulators, designing the identifiers as symmetric counterparts of the model may not work. This may be the case with equilibria incompatibility between the model and the manipulator, say if the $\Pi_{mi}(\cdot)$ are quite different functions from $\Pi_i(\cdot)$ and (4.7.35) does not hold. Since the identifier must be compatible with the manipulator, the symmetry does not apply and we have to verify the identifier with Theorem 6.1.1 independently, perhaps using different adaptive laws. Then a second control condition must also be added to (6.3.13), without contradicting the (6.3.13).

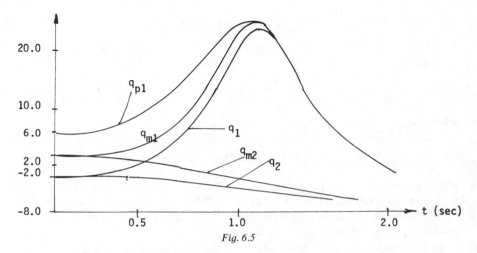

Fig. 6.5

As mentioned many times in this book the speed of on-line calculations is an essential feature in the work of present manipulators. In particular, digital solving of either inverse kinematics or, especially, differential equations of motion is a very unwelcomed procedure. Unfortunately the state and parameter information required for feedback signal adaptive controllers need such solutions. Hence we see so many various inventions of better or worse approximative methods attempting to provide the computer with analytic, closed type algorithms that save time. In this sense our nonlinear tracking method that multiplies dimensions of the system making the calculations more difficult may look odd at the first glance. Observe however that with the outlined adaptive identification method both the states and parameters of the manipulator may be replaced by those of the predictor—and the predictor may be designed to be simple enough to have exact solutions. The same goes for solving the linear adaptive laws. Hence, adopting this approach we may in fact avoid all together solving on-line any differential equations. The case has been illustrated in Skowronski [30]. It is a result of the consequent use of the Liapunov theory.

References

Abele, E., and Sturz, W.
- [1] Sensors for adaptive control of settling tasks with industrial robots, *Proc. 2nd Internat. Conf. Robot Vision and Sensory Controls, Stuttgart, 1982.*

Aggarwal, R., and Leitmann, G.
- [1] A max–min distance problem, *Trans. ASME Ser. G J. Dynamic Systems Measurement Control* and **94** (1972), 155–158.

Aksenov, G.S., and Formin, V.H.
- [1] On the problem of adaptive control of a manipulator, in *Kybernetic Problems, Adaptive Systems*, pp. 164–168. Nauka, Moscow, 1976.

Antosiewicz, H.A.
- [1] A survey of Liapunov second method, *Contributions to Theory of Nonlinear Oscillations*, Vol. 4, pp. 141–166. Princeton Univ. Press, Princeton, New Jersey, 1958.

Ardafio, D.D.
- [1] Model formulation of flexible manipulator arms, in J.R. Avula and R. Kalmann (eds.), *Mathematical Modelling in Science and Technology*, pp. 891–896. Pergamon, Oxford, 1984.

Aronsson, G.
- [1] Global controllability and bang–bang steering of certain nonlinear systems, *SIAM J. Control* **11** (1973), 607–619.

Asher, R.B., and Matuszewski, J.P.
- [1] Optimal guidance with maneuvering targets, *J. Spacecraft Rockets* **11** (1974), 204–206.

Balestrino, A., Demaria, G., and Sciavicco, L.
- [1] An adaptive model following control for robotic manipulators, *Trans. ASME Ser. G J. Dynamic Systems Measurement Control* **105** (1983), 143–151.

Balestrino, A., Demaria, G., and Zinober, A.S.I.
- [1] Nonlinear adaptive model following control, *Automatica* **20** (1984), 559–568.

Barmish, B.R., Corless, M., and Leitmann, G.
- [1] A new class of stabilizing controllers for uncertain dynamical systems, *SIAM J. Control Optim.* **21** (1983), 246–252.

Barmish, B.R., and Feuer, A.
- [1] Instability in optimal aim control, *IEEE Trans. Automat. Control* **AC-25** (1980), 1250–1252.

Barmish, B.R., and Leitmann, G.
- [1] On ultimate boundedness control of uncertain systems in the absence of matching conditions, *IEEE Trans. Automat. Control* **AC-27** (1982), 153–155.

Barmish, B.R., Petersen, I.R., and Feuer, A.
- [1] Linear ultimate boundedness control of uncertain dynamical systems, *Automatica* **19** (1983), 523–528.

Barmish, B.R., Thomas, R.J., and Lin, Y.H.
- [1] Convergence properties of a class of point-wise strategies, *IEEE Trans. Automat. Control* **AC-23** (1978), 954–956.

Barnard, R.D.
[1] Optimal aim-control strategies applied to large scale nonlinear regulation and tracking system, *IEEE Trans. Circuits and Systems* **CAS-23** (1976), 800–806.

Barnett, S.
[1] *Introduction to Mathematical Control Theory*. Oxford Univ. Press (Clarendon), London and New York, 1975.

Bejczy, A.K.
[1] Robot arm dynamics and control, Tech. Memo 33-669, Jet Propul. Lab., Pasadena, California, Feb. 1974.
[2] Algorithmic formulation of control problems in manipulation, *Proc. Internat. Conf. Cybernet. and Soc.* (1975), 135–142.

Bejczy, A.K., and Zawacki, R.L.
[1] Computer aided manipulator control, *Proc. 1st Int. Sympos. Micro- and Macrocomputs. in Control, San Diego* (1979), 129–143.

Beletskii, V.V., and Chudinov, P.S.
[1] Parametric optimization in the task of biped locomotion, *Izv. Akad. Nauk SSSR, Meh. Tverd. Tela* (1977), No. 1.

Belusov, A.P., and Furasov, V.D.
[1] Stabilization of nonlinear control systems with incomplete information on object's state, *Automat. Remote Control* **36** (1975), 1385–1392.

Benedict, C.E., and Tesar, D.
[1] Model formulation of complex mechanisms with multiple inputs, part I and II, *ASME J. Mech. Design* **100** (1978), 747–750, 755–760.

Blaquiere, A., Gerard, F., and Leitmann, G.
[1] *Quantitative and Qualitative Games*. Academic Press, New York, 1969.

Bogusz, W., and Skowronski, J.M.
[1] Kinetic synthesis of lumped mechanical systems, *Mech. Teoret. Stos.* **1** (1965), 13–27.
[2] Kinetic synthesis of general mechanical systems, *Bull. Polytechnic Inst. Jassi, Romania* **13** (1967), 409–413.

Bradshaw, A., and Porter, B.
[1] Design of linear multivariable discrete time tracking systems for plants with inaccessible states, *Int. J. Control* **24** (1976), 275–281.

Brady, M., Hollerback, J., Johnson, T., Lozano-Perez, T., and Mason, M. (Eds.)
[1] *Robot Motion Planning and Control*. MIT Press, Cambridge, Massachusetts, 1983.

Breakwell, J.V.
[1] Pursuit of a faster evader, in J.D. Grote (ed.), *Theory and Application of Differential Games*, pp. 243–256. Reidel, Dordrecht, 1975.
[2] A differential game with two pursuers and one evader, *J. Optim. Theory Appl.* **18** (1976), 15–29.

Brooks, R.A.
[1] Solving the find-path problem by representing free space as generalized cones, AI Memo No. 674, MIT Artificial Intelligence Lab., May 1982.
[2] Solving the find-path problem by good representation of free space, *IEEE Trans. Systems Man Cybernet.* **SMC-13** (1983), 190–197.

Butchard, R.L., and Shackloth, B.
[1] Synthesis of model reference adaptive control systems by Liapunov's Second Method, *Proc. 2nd IFAC Sympos. Theory of Self-Adaptive Control Systems, Teddington* (1966), 145–152.

Calm, D.F., and Phillips, S.K.
[1] Robnav-range based robot navigation and obstacle avoidance algorithm, *IEEE Trans. Systems Man Cybernet.* **SMC-5** (1975), 544–551.

REFERENCES

Carrol, R.L.
- [1] New adaptive algorithms in Liapunov synthesis, *IEEE Trans. Automat. Control* **AC-21** (1976), 246–249.

Chikriy, A.A.
- [1] Sufficient conditions for avoidance in nonlinear differential games, *Dokl. Akad. Nauk* **241** (1978), 547–550.
- [2] Nonlinear differential evasion games, *Soviet Math. Dokl.* **20** (1979), 591–595.

Choe, H.H., and Nikiforuk, P.N.
- [1] Inherently stable feedback control of a class of unknown plants, *Automatica* **7** (1971), 607–625.

Christensen, R.M.
- [1] *Theory of Visco-Elasticity*. Academic Press, New York, 1971.

Chukwu, E.N.
- [1] Finite time controllability of nonlinear control processes, *SIAM J. Control* **13** (1975), 807–876.

Coiffet, P.
- [1] *Modelling and Control*. Kogan Page, London, 1983.

Corless, M., Goodall, D.P., Leitmann, G., and Ryan, E.P.
- [1] Model following controls for a class of uncertain dynamical systems, *7th IFAC Sympos. Identif. & System Parameter Estimation, London, 1985*, 38–50.

Corless, M., and Leitmann, G.
- [1] Continuous state feedback guaranteed ultimate boundedness for uncertain dynamical systems, *IEEE Trans. Automat. Control* **AC-26** (1981), 1139–1142.
- [2] Adaptive control of systems containing uncertain functions and unknown functions with uncertain bounds, *J. Optim. Theory Appl.* **41** (1983), 155–168.
- [3] Adaptive control for uncertain dynamical systems, in A. Blaquière and G. Leitmann (eds.), *Dynamical Systems and Microphysics: Control Theory and Mechanics*. Academic Press, New York, 1983.
- [4] Adaptive controllers for a class of uncertain systems, *Ann. Found. Louis de Broglie* **9** (1984), 65–95.
- [5] Memoryless controllers for uncertain systems, in F. Gagliardi (ed.), *Mathematical Methods for Optimization in Engineering*. Univ. of Cassino Press, Italy, 1984.

Corless, M., Leitmann, G., and Ryan, E.P.
- [1] Tracking in the presence of bounded uncertainties, *Proc. IV Int. Conf. Control Theory, Cambridge Univ.*, Sept. 1984, 87–92.

Corless, M., Leitmann, G., and Skowronski, J.M.
- [1] Adaptive control for avoidance or evasion in an uncertain environment, in Y. Yavin (ed.), *Pursuit–Evasion Differential Games, Comput. Math. Appl.*, to appear (1986).

Cvetkovič, V., and Vukobratovič, M.
- [1] One robust dynamic control algorithm for manipulation systems, *Internat. J. Robotic Res.* **1** (1982), 15–28.

Dailly, C.
- [1] Optimal workpiece positioning in a robot workspace, *Tech. Notes, Engng. Dept., Univ. of Cambridge*, 1985.

Dauer, J.P.
- [1] A controllability technique for nonlinear systems, *J. Optim. Theory Appl.* **37** (1972), 442–451.
- [2] A note on bounded perturbations of controllable systems, *J. Optim. Theory Appl.* **42** (1973), 221–225.
- [3] Controllability of nonlinear systems with restrained control, *J. Optim. Theory Appl.* **44** (1974), 251–261.

[4] Bounded perturbations of controllable systems, *J. Math. Anal. Appl.* **48** (1974), 61–69.

Davy, I.L.
[1] Properties of solutions set of generalized differential equations, *Bull. Austral. Math. Soc.* **6** (1972), 379–398.

Desa, S., and Roth, B.
[1] Mechanics: Kinematics & dynamics, in G. Beni and S. Hackwood (eds.), *Recent Advances in Robotics*, Wiley, New York, 1985, pp. 71–130.

Donaldson, D.D., and Leondes, C.T.
[1] Model referenced parameter tracking technique for adaptive control systems, *Trans. IEEE Appl. and Industry* **82** (1963), 241–262.

Drenick, R.F.
[1] Optimization under uncertainty, in J. Stoer (ed.), *Optimization Techniques*, Lecture Notes on Control and Information Sciences, Vol. 6. Springer-Verlag, Berlin and New York, 1978, pp. 40–58.
[2] Feedback control of partly unknown systems, *SIAM J. Control Optim.* **15** (1977), 506–509.

Dubovsky, S., and Desforges, D.T.
[1] Application of model referenced adaptive control to robotic manipulators, *Trans. ASME Ser. G J. Dynamic Systems Measurement Control* **101** (1979), 193–200.

Dubovsky, S., and Gardner, T.N.
[1] Dynamic interactions of link elasticity and clearance connections in planar mechanical systems, *Trans. ASME J. Engrg. Industry* **97** (1975), 652–661.

Duffy, J.
[1] *Analysis of Mechanisms and Robot Manipulators.* Wiley, New York, 1980.

Duinker, S.
[1] Traditors, a new class of nonenergetic network elements, *Philips Res. Rep.* **14** (1959), 29–51.

Erzberger, H.
[1] Analysis and design of model following control systems by state space techniques, *Proc. JACC, Ann Arbor* (1968), 572–581.

Eykhoff, P.
[1] *System Identification: Parameter and State Estimation.* Wiley, New York, 1974.

Filippov, A.F.
[1] Existence of solutions of generalized differential equations, *Math. Notes* **10** (1971), 608–611.

Filippov, S.D.
[1] Avoidance problem with incomplete information, *Differential Equations* **13** (1977), 1267–1272.

Flashner, H., and Skowronski, J.
[1] Model referenced adaptive control of flexible spacecraft links, to be published.

Foley, M.A., and Schmitendorf, W.E.
[1] A class of differential games with two pursuers versus one evader, *IEEE Trans. Automat. Control* **AC-19** (1974), No. 3.

Frederick, C.
[1] *Modelling and Analogies of Dynamical Systems.* Houghton, Boston, Massachusetts, 1978.

Furasov, W.D.
[1] *Stability of Motion.* Nauka, Moscow, 1977.

Galperin, E.A.
[1] Asymptotic observers for nonlinear control systems, *Proc. IEEE Conf. Decision and Control (CDC)*, New York (1976), 1929–1300.

REFERENCES

Galperin, E.A., and Skowronski, J.M.
 [1] Playable asymptotic observers for differential games with incomplete information—the user's guide, *Proc. 23rd IEEE Conf. Decision and Control (CDC)* Las Vegas, (1984), 1201–1206.

Gayek, J., and Vincent, T.L.
 [1] Manoeuverable sets, *J. Optim. Theory Appl.*, to appear (1986).

Gershwin, S.B., and Jacobson, D.H.
 [1] Controllability theory for nonlinear systems, *IEEE Trans. Automat. Control* **AC-16** (1971), 37–50.

Getz, W.M., and Leitmann, G.
 [1] Qualitative differential games with two targets, *J. Optim. Theory Appl.* **68** (1979), 421–430.

Gevarter, W.B.
 [1] Basic relations for control of flexible vehicles, *AIAA J.* **8** (1970), 666–678.

Grantham, W.J.
 [1] A controllability Minimum Principle, Ph.D. Thesis, Aerospace Engrg., Univ. of Arizona, Tucson, 1973.

Grantham, W.J., and Vincent, T.L.
 [1] Controllability Minimum Principle, *J. Optim. Theory Appl.* **17** (1975), 93–114.

Grayson, L.P.
 [1] Two theorems on the second method, *IEEE Trans. Automat. Control* **AC-9** (1964), 587–562.
 [2] The status of synthesis using Liapunov's method, *Automatica* **3** (1965), 91–121.

Griffith, E.W., and Kumar, E.S.P.
 [1] On observability of nonlinear systems, *J. Math. Anal. Appl.* **35** (1971), 61–69.

Gusev, S.V., and Yakubovich, V.A.
 [1] Adaptive control algorithm for a manipulator, *Automation and Remote Control* **41** (9), Part 2 (1980), 1268–1277.

Gusev, S.V., Timofeev, A.V., and Yakubovich, V.A.
 [1] Algorithms of adaptive control of robot movements, *Mechanism and Machine Theory* **18** (1983), 279–281.

Gutman, S.
 [1] On optimal guidance for homing missiles, *Proc. 20th Israel Conf. Aviation and Astronaut.* (1978), 75–81.
 [2] Uncertain dynamical systems—a Lyapunov min-max approach, *IEEE Trans. Automat. Control* **AC-24** Haifa, Israel, (1979), 437–443.
 [3] Remarks on capture-avoidance games, *J. Optim. Theory Appl.* **32** (1980), 365–377.

Gutman, S., and Leitmann, G.
 [1] Stabilizing feedback control for dynamical systems with bounded uncertainty, *Proc. IEEE Conf. Decision and Control (CDC)*, New York *1976*.

Gutman, S., and Palmor, Z.
 [1] Properties of min-max controllers in uncertain dynamical systems, *SIAM J. Control Optim.* **20** (1982), 850–861.

Haas, U.L., Kuntze, H.B., and Schill, W.
 [1] A surveillance system of obstacle recognition and collision avoidance control, *Proc. 2nd Internat. Conf. Robot Vision and Sensory Controls*, Stuttgart (1982), 357–366.

Hagedorn, P., and Breakwell, J.V.
 [1] A differential game with two pursuers and one evader, *J. Optim. Theory Appl.* **18** (1976), 15–29.
 [2] A differential game with two pursuers and one evader, *in* G. Leitmann (ed.), *Multicriteria Decision Making and Differential Games*. Plenum, New York, 1976, 80–88.

Hajek, O.
 [1] *Pursuit Games*. Academic Press, New York, 1975.

Hanafi, A., Wright, F.W., and Hewitt, J.R.
[1] Optimal trajectory control of robotic manipulators, *Mechanism and Machine Theory* **19** (1984), 267–273.

Hewitt, J.R., and Burdess, J.S.
[1] Fast dynamic decoupled control for robotics using active force control, *Mechanism and Machine Theory* **16** (1981), 535–542.

Hollerbach, J.M.
[1] A recursive formation of Lagrangian manipulator dynamics, *IEEE Trans. Sys. Man. Cybern.* **SMC-10** (1980) 730–736.

Hooker, W.W.
[1] A set of r dynamical attitude equations for an arbitrary n-body satellite having r rotational degrees of freedom, *AIAA J.* **8** (1970), No. 7.

Hooker, W.W., and Margulies, G.
[1] The dynamical attitude equations for a n-body satellite, *J. Astronaut. Sci.* **12** (1965), 123–128.

Horowitz, R., and Tomizuka, M.
[1] An adaptive control scheme for mechanical manipulators, *ASME Paper* **80-WA/DSC-6** (1982).

Hwang, M., and Seinfeld, J.H.
[1] Observability of nonlinear systems, *J. Optim. Theory Appl.* **10** (1972), 67–77.

Isaacs, R.
[1] *Differential Games*. Wiley, New York, 1964.

Jumarie, G.
[1] Structural differential games, *C. R. Hebd. Seances Acad. Sci. Sér. A* **280** (1975), 969–972.

Kalman, R.E.
[1] On general theory of control systems, in *Automatic and Remote Control*, pp. 481–491. Butterworth, London, 1961.

Kalman, R.E., and Bertram, J.E.
[1] Control system analysis and design via the second method of Liapunov, *Trans. ASME J. Basic Eng.* **1** (1960), 371–393.

Kane, T.R., and Levinson, D.A.
[1] The use of Kane's dynamical equations in robotics, *Internat. J. Robotic Res.* **2** (1983), No. 3.

Kane, T.R., Likins, P.W., and Levinson, D.A.
[1] *Spacecraft Dynamics*. McGraw-Hill, New York, 1982.

Kaskosz, B.
[1] A sufficient condition for evasion in a nonlinear game, *Control Cybernet.* **7** (1978), 515–521.
[2] On a nonlinear evasion problem, *SIAM J. Control Optim.* **15** (1977), 661–679.

Kersten, L.
[1] The Lemma concept: a new manipulator, *Mechanism and Machine Theory* **12** (1977), 77–84.

Khatib, O., and Lemaitre, J.F.
[1] Dynamic control of manipulators operating in complex environment, *Proc. CISM Sympos. Theory and Practice of Robots*, Udine, Italy *1980*, 267–282.

Klamka, J.
[1] On global controllability of perturbed nonlinear systems, *IEEE Trans. Automat. Control* **AC-20** (1975), 170–172.
[2] On local controllability of perturbed nonlinear systems, *IEEE Trans. Automat. Control* **AC-20** (1975), 289–291.

REFERENCES

Klein, A.
- [1] Energy consumption of 3 link manipulators, *Robotica* **1** (1983), 79–83.

Kobrinski, A.A., and Kobrinski, A.E.
- [1] Motion design for manipulator systems in environment with obstacles, *Dokl. Akad. Nauk SSSR* **224** (1975), 1279–1285.

Koivo, A.J., and Paul, R.P.
- [1] Manipulator with self-tunning controller, *IEEE Conf. Cybernet. and Soc. 1980*.

Koivo, A.J., and Repperger, D.W.
- [1] Optimization of terminal rendezvous as a cooperative game, *Proc. 12th JACC, St. Louis* (1971), 508–516.

Kononenko, V.O.
- [1] Some autonomous problems of theory of nonlinear vibrations, *Proc. IUTAM Sympos., Kiev, 1961* (1962), 151–179.
- [2] *Vibrating Systems with Limited Power Supply.* Iliffe Books, London, 1969.

Konstantinov, M.S.
- [1] Inertia forces of robots and manipulators, *Mechanism and Machine Theory* **12** (1977), 387–401.

Kooleshov, V.S., and Lakota, N.A.
- [1] *Manipulators Control System Dynamics.* Energia, Moscow, 1971.

Kostiukovski, Y.M.L.
- [1] Observability of nonlinear control systems, *Automat. Remote Control* (1968), 1384–1396.

Kon, S.R., Elliot, D.L., and Tarn, T.J.
- [1] Exponential observers for nonlinear dynamic systems, *Inform. and Control* **29** (1975), 204–216.

Krassovski, N.N.
- [1] *Control Theory.* Nauka, Moscow, 1968.
- [2] Differential game of approach–evasion, *Izv. Akad. Nauk SSSR, Tech. Cybernet.* (1973), Nos. 2 and 3.
- [3] Game theoretic control and problems of stability, *Problems Control Inform. Theory* **3**, (1974), No. 3, 171–182.
- [4] Game theoretic control under incomplete phase-state information, *Problems Control Inform. Theory* **5** (1976), 291–302.

Krassovski, N.N., and Subbotin, A.I.
- [1] *Positional Differential Games.* Nauka, Moscow, 1974.

Kroc, L.
- [1] On synthesis of adaptive multiparameter control systems by Liapunov method, *Kibernetica* **2** (1975), 277–286.

Krogh, B.H.
- [1] Feedback obstacle avoidance control, *Proc. XXI Allerton Conf., Univ. of Illinois* (1983), 325–334.
- [2] Guaranteed steering control, *Proc. 1984 ACC*.

Kulinich, A.S., and Penev, G.P.
- [1] Parametric optimization of the equations of motion of multi-link system and adaptive control algorithms, *Automation and Remote Control* **40** (1979), 1793–1803.

Kumar, A., and Waldron, K.J.
- [1] The workspace of a mechanical manipulator, *J. Mech. Design* **103** (1981), 665–672.

Kuntze, H.B., and Schill, W.
- [1] Methods for collision avoidance in computer controlled industrial robots, *Proc. Internat. Sympos. Indust. Robots, Paris* (1982), 519–530.

Leborgue, M., Ibarra, J.M., and Espiau, B.
[1] Adaptive control of high velocity manipulators, *Proc. XI Internat. Sympos. Indust. Robots, Tokyo, 1981.*
Landau, I.D.
[1] A survey of model reference adaptive techniques, *Automatica* **10** (1974), 313–379.
[2] *Adaptive Control: The Model Reference Approach.* Dekker, New York, 1979.
Lee, C.S.G., and Chung, M.J.
[1] Adaptive perturbation control with feed forward compensation for robot manipulators, *Simulation* **44** (1985), 127–136.
Leitmann, G.
[1] Sufficiency theorems for optimal control, *J. Optim. Theory Appl.* **2** (1968), 285–289.
[2] Stabilization of dynamical systems under bounded input disturbances and parameter uncertainties, in E.O. Roxin (P.T. Liu and R.L. Sternbey, eds.) "Differential Games and Control Theory" *Proc. II Kingston Conf. Differential Games and Control, 1976,* 47–52.
[3] Guaranteed asymptotic stability for a class of uncertain linear dynamical systems, *J. Optim. Theory Appl.* **27** (1979), 99–104.
[4] Guaranteed avoidance strategies, *J. Optim. Theory Appl.* **32** (1980), No. 4, 569–576.
[5] Deterministic control of uncertain systems, *Acta Astronaut.* **7** (1980), 1457–1461.
[6] On efficacy of nonlinear control in uncertain linear systems, *Trans. ASME Ser. G J. Dynamic Systems Measurement Control* **103** (1981), 95–102.
[7] An introduction to optimal control, McGraw-Hill, New York, 1966.
Leitmann, G., and Liu, H.S.
[1] Evasion in the plane, *8th IFIP Sympos. Optim. Tech., Würzburg, 1977.*
Leitmann, G., and Skowronski, J.
[1] Avoidance control, *J. Optim. Theory Appl.* **23** (1977), 581–591.
[2] A note on avoidance control, *Optimal Control Appl. Methods* **4** (1983), 335–342.
Lenarcic, J.
[1] A new method for calculating the Jacobian for a robot manipulator, *Robotica* **1** (1983), 205–209.
Lewandowska, A.
[1] Controller synthesis for nonlinear plants with conditions on stability region, *Control Cybernet.* **5** (1976), 5–19.
Lewis, R.A.
[1] Adaptive control of robotic manipulators, *Proc. IEEE-CDC* (1977), 743–748.
Liegeois, A., Fournier, A., and Aldon, M.J.
[1] Model reference control of high velocity industrial robots, *Proc. JACC, San Francisco* (1980), TP10-D.
Liegeois, A., Khalil, W., Dumas, J.M., and Renaud, M.
[1] Mathematical and computer models of inter-connected mechanical systems, in A. Morecki and K. Kedzior (eds.), *Theory and Practice of Robot Manipulators.* Elsevier, Amsterdam, 1977, 23–29.
Lindorf, D.P., and Carrol, R.L.
[1] Survey of adaptive control using Liapunov design, *Int. J. Control* **18** (1973), 897–914.
Liu, R.W., and Leake, R.J.
[1] Inverse Liapunov problems, *Proc. Internat. Sympos. Differential Equations Dynamic Systems, Puerto Rico* (1965), 75–80.
[2] Exhaustive equivalence classes of optimal systems with separable controls, *SIAM J. Control* **4** (1966), 678–685.
Lozano-Perez, T.
[1] Automatic planning of manipulator transfer movements, AI Memo No. 606, MIT Artificial Intelligence Lab., Dec. 1980.

REFERENCES

[2] *Spatial planning: a configuration space approach*, AI Memo No. 605, MIT Artificial Intelligence Lab., Dec. 1980.

Lozano-Perez, T., and Wesley, M.S.
[1] An algorithm for planning collision free paths among polyhedral obstacles, *Comm. ACM* **44** (1979), 560–570.

Luenberger, D.G.
[1] An introduction to observers, *IEEE Trans. Automat. Control* **AC-16** (1971), 596–602.
[2] Dynamic systems in descriptor form, *IEEE Trans. Automat. Control* **AC-22** (1977), 312–331.

Luh, J.Y.S.
[1] Anatomy of industrial robots and their controls, *IEEE Trans. Automat. Control* **AC-28** (1983), 133–153.
[2] Conventional controller design for industrial robots—a tutorial, *IEEE Trans. Systems Man Cybernet.* **SMC-13** (1983), 298–316.

Luh, J.Y.S., and Lin, C.S.
[1] Optimum path planning for mechanical manipulators, *Trans. ASME Ser. G. J. Dynamic Systems Measurement Control* **102** (1981), 142–151.

Lukas, D.L.
[1] Global controllability nonlinear systems, *SIAM J. Control* **10** (1972), 112–126.
[2] Global controllability of disturbed nonlinear equations, *SIAM J. Control* **12** (1974), 695–704.

Mahil, S.S.
[1] On application of Lagrange's Method to description of mechanical systems, *IEEE Trans. Systems Man Cybernet.* **SMC-12** (1982), No. 6.

Mansour, M.
[1] Generalized Lyapunov function for power systems, *IEEE Trans. Automat. Control* **AC-19** (1974), 247–248.

Markov, M.D., Zamanov, V.B., and Nenchev, D.N.
[1] Trajectory modelling and teaching of robots under Cartesian path control, *Internat. J. Prod. Res.* **21** (1983), 173–182.

Markus, L.
[1] Controllability of nonlinear processes, *SIAM J. Control* **3** (1965), 78–90.

Medvedev, V.S., Leskov, A.G., and Yushchenko, A.S.
[1] *Control System for Manipulator Robots*. Nauka, Moscow, 1978.

Meirovitch, L.M.
[1] *Methods of Analytic Dynamics*. McGraw-Hill, New York, 1970.

Merz, A.W.
[1] Optimal evasive manoeuvers in maritime collision avoidance, *Navigation* **20** (1973), No.2.
[2] The game of two identical cars, in G. Leitmann (ed.), *Multicriteria Decision Making*, pp. 421–442. Plenum, New York, 1976.

Moravec, H.P.
[1] Obstacle avoidance and navigation in the real world by a seeing robot rover, Tech. Rep. AIM 340, Stanford Univ., Sept. 1980.

Narendra, K.S.
[1] Stable identification schemes, in R.K. Mehra and D.G. Lainiotis (eds.), *System Identification*, pp. 165–209. Academic Press, New York, 1976.

Narendra, K.S., and Tripathi, S.S.
[1] Identification and optimization of aircraft dynamics, *J. Aircraft* **10** (1973), 193–199.

Nemitcky, V.V., and Stepanov, V.V.
[1] *Qualitative Theory of Differential Equations*. Academic Press, New York, 1960.

Nguen, T.L.
[1] Synthesis of adaptive control systems using standard models, *Engrg. Cybernet.* **9** (1971), 386–394.

Nikolsky, M.S.
[1] On some differential games with fixed time, *Soviet Math. Dokl.* **19** (1978), 587–591.

Nikonov, O.I.
[1] On combination of control and observation processes in game problems of evading motion, *Differential Equations* **13** (1977), 1053–1060.

Okhocimski, D.E., and Platonov, A.K.
[1] Control algorithms for a stepping machine capable of avoiding obstacles, *Tech. Kibernet.* (1973), 3–10.

Olas, A., Ryan, P.W., and Skowronski, J.M.
[1] Linear damping of free motion of strongly nonlinear double mechanical system, *Bull. Acad. Polon. Sci. Sér. Sci. Tech.* **15** (1967), 97–100.

Olsder, G.J., and Breakwell, J.V.
[1] Role determination in an aerial dogfight, *Internat. J. Game Theory* **3** (1974), No. 1.

Olsder, G.J., and Walter, J.L.
[1] A differential game approach to collision avoidance of ships, *Lecture Notes in Control and information Sci.* **6** (1978), 264–271.

Ostapenko, V.V.
[1] On linear avoidance problem, *Kibernetika* (1978), 106–112.

Paskov, A.G.
[1] In sufficient conditions for nonlinear positional games of encounter, *J. Appl. Math. Mech.* **40** (1976), 148–151.

Paul, R.P.
[1] *Robot Manipulators: Mathematics, Programming and Control.* MIT Press, Cambridge, Massachusetts, 1981.
[2] Modelling trajectory calculation and surveying of a computer controlled arm, Memo No. 177, Stanford Artificial Intelligence Lab., Sept. 1972.

Pearson, J.O.
[1] Worst case design subject to linear parameter uncertainties, *IEEE Trans. Automat. Control* **AC-20** (1975), 167–169.

Peng, W.Y., and Vincent, T.L.
[1] Some aspects of aerial combat, *AIAA J.* **13** (1975), 7–11.

Petrov, A.A., and Sirota, I.A.
[1] Obstacle avoidance by a robot manipulator under limited information about environment, *Automat. i Telemeh.* (1983), 29–40.

Popov, E.P., Vereshchagin, A.F., and Zinkevich, S.L.
[1] *Manipulational Robots: The Dynamics and Algorithms.* Nauka, Moscow, 1978.

Porter, B.
[1] *Synthesis of Dynamical Systems.* Nelson, New York, 1969.

Pozaritskii, G.K.
[1] Game problem of impulse hard contact in a position attraction field with an opponent who realizes a bounded thrust, *J. Appl. Math. Mech.* **39** (1975), 185–195.
[2] Game problem of impulse encounter with an opponent limited in energy, *J. Appl. Math. Mech.* **39** (1975), 555–565.

Pshenichnyi, B.N.
[1] On the problem of avoidance, *Kibernetika* (1975), No. 4.

Pszczel, M., and Skowronski, J.M.
[1] Obstacles avoidance control for robot manipulators under uncertainty, to be published (1986).

Rajbman, N.S.
- [1] An application of identification methods in USSR—a survey, *Automatica* **12** (1976), 73–95.

Rang, E.R.
- [1] Adaptive controllers derived by stability considerations, Memo No. MR 7905, Minneapolis-Honeywell Regulator Co., 1962.

Red, W.E.
- [1] Minimum distances for robot task simulation, *Robotica* **1** (1983), 231–238.

Red, W.E., and Truong-Cao, H.V.
- [1] Configuration space approach to robot path planning, *Proc. ACC, San Diego* (1984), 817–821.
- [1] Vibrations due to motion discontinuities in hydraulically actuated robots, *Robotica* **1** (1983), 211–215.

Reibert, M.H., and Horn, B.K.P.
- [1] Manipulator control using the configuration space method, *Indust. Robot* (1978), 69–73.

Roitenberg, Y.Y.
- [1] Observability of nonlinear systems, *SIAM J. Control* **8** (1970), No. 3.

Roxin, E.O.
- [1] Stability in general control systems, *J. Differential Equations* **1** (1965), 115–150.
- [2] On generalized dynamical systems defined by contingent equations, *J. Differential Equations* **1** (1965), 188–205.
- [3] *Ordinary Differential Equations.* Wadsworth, Belmont, California, 1972.

Ryan, E.P., and Corless, M.
- [1] Ultimate boundedness and symmetric stability of a class of uncertain dynamical systems via continuous and discontinuous feedback control, *IMA J. Methods Control Inform.* **1** (1984), 223–242.

Ryan, E.P., Leitmann, G., and Corless, M.
- [1] Practical stabilizability of uncertain dynamical systems—application to robotic tracking, *J. Optim. Theory Appl.*, to appear (1986).

Saridis, G.N., and Lobbia, R.N.
- [1] Parameter identification and control of linear discrete time systems, *IEEE Trans. Automat. Control* **AC-17** (1972), 52–60.

Schmitendorf, W.E.
- [1] Min–max control of systems with uncertainty in the initial state and in state equations, *IEEE Trans. Automat. Control* **AC-22** (1977), 439–443.

Shaked, U.
- [1] Design of general model following control system, *Int. J. Control* **25** (1977), 57–79.

Shestakov, A.A.
- [1] Alout distribution of critical points of a system of differential equations, Trudy Kazanskoho Aviationnogo Instituta, **27**, 1953, 41–50.

Shinar, J., and Gutman, S.
- [1] Recent advances in optimal pursuit and evasion, *Proc. IEEE CDC, San Diego* (1979), pp. 960–965.

Silver, W.M.
- [1] An equivalence of the Lagrangian and Newton–Euler Methods, *Internat. J. Robotics Res.* **1** (1982), 60–70.

Skowronski, J.M.
- [1] Some remarks about the character of motion of mechanical systems, *Proc. IV Nat. U.S. Congr. Appl. Mech., Berkeley* (1962), pp. 356–364.
- [2] Structural investigation of vibrating nonlinear mechanical systems, *Nonlinear Vibr. Problems* **6** (1964), 253–310.

[3] Synthesizable stability of limit domain for response of general mechanical systems, *Abh. Deutsch. Akad. Wiss. Berlin Kl. Math. Phys. Tech.* (1965), 344–346.
[4] Nonlinear mechanical lumped systems, *Nonlinear Vibr. Problems* **7** (1965), 7–224.
[5] Sufficient criterion for synthesizable stability of general physical lumped systems, *Bull. Acad. Polon. Sci. Sér. Sci. Tech.* **14** (1966), 425–428.
[6] *Elements of Geometric Dynamics.* Sci. & Tech. Publishers (PWN), Warsaw, 1967.
[7] Geometric aspects of kinetic synthesis of multiple lumped systems, *Proc. IV Internat. Conf. Nonlinear Oscillations, Prague, 1968*, 257–268.
[8] Sufficient conditions for optimal synthesizability, *Proc. Internat. Congr. Math., Moscow* **12** (1968), 385–386.
[9] Asymptotic equivalence method for kinetic synthesis of shape, *Bull. Acad. Polon. Sci. Sér. Sci. Tech.* **16** (1968), 623–628.
[10] System asymptotic equivalence as a method for kinetic synthesis of a set, *Nonlinear Vibr. Problems* **10** (1969), 31–35.
[11] Inverse delta method for synthesis of nonlinear lumped systems, *Nonlinear Vibr. Problems* **10** (1969), 27–30.
[12] *Multiple Nonlinear Lumped Systems.* Polish Sci. Publishers (PWN), Warsaw, 1970.
[13] Space delta method for analysis and synthesis of nonlinear lumped systems, *Int. J. Control* **12** (1970), 109–120.
[14] An attempt to design qualitative differential games, *Int. J. Control* **12** (1970), 121–127.
[15] Conjectures on qualitative Liapunov games, *Int. J. Control* **16** (1972), 501–507.
[16] On qualitative competitive Liapunov game, *Proc. V Hawaii Internat. Conf. Systems Sci.* (1972), pp. 152–154.
[17] Game modelled lumped physical system, *Proc. IX Hawaii Internat. Conf. Systems Sci.* (1976).
[18] Liapunov type playability for adaptive physical systems, *Proc. Nat. Systems Conf., Combaitore, PSG College of Techn., India* (1977), pp. Q11, 1–5.
[19] Note on Liapunov design of systems in conflict with environment, *Proc. IFAC Sympos. Environmental Systems Planning, Design & Control, Kyoto, 1977*, 28–35.
[20] Adaptive identification of models stabilizing under uncertainty, *Lecture Notes in Biomath*, **40** (1981), 64–78.
[21] Collision with capture and escape, *Israel J. Tech.* **18** (1981), 70–75.
[22] Deterministic identification of turbulent loads for general mechanical structures with uncertain state measurements, *Mech. Res. Comm.* **10** (1983), No. 6, 345–350.
[23] Parameter and state identification in nonlinearizable uncertain systems, *Internat. J. Nonlinear Mech.* **19** (1984), 345–353.
[24] *Applied Liapunov Dynamics*, Systems & Control Engineering Consultants Publ., Brisbane, 1984.
[25] Adaptive identification of opposition in nonlinear differential game, *Internat. J. Nonlinear Mech.* **21** (1986), 83–94.
[26] Nonlinear model reference adaptive control, *J. Austral. Math. Soc. Ser. B*, **28**, 1986, 23–35.
[27] Model reference control and identification of robot manipulators under certainty, X. Arula et al. (eds.), *Proc. Internat. Conf. Math Modelling, Berkeley, 1985*, to appear.
[28] Adaptive control of robotic manipulators under uncertain pay-load, in M. Hamza (ed.), *Advances in Robotics*, pp. 40–48. Acta Press, Anaheim, California, 1985.
[29] Control for collision with rendezvous or capture, *J. Optim. Theory Appl.*, to appear.
[30] Model reference adaptive control under uncertainty of nonlinear flexible manipulators, *Proc. 1986 AIAA Guidance, Navigation and Control Conf.*, Williamsburg, Virginia, paper no. 86-1976-CP.

REFERENCES

Skowronski, J.M., and Pszczel, M.
 [1]   Control of robotic manipulators under adaptively identified uncertainty, *Proc. Control '85, IEE Conf., London* (1985), 46–47.

Skowronski, J.M., and Stonier, R.J.
 [1]   The barrier in a pursuit–evasion game with two targets, in Y. Yavin (ed.), *Pursuit–Evasion Differential Games, Comput. Math. Appl.*, to appear.
 [2]   Liapunov barrier in qualitative dynamic games, *J. Optim. Theory Appl.*, to appear, 1986.

Skowronski, J.M., and Vincent, T.L.
 [1]   Playability with and without capture, *J. Optim. Theory Appl.* 36 (1982), 111–128.

Skowronski, J.M., and Ziemba, S.
 [1]   Certain properties of mechanical models of structures, *Arch. Mech. (Arch. Mech. Stos.)* 11 (1959), 193–209.
 [2]   Boundedness of motions and the existence and stability of limit regimes in strongly nonlinear nonautonomous mechanical systems, in *Automatic and Remote Control*, pp. 906–912. Butterworth, London, 1961.
 [3]   Domain of boundedness of motions of strongly nonlinear nonautonomous mechanical systems with partially negative damping, *Proc. IUTAM Sympos. Nonlinear Oscillations, Kiev, 1961* 2 (1963), 356–364.
 [4]   Abstract machine applied to physical systems, *Bull. Acad. Polon. Sci. Sér. Sci. Tech.* 15 (1967), 1–7.

Slotine, J.J., and Sastry, S.S.
 [1]   Tracking control of nonlinear systems using sliding surfaces with application to robotic manipulators, *Int. J. Control* 38 (1983), 465–492.

Snyder, W.E.
 [1]   *Industrial Robots*. Prentice-Hall, Englewood Cliffs, New Jersey, 1985.

Spong, M.W., Thorp, J.S., and Kheradpir, S.
 [1]   Control of robot manipulators using an optimal decision strategy, *Proc. XXI Allerton Conf.*, Univ. of Illinois (1983), 303–311.

Stalford, H.
 [1]   Sufficient conditions for optimal control with state and control constraints, *J. Optim. Theory Appl.* 7 (1971), No. 2.

Stalford, H., and Leitmann, G.
 [1]   Sufficient conditions for optimality in two person zero-sum differential game, *J. Math. Anal. Appl.* 33 (1971), 650–654.

Stensloff, H.
 [1]   *Wege zu sehr fortgeschnitten Handhabungs Systemen, Fachberichte Messen, Steuern, Regeln*, Vol. 4. Springer-Verlag, Berlin and New York, 1980.

Stepanenko, Y., and Vukobratovic, M.
 [1]   Dynamics of articulated open chain active mechanism, *Math. Biosci.* 28 (1976), 137–170.

Sticht, D.J., Vincent, T.L., and Schultz, D.G.
 [1]   Sufficiency theorems for target capture, *J. Optim. Theory Appl.* 17 (1975), pp. 523–543.

Stonier, R.J.
 [1]   Liapunov reachability and optimization in control, *J. Optim. Theory Appl.* 39 (1983), 403–416.
 [2]   On qualitative differential games with two targets, *J. Optim. Theory Appl.* 41 (1983), 587–598.

Stoten, D.P.
 [1]   Adaptive control of manipulator arms, *Mechanism and Machine Theory* 18 (1983), 283–288.

Takegaki, M., and Arimoto, S.
 [1]   Adaptive trajectory control of manipulators, *Int. J. Control* 34 (1981), 219–230.

[2] A new feedback method for dynamic control of manipulators, *Trans. ASME Ser. G J. Dynamic Systems Measurement Control* **102** (1981), 119–125.

Timofeev, A.V.
[1] *Robots and Artificial Intelligence*. Nauka, Moscow, 1978.
[2] *Design of Adaptive Systems of Control for Programmed Motion*. Energya, Leningrad, 1980.

Timofeev, A.V., and Ekalo, Y.V.
[1] Stability and stabilization of programmed motion of a manipulation robot, *Automat. i Telemeh.* (1976), 143–156.

Tkachenko, A.N., Brovinskaya, N.M., and Kondratenko, Y.P.
[1] Evolutionary adaptation of control processes in robots operating in nonstationary environments, *Mechanism and Machine Theory* **18** (1983), 275–278.

Tomizuka, M., and Horowitz, R.
[1] Model reference adaptive control of mechanical manipulators, *IFAC, Adaptive Systems in Control and Signal Processing*, San Francisco (1983), 23–32.

Totani, T., and Miyakawa, S.
[1] Mathematical model of hand transfer motion for application to manipulator control, *Trans. ASME Ser. G J. Dynamic Systems Measurement Control* **102** (1981), 152–157.

Udupa, S.M.
[1] Collision detection and avoidance in computer controlled manipulators, Ph.D. Thesis, California Inst. of Technol., Pasadena, 1977.

Uicker, J.T.
[1] Dynamic behaviour of spatial linkages, *Trans. ASME J. Engrg. Industry* **91** (1969), 251–258.

Vincent, T.L.
[1] Avoidance of guided projectiles, in J.D. Grote (ed.), *Theory and Application of Differential Games*, pp. 267–279. Reidel, Dordrecht, 1975.
[2] Collision avoidance at sea, *Lecture Notes in Control and Information Sci.* **3** (1977).

Vincent, T.L., and Skowronski, J.M.
[1] Controllability with capture, *J. Optim. Theory Appl.* **29** (1979), 77–86.

Voroneckaya, D.K., and Fomin, V.H.
[1] To the problem of path tracking by a manipulator, *Vestnik Leningrad. Univ.* **13** (1977), 132–136.

Vukobratovič, M., and Kirčanski, M.
[1] *Real Time Dynamics of Manipulation Robots*. Springer-Verlag, Berlin and New York, 1984.

Vukobratovič, M., and Potkonjak, V.
[1] *Dynamics of Manipulation Robots*. Springer-Verlag, Berlin and New York, 1982.

Vukobratovič, M., Potkonjak, V., and Nikolič, I.
[1] *Computer Aided Design Manipulation Robots*. Springer-Verlag, Berlin and New York, 1986.

Vukobratovič, M., and Stokič, D.
[1] *Control of Manipulation Robots*. Springer-Verlag, Berlin and New York, 1983.

Vukobratovič, M., Stokič, D., and Kirčanski, M.
[1] *Nonadaptive and Adaptive Control of Manipulation Robots*. Springer-Verlag, Berlin and New York, 1985.

Walker, M.W., and Orin, D.E.
[1] Efficient dynamic computer simulation of robotic mechanisms. *J. Dyn. Sys. Meas. Contr.* **104** (1982), 205–211.

Whittaker, E.T.
[1] *A Treatise on Analytical Dynamics of Particles and Rigid bodies*. Cambridge Univ. Press, London and New York, 1964.

REFERENCES

Willems, J.C.
- [1] Generation of Liapunov functions for input-output stable systems, *SIAM J. Control* **9** (1971), 105–133.
- [2] Dissipative dynamical systems, *Arch. Rational Mech. Anal.* **45** (1972), 321–351.
- [3] Consequences of a dissipation inequality in the theory of dynamical systems, in Dixhorn-Evans (eds.), *Physical Structure in Systems Theory*. Academic Press, New York, 1974, 193–218.

Wilson, D.J., and Leitmann, G.
- [1] Min–max control of systems with uncertain state measurements, *Appl. Math. Optim.* **2** (1976), 315–336.

Winsor, C.A., and Roy, R.J.
- [1] Design of model reference adaptive control systems by Liapunov second method, *IEEE Trans. Automat. Control* **AC-13** (1968), 204–210.

Wittenberg, J.
- [1] *Dynamics of Systems of Rigid Bodies*. Teubner, Stuttgart, 1977.

Yoshizawa, T.
- [1] Stability theory by Liapunov second method, *Publ. Math. Soc. Japan* Tokyo, (1966).

Math. Soc. Japan Tokyo, (1966).

Zinober, A.S., El-Ghezawi, O.M.E., and Billings, S.A.
- [1] Multivariable structure adaptive model following control systems, *Proc. IEE-D* **129** (1982), 6–12.

Index

A

Accumulation, axiom of, 63
Adaptive, 23, 31
   control, 23, 72, 73, 98, 115
   controllers, 117, 126
   identification, 98, 231
   law, 115, 117, 188, 223
   model reference, control, 185
   parameters, 31
   stabilization, 190
Admissible, 35
   control, 35
   program, 35
   subset of space, R^N, 24
Association table, 11
Attainable set, 100
Attraction, 85
   equi, 85
   finite time, 90, 91
   region of, 85, 89
   uniform, 85
Augmented body approach, 4
Avoidance, 131, 198
   adaptive, 221, 228
   band, 6
   controllability for, 131
     region of, 131
   of obstacles, 198, 201, 217, 225, 228
   real time, 210, 216, 221, 222, 228
   set, 7, 30
   with stipulated handling time, 209, 210, 211
   strong controllability for, 200, 205, 207
   ultimate, 214, 215

B

Barrier, 107
   semibarrier, 113, 122
     for property Q, 105, 106
Boundedness of motions, 89
   conditions for, 90
   equi, 89, 90
   uniform, 89
   ultimate, 89
     equi, 89
     uniform, 90, 91, 92

C

Capture, 80, 83, 156, 157
   with avoidance, 219
   controllability for, 80
   optimal, 156, 167
   strong controllability for, 160
     optimal, 166
   subtarget, 159
Chain, 9
   with colocated joints, 11
   kinetic, 17
   simple, 11
   structure, 9
Characteristic
   centrifugal, 21
   coriolis, 21
   damping, 9, 21
   elastic, 8
   gravity, 8, 22
   input, 22

Characteristic, *(cont.)*
  kinetic, 9, 23
  of potential forces, 8
  potential, 22
  restoring force, 22
  spring, 22
Compensation, 14, 125, 128
  control force, 126
  gravity, 8, 14, 127, 194
  potential, 127, 194,
    force, 125
  spring, 127, 128, 194
Configuration, 5
  of all joints, 5
  of manipulator arm, 1
  representing point, 26
  space, 5, 17
Control, 1, 30, 70
  adaptive, 23, 72, 73, 98, 115, 117, 185
  condition, 113
  direct method of, 70
  feedback, 23, 72, 108, 181
  forward, 23, 72
  on-line, 72
  open loop, 23, 71
  optimal, 23
  program, 31, 35, 83, 108
    admissible, 35
    memoryless, 72
    optimal, 118
    set valued, 100
  torque, 142
  vector, 30
Controllability, 70, 77
  for capture, 80, 81, 82, 108
  optimal, 117, 119
  for reaching, 75, 76, 77, 137
    in stipulated time, 78
  region of, 75, 84, 131
  for rejection, 202
  for qualitative property Q, 103
    region of, 103
  problem, 83
  strong
    for avoidance, 205, 228
    for capture, 160, 166
    for penetration, 146, 147
    for reaching, 137, 138, 171
    for rendezvous, 171, 172, 174
    for qualitative property Q, 104
      degree of, 104
      region of, 104
      uniform, 104
Coordinates, 1, 17
  base, 1, 3
  Cartesian, 1, 18, 19
  generalized, 17, 19
  inertial, 1
  joint, 5
  Lagrangian, 1, 5, 17, 18, 19
  link, 1
  relative, 85, 175, 225
  target, 180
  world, 1

D

Damping, 15, 61
  at rest, 61
  characteristic, 21
  control, 143
  forces, 42, 62
  negative, 61, 62
  positive, 61
  structural, 15
Dirichlet
  instability, 92
  stable equilibrium, 50, 51, 52, 97, 164

E

Energy, 32, 44, 60
  basic, level of, 55
  conservative reference frame, 48
  cup, 55, 56
    local, 56, 76, 144
    in-the-large, 56, 57, 94, 95, 111
  flow, 60, 64, 69, 76
    autonomous, 64
    accumulative, 66, 69
    dissipative, 66, 69, 141
    monotone, 66
  flux, 64, 142, 155
  kinetic, 8, 18, 20, 23, 44
  influx, 65, 142, 155, 165
  iso, level, 55
  neutral, 49
  outflux, 67, 142, 165
  potential, 8, 23, 44, 91, 125
  state, 156

INDEX

trajectories, 63, 64
motions, 69
surface, 48
threshold, 53, 55, 92, 93
usage, 169
zones, 66
Equations of motion, 8, 15, 23
canonical, 42
contingent, 100, 102, 234
cost dissipative, 121
Lagrange, 8, 19, 20, 23, 30
Hamiltonian state, 33, 41, 99, 128
main, 123
Newtonian form, 15, 20, 22, 128
normal form, 30
phase-space, 32
Equilibrium, 50, 53, 57, 86, 97, 108, 148, 232
basic, 51, 68, 240
isolated, 36
level, 76
equistable, 86
stable, 53
uniform asymptotically stable, 87, 88
uniform stable, 86
unstable, 53, 55, 56, 89
zero, 36

F

Force, 15
accumulative, 60
applied, 15, 16, 19, 23
centrifugal, 8, 9, 20, 21
compensation, 125
conservative, 49
coriolis, 8, 9, 14, 20, 21
damping, 9, 15, 42, 61, 63, 64
dissipative, 60
elastic, 9
fictitious, 15
generalized, 19
gravity, 23, 30, 45
input, 15,
transmission, 7, 96
kinetic, 15, 18, 23
nonpotential, 60
neutral, 49, 60
perturbation, 37
potential, 9, 44, 49
per-mass, 44

reaction, 15
spring, 45

G

Gravity, 8, 14
characteristic, 8, 22
compensation, 125
equilibria design, 124
forces, 14, 22

I

Identification, 231
adaptive, 98, 231, 238, 245
of states, 231
of parameters, 231
of payloads, 239
constant parameters, 242
model reference, 231, 232
Inertia, 15
coefficient, rotary actuator, 40
coupled, 22
decoupling, 32, 41, 47
force, 15
translative, 15
generalized, matrix, 18, 21
moment, 23
Invariant set, 35, 85
minimal, 85
negatively, 85
positively, 81, 85, 134
strongly, 149
cost, 122

J

Joints
configuration of, 5
collocated, 11
coordinates, 5, 24
dummy, 16
k-multiple, 17
prismatic, 1
revolute, 1
variable, 17

K

kinematics, 13
  forward, 13, 30
  independence of, 17
  inverse, 13, 14, 30
  only approach, 9
Kinetic
  chain, 17
  energy, 8, 18, 20, 32, 39, 40
  potential, 20

L

Legendre's transformation, 33
Leitmann
  –Gutman controller, 110, 111, 126, 142, 163
  school, 95
  sufficient conditions, 120, 135, 166
Liapunov
  direct method, 73
  derivative, 110, 233
  function, 97, 139, 187, 189, 244
  formalism, 124, 198
  levels, 97
  matrix equation, 88, 126, 187, 204
  stability, 85
Link, 1
  flexible, 16
  modular, 14
  moving, 10
  optional, 11
  rotating, 10
  single DOF, 142
Lipschitz
  condition, 35
  constant, 35, 39
  continuous, 39
    locally, 38, 99

M

Maneuvereability, 70
Maneuvereable set, 134
Manipulator, 1
  horizontal, 181
  modular, 1, 39
  overhead, 142
  PR-type, 39
  recent, 9
  with several chain structures, 9
  unit, 1
Model, 29
  Cartesian, 9
  dynamical
    tracked, 29
    following, 185
  Lagrangian, 17
  mechanical, 1
  reference, 185, 187, 290
    adaptive control, 185, 231, 232
  tracking, 73, 188, 190, 242, 245
Motion
  equibounded, 89
  equiultimately bounded, 89
  equation
    Lagrange, 8, 19, 35
    Hamilton, 33, 35
  representing vector, 17
  uniformly bounded, 89

P

Path
  avoidance, 225, 228
  collision-free, 28
  following, 80
  planned, 28, 70, 175
  stipulated, 70
  Cartesian, 70
  target, 175, 176
Penetration, 130
  strong controllability for, 146, 147
    region of, 147
  usable for, 149
Potential, 16, 44
  of conservative forces, 16
  cost, force, 122
  game, 120
  kinetic, 20
  push-off, 198, 201
Pontriagin, min–max principle, 124
Power, 44, 60
  accumulation, 96
  balance, 60, 63, 89, 113
  characteristic, 63
  damping, 60
  limited, 143
  input, 60
  zero, contour, 141
PUMA, 23

INDEX

R

Reachable, 100
  cone, 100
  point, 75, 133
  set, 100, 133
Reaching, 75, 76, 78, 129, 130
  with capture, 156, 158, 160, 151, 162, 166
    region of
  without capture 170
  with escape 157
  with penetration, 146
  with rejection, 149
  sequential, 151, 161
    in stipulated time, 161
  soft, 129, 146
  stipulated target, 130
  in stipulated time, 151, 152, 153
  strong controllability for, 137, 138, 171
    region of, 171
  ultimate, 153
  under water, 145
Rendezvous, 83, 158
  during T_z, 158
  reaching with, 171, 172, 174
    region of, 171

S

Space, 30
  configuration, 5
  of events, 38
  phase, 5, 17
  product, method of, 29, 179, 227, 234
  state, 5, 30
  task, 28
  work, 25, 140
Stability, 85
  asymptotic, 85, 86, 88, 233
    quasi, 85
    region of, 97, 98
    uniform, 85, 86, 187
  Dirichlet, 50, 51
  equi, 86
  Liapunov, 85
  uniform, 85, 86
    asymptotic, 85, 88
  unstable
    equilibrium, 89

  set, 85
Stabilization, 93, 96
  adaptive, 190
  asymptotic, 94, 95, 97, 179
    region of, 94
  practical, 94, 97
    region of, 94
  stabilizable system, 93, 204
  strong, 109, 111
    optimal, 123
  synthesis, 124
  under given level, 94
State, 30
  force, 32
  relative, vector, 175
  space, 5
    form, 97
  vector, 5, 17, 31
System
  autonomous, 36, 39, 74, 91
  conservative, 49
  dissipative, 68
  multiple mechanical, 9
  nonautonomous, 37, 69
  programs, 37
  reference, 55
  stabilizable, 93, 204

T

Target, 28
  Cartesian, 28
  configuration, 6, 7
  capture sub, 81
  in-the-large, 141
  path, 175, 176
  successive, 140
  terminal, 139
Tracking, 73, 188, 245
  adaptive, 244
  model, 73, 188, 190, 242, 245
  path, 175, 176
Trajectory, 35
  minimal, 36
  zero, 36
Transformation, 3, 4
  homogeneous, 3, 4, 23
  matrix, 3

U

Uncertain, 98, 99, 116, 231, 234, 239
  environmental perturbation, 98, 114
  known bound, 116
  parameters, 98
  system, 224
  unknown bound, 115

W

Work
  done, potential forces, 15
  region, 5, 26, 27

£33.00
ML

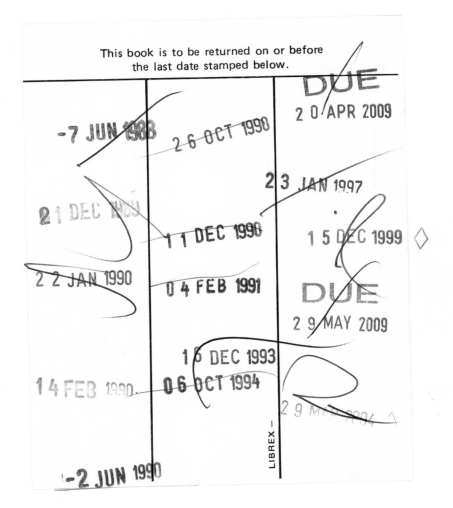